Lecture Notes in Physics

T0238602

The Editorial Policy for Edited Volumes

The series *Lecture Notes in Physics* (LNP), founded in 1969, reports new developments in physics research and teaching - quickly, informally but with a high degree of quality. Manuscripts to be considered for publication are topical volumes consisting of a limited number of contributions, carefully edited and closely related to each other. Each contribution should contain at least partly original and previously unpublished material, be written in a clear, pedagogical style and aimed at a broader readership, especially graduate students and nonspecialist researchers wishing to familiarize themselves with the topic concerned. For this reason, traditional proceedings cannot be considered for this series though volumes to appear in this series are often based on material presented at conferences, workshops and schools.

Acceptance

A project can only be accepted tentatively for publication, by both the editorial board and the publisher, following thorough examination of the material submitted. The book proposal sent to the publisher should consist at least of a preliminary table of contents outlining the structure of the book together with abstracts of all contributions to be included. Final acceptance is issued by the series editor in charge, in consultation with the publisher, only after receiving the complete manuscript. Final acceptance, possibly requiring minor corrections, usually follows the tentative acceptance unless the final manuscript differs significantly from expectations (project outline). In particular, the series editors are entitled to reject individual contributions if they do not meet the high quality standards of this series. The final manuscript must be ready to print, and should include both an informative introduction and a sufficiently detailed subject index.

Contractual Aspects

Publication in LNP is free of charge. There is no formal contract, no royalties are paid, and no bulk orders are required, although special discounts are offered in this case. The volume editors receive jointly 30 free copies for their personal use and are entitled, as are the contributing authors, to purchase Springer books at a reduced rate. The publisher secures the copyright for each volume. As a rule, no reprints of individual contributions can be supplied.

Manuscript Submission

The manuscript in its final and approved version must be submitted in ready to print form. The corresponding electronic source files are also required for the production process, in particular the online version. Technical assistance in compiling the final manuscript can be provided by the publisher's production editor(s), especially with regard to the publisher's own LaTeX macro package which has been specially designed for this series.

LNP Homepage (springerlink.com)

On the LNP homepage you will find:
−The LNP online archive. It contains the full texts (PDF) of all volumes published since 2000. Abstracts, table of contents and prefaces are accessible free of charge to everyone. Information about the availability of printed volumes can be obtained.
−The subscription information. The online archive is free of charge to all subscribers of the printed volumes.
−The editorial contacts, with respect to both scientific and technical matters.
−The author's / editor's instructions.

U. Carow-Watamura Y. Maeda S. Watamura (Eds.)

Quantum Field Theory and Noncommutative Geometry

Editors

Ursula Carow-Watamura
Tohoku University
Department of Physics
Graduate School of Science
Aoba-ku
Sendai 980-8578
Japan

Satoshi Watamura
Tohoku University
Department of Physics
Graduate School of Science
Aoba-ku
Sendai 980-8578
Japan

Yoshiaki Maeda
Keio University
Department of Mathematics
Faculty of Science and Technology
Hiyoshi Campus
4-1-1 Hiyoshi, Kohoku-ku
Yokohama 223-8825
Japan

U. Carow-Watamura Y. Maeda S. Watamura (Eds.),*Quantum Field Theory and Noncommutative Geometry*, Lect. Notes Phys. **662** (Springer, Berlin Heidelberg 2005), DOI 10.1007/b102320

ISBN 978-3-642-06287-2   e-ISBN 978-3-540-31526-1

ISSN 0075-8450

Springer is a part of Springer Science+Business Media

springeronline.com

Cover design: *design & production*, Heidelberg

Printed on acid-free paper
54/3141/jl - 5 4 3 2 1 0

Preface

This book is based on the workshop "Quantum Field Theory and Noncommutative Geometry" held in November 2002 at Tohoku University, Sendai, Japan. This workshop was the third in a series, the first one having been held at the Shonan International Village at Hayama in Kanagawa-ken in 1999, and the second one at Keio University, Yokohama in 2001. The main aim of these meetings is to enhance the discussion and cooperation between mathematicians and physicists working on various problems in deformation quantization, noncommutative geometry and related fields.

The workshop held in Sendai was focused on the topics of noncommutative geometry and an algebraic approaches to quantum field theory, which includes the deformation quantization, symplectic geometry and applications to physics as well as topological field theories.

The idea to treat quantized theories by using an algebraic language can be traced back to the early days of quantum mechanics, when Heisenberg, Born and Jordan formulated quantum theory in terms of matrices (matrix mechanics). Since then, a continuous effort has been made to develop an algebraic language and tools which would also allow the inclusion gravity. Among the physicist is point of view, the concept of a minimum length is discussed many times in various theories, especially in the theories of quantum gravity. Since the string is an extended object, string theory strongly suggests the existence of a minimum length, and this brought the discussion on the quantization of space into this field. However, this discussion raised several problems, in particular, how such a geometry with minimum length should be formulated and how a quantization should be performed in a systematic way.

A hint in this direction came from the theory of quantum groups, which had been developed in the 1980s and which gave a method to deform an algebra to become noncommutative, thereby preserving its symmetry as a q-deformed structure. Nearly at the same time A. Connes published his work on noncommutative differential geometry. It was the impact from these two new fields, that put forward the research on quantized spaces, and drew more and more the physicists' attention towards this field.

Noncommutative differential geometry (NCDG) led to striking extensions of the Atiyah-Singer index theorem and it also shows several common points with deformation quantization. Another result is the development of noncommutative gauge theory, which became a very promising candidate as an the effective theory of the so-called D-brane; a D-brane is a configuration which evolved in the course of the development of string theory, leading to solutions of nonperturbative configurations of the string in the D-brane background. Inspired by the possibilities opened by NCDG; there is now a number of physicists developing the "matrix theory", about 80 years after the "matrix mechanics".

Deformation quantization is a quantization scheme which has been introduced by Bayen, Flato, Fronsdal, Lichnerowicz and Sternheimer. In this approach the algebras of quantum observables are defined by a formal deformation of the classical observables as formal power series. The expansion parameter is $\hbar$ and the product of these deformed algebras is the star product. Symplectic geometry and Poisson geometry fit very well to this quantization scheme since they possess a Poisson structure, and thus deformation quantization is regarded as a quantization from an algebraic point of view. As we know from the theorem of Gel'fand and Naimark, we can often realize a classical space from a suitable algebra of the classical observables. From this point of view, we expect the deformation quantization may give a reasonable quantum space, whose investigation will contribute a development to noncommutative geometry.

We collected here the lectures and talks presented in the meeting. When preparing this proceedings we made effort to make this book interesting for a wider community of readers. Therefore, the introductions to the lectures and talks are more detailed than in the workshop. Also some derivations of results are given more explicitly than in the original lecture, such that this volume becomes accessible to researchers and graduate students who did not join the workshop. A large number of contributions are devoted to presentations of new results which have not appeared previously in professional journals, or to comprehensive reviews (including an original part) of recent developments in those topics.

Now we would like to thank all speakers for their continuous effort to prepare these articles. Also we would like to thank all participants of the workshop for sticking together until the end of the last talk, thus creating a good atmosphere and the basis for many fruitful discussions during this workshop. We also greatly acknowledge the Ministry of Education, Culture, Sports, Science and Technology, Japan, who supported this workshop by a Grant-in-Aid for Scientific Research (No. 13135202).

Sendai and Yokohama   *Ursula Carow-Watamura*
January 2005   *Yoshiaki Maeda*
                                                 Satoshi Watamura

Contents

Part III Applications in Physics

Part IV Topological Quantum Field Theory

List of Contributors

B. Jurčo
Theoretische Physik, Universität München, Theresienstr. 37, 80333 München, Germany
jurco@theorie.physik. uni-muenchen.de

G. Landi
Dipartimento di Scienze Matematiche, Università di Trieste, Via Valerio 12/b, 34127 Trieste, Italia, and INFN, Sezione di Napoli, Napoli, Italia
landi@univ.trieste.it

H. Bursztyn
Department of Mathematics, University of Toronto, Toronto, Ontario M5S 3G3, Canada
henrique@math.toronto.edu

J. Wess
Sektion Physik der Ludwig-Maximillians-Universität, Theresienstr. 37, 80333 München, Germany, and Max-Planck-Institut für Physik (Werner-Heisenberg-Institut), Föhringer Ring 6, 80805 München, Germany

K. Nakada
Department of Pure and Applied Mathematics, General School of Information Science and Technology, Osaka University, Japan
smv182nk@ecs.cmc.osaka-u.ac.jp

L.D. Paniak
Michigan Center for Theoretical Physics, University of Michigan, Ann Arbor, Michigan 48109-1120, U.S.A.
paniak@umich.edu

M. Cahen
Université Libre de Bruxelles, Campus Plaine, CP 218, 1050 Brussels, Belgium
mcahen@ulb.ac.be

M. Crainic
Depart. of Math., Utrecht University 3508 TA Utrecht, The Netherlands
crainic@math.uu.nl

M. Iwami
Graduate School of Pure and Applied Sciences, University of Tsukuba, Japan
maki@math.tsukuba.ac.jp

N. Iiyori
Unit of Mathematics and Information Science, Yamaguchi University, Japan
iiyori@yamaguchi-u.ac.jp

N. Miyazaki
Department of Mathematics, Faculty of Economics, Keio University, 4-1-1, Hiyoshi, Yokohama, 223-8521, JAPAN
miyazaki@math.hc.keio.ac.jp

P. Bieliavsky
Université Libre de Bruxelles,
Belgium
pbiel@ulb.ac.be

P. Bonneau
Université de Bourgogne, France
bonneau@u-bourgogne.fr

R.J. Szabo
Department of Mathematics,
Heriot-Watt University,
Scott Russell Building, Riccarton,
Edinburgh EH14 4AS, U.K.
R.J.Szabo@ma.hw.ac.uk

R.L. Fernandes
Depart. de Matem., Instituto
Superior Técnico
1049-001 Lisbon, Portugal
rfern@math.ist.utl.pt

S. Kamimura
Department of Mathematics,
Keio University, Yokohama
223-8522, Japan
kamimura@math.keio.ac.jp.

S. Waldmann
Fakultät für Mathematik und
Physik, Physikalisches
Institut, Albert-Ludwigs-Universität
Freiburg, Hermann Herder Strasse 3,
79104 Freiburg, Germany
Stefan.Waldmann@physik.
uni-freiburg.de

T. Itoh
Department of General Education,
Kinki University Technical College,
Japan
titoh@ktc.ac.jp

T. Kimura
Department of Mathematics and
Statistics, Boston University,
111 Cummington Street, Boston,
MA 02215, USA, and Institut des
Hautes Études Scientifiques, Le
Bois-Marie, 35, routes de
Chartres, 91440 Bures-sur-Yvette,
France

T. Masuda
Institute of Mathematics,
University of Tsukuba, Japan
tetsuya@math.tsukuba.ac.jp

T. Natsume
Division of Mathematics,
Nagoya Institute of Technology,
Showa-ku, Nagoya 466-8555, Japan

Y. Kawahigashi
Department of Mathematical
Sciences, University of Tokyo,
Komaba, Tokyo, 153-8914, JAPAN
yasuyuki@ms.u-tokyo.ac.jp

Y. Maeda
Keio University, Japan
maeda@math.keio.ac.jp

Part I

Noncommutative Geometry

Since the Gel'fand-Naimark theorem has been formulated, mathematicians and physicists are investigating the various possibilities to generalize the geometry of spaces and of space-time. The theorem states that geometry may be constructed from a function algebra. To express geometry in algebraic terms gives us the possibility to replace such a function algebra by a more general one, including the noncommutative algebra. This leads to our subject "Noncommutative Geometry".

The first lecture in Part I gives a general and kind introduction to noncommutative geometry, following Connes' approach. This is done in such a way that the reader can get some orientation about the developments and the present status of research in this field. Then, the recent developments on the non-commutative spheres in higher dimensinons are described.

The second contribution stresses the C^* algebra aspect of noncommutative spheres, and the deformation of spheres in 2, 3 and 4 dimensions is studied. Those include noncommutative spheres due to Bratteli, Eliott, Evans and Kishimoto, and also the ones due to Woronowicz, Podleś and Matsumoto, as well as the types studied by Connes and Landi, and the spheres investigated by Natsume and Olsen.

In the third contribution, the so-called θ deformation due to Connes and Dubois-Violette is applied in order to construct quantum homogeneous spaces as quantum groups. After the discussion about the θ-deformation, the theory of the quantum projective spaces as quantum homogeneous spaces based on quantum unitary group and the consequences are discussed.

Noncommutative Spheres and Instantons

G. Landi

Dipartimento di Scienze Matematiche, Università di Trieste, Via Valerio 12/b, 34127 Trieste, Italia, and INFN, Sezione di Napoli, Napoli, Italia
landi@univ.trieste.it

Summary. We report on some recent work on deformation of spaces, notably deformation of spheres, describing two classes of examples.

The first class of examples consists of noncommutative manifolds associated with the so called θ-deformations which were introduced in [17] out of a simple analysis in terms of cycles in the (b, B)-complex of cyclic homology. These examples have non-trivial global features and can be endowed with a structure of noncommutative manifolds, in terms of a spectral triple $(\mathcal{A}, \mathcal{H}, D)$. In particular, noncommutative spheres S_θ^N are isospectral deformations of usual spherical geometries. For the corresponding spectral triple $(C^\infty(S_\theta^N), \mathcal{H}, D)$, both the Hilbert space of spinors $\mathcal{H} = L^2(S^N, \mathcal{S})$ and the Dirac operator D are the usual ones on the commutative N-dimensional sphere S^N and only the algebra and its action on $\mathcal{H}$ are deformed. The second class of examples is made of the so called quantum spheres S_q^N which are homogeneous spaces of quantum orthogonal and quantum unitary groups. For these spheres, there is a complete description of K-theory, in terms of nontrivial self-adjoint idempotents (projections) and unitaries, and of the K-homology, in term of nontrivial Fredholm modules, as well as of the corresponding Chern characters in cyclic homology and cohomology [35].

These notes are based on invited lectures given at the *International Workshop on Quantum Field Theory and Noncommutative Geometry*, November 26–30 2002, Tohoku University, Sendai, Japan.

1 Introduction

We shall describe two classes of deformation of spaces with particular emphasis on spheres.

The first class of examples are noncommutative manifolds associated with the so called θ-deformations and which are constructed naturally [17] by a simple analysis in terms of cycles in the (b, B)-complex of cyclic homology. These examples have non-trivial global features and can be endowed with a structure of noncommutative manifolds, in terms of a spectral triple $(\mathcal{A}, \mathcal{H}, D)$ [10, 12]. In particular we shall describe noncommutative spheres S_θ^N which are isospectral deformations of usual spherical geometries; and we shall also show quite generally that any compact Riemannian spin manifold whose isometry group has rank $r \geq 2$ admits isospectral deformations to noncommutative geometries.

The second class of examples is made of the so called quantum orthogonal spheres S_q^N, which have been constructed as homogeneous spaces [30] of quantum orthogonal groups $\mathrm{SO}_q(N+1)$ and quantum unitary spheres S_q^{2n+1} which are homogeneous spaces of quantum unitary groups $\mathrm{SU}_q(n+1)$ [61]. The quantum groups $\mathrm{SO}_q(N+1)$ and $\mathrm{SU}_q(n+1)$ are R-matrix deformations of the usual orthogonal and unitary groups $\mathrm{SO}(N+1)$ and $\mathrm{SU}_q(n+1)$ respectively. In fact, it has been remarked in [35] that "odd" quantum orthogonal spheres are the same as "odd" quantum unitary ones, as it happens for undeformed spheres.

It is not yet clear if and to which extend these quantum spheres can be endowed with the structure of a noncommutative geometry via a spectral triple. There has been some interesting work in this direction recently. In [6] a 3-summable spectral triple was constructed for $\mathrm{SU}_q(2)$; this has been thoroughly analyzed in [14] in the context of the noncommutative local index formula of [18]. A 2-summable spectral triple on $\mathrm{SU}_q(2)$ was constructed in [7] together with a spectral triple on the spheres S_{qc}^2 of Podleś [51]. Also, a '0-summable' spectral triple on the so called standard spheres S_{q0}^2 has been given in [25, 39, 57]. Instead, on these spheres one can construct Fredholm modules, which provide a structure which is somewhat weaker that the one given by spectral triples. Indeed, a Fredholm module can be though of as a noncommutative conformal structure [20]. This construction for the quantum spheres S_q^N will be described in Sect. 6 closely following [35].

All our spaces can be regard as "noncommutative real affine varieties". For such an object, X, the algebra $A(X)$ is a finitely presented $*$-algebra in terms of generators and relations. In contrast with classical algebraic geometry, there does not in general exist a topological point set X. Nevertheless, we regard X as a noncommutative space and $A(X)$ as the algebra of polynomial functions on X. In the classical case, one can consider the algebra of continuous functions on the underlying topological space of an affine variety. If X is bounded, then this is a C*-algebra and is the completion of $A(X)$. In general, one defines $\mathcal{C}(X)$ to be the C*-algebraic completion of the $*$-algebra $A(X)$. To construct this, one first considers the free algebra $F(X)$ on the same generators of the algebra $A(X)$. Then, one takes all possible $*$-representations π of $F(X)$ as bounded operators on a countably infinite-dimensional Hilbert space $\mathcal{H}$. The representations are taken to be *admissible*, that is in $\mathcal{B}(\mathcal{H})$ the images of the generators of $F(X)$ satisfy the same defining relations as in $A(X)$. For $a \in F(X)$ one defines $\|a\| = \mathrm{Sup} \|\pi(a)\|$ with π ranging through all admissible representations of $F(X)$. It turns out that $\|a\|$ is finite for $a \in F(X)$ and $\| \cdot \|$ is a seminorm. Then $\mathcal{I} := \{a \in F(X) = 0\}$ is a two-sided ideal and one obtains a C^*-norm on $F(X)/\mathcal{I}$. The C^*-algebra $\mathcal{C}(X)$ is the completion of $F(X)/\mathcal{I}$ with respect to this norm. The C^*-algebra $\mathcal{C}(X)$ has the universal property that any $*$-morphism from $A(X)$ to a separable C^*-algebra factors through $\mathcal{C}(X)$. In particular, any $*$-representation of $A(X)$ extends to a representation of $\mathcal{C}(X)$.

The word instanton in the title refers to the fact that all (in particular even) spheres come equipped with a projection $e \in \mathrm{Mat}_r(A(X))$, $e^2 = e = e^*$, for $X = S_\theta^N$ and $X = S_q^N$. These projection determines the module of sections of a vector bundle which deforms the usual monopole bundle and instanton bundle in two and four dimensions respectively, and generalizes them in all dimensions.

In particular on the four dimensional S_θ, one can develop Yang-Mills theory, since there are all the required structures, namely the algebra, the calculus and the "vector bundle" e (naturally endowed, in addition, with a preferred connection ∇). Among other things there is a basic inequality showing that the Yang-Mills action, $YM(\nabla) = \!\!\int \theta^2 \, ds^4$, (where $\theta = \nabla^2$ is the curvature, and $ds = D^{-1}$) has a strictly positive lower bound given by the topological invariant $\varphi(e) = \!\!\int \gamma(e - \frac{1}{2})[D, e]^4 \, ds^4$ which, for the canonical projections turns out to be just 1: $\varphi(e) = 1$.

In general, the projection e for the spheres S_θ^{2n} satisfies self-duality equations

$$*_H \, e(de)^n = \mathrm{i}^n e(de)^n \,, \tag{1}$$

with a suitably defined Hodge operator $*_H$ [15] (see also [1] and [42]). An important problem is the construction and the classification of Yang-Mills connections in the noncommutative situation along the line of the ADHM construction [3]. This was done in [19] for the noncommutative torus and in [50] for a noncommutative $\mathbb{R}^4$.

It is not yet clear if a construction of gauge theories along similar lines can be done for the quantum spheres $X = S_q^N$.

There has been recently an explosion of work on deformed spheres from many points of view. The best I can do here is to refer to [22] for an overview of noncommutative and quantum spheres in dimensions up to four. In [60] there is a family of noncommutative 4-spheres which satisfy the Chern character conditions of [17] up to cohomology classes (and not just representatives). Additional quantum 4-dimensional spheres together with a construction of quantum instantons on them is in [32]. A different class of spheres in any even dimension was proposed in [4]. At this workshop T. Natsume presented an example in two dimensions [49].

2 Instanton Algebras

In this section we shall describe how to obtain in a natural way noncommutative spaces (i.e. algebras) out of the Chern characters of idempotents and unitaries in cyclic homology. For this we shall give a brief overview of the needed fundamentals of the theory, following [5]. For later use we shall also describe the dual cohomological theories.

2.1 Hochschild and Cyclic Homology and Cohomology

Given an algebra $\mathcal{A}$, consider the chain complex $(C_*(\mathcal{A}) = \bigoplus_n C_n(\mathcal{A}), b)$ with $C_n(\mathcal{A}) = \mathcal{A}^{\otimes(n+1)}$ and the boundary map b defined by

$$b : C_n(\mathcal{A}) \to C_{n-1}(\mathcal{A}) ,$$

$$b(a_0 \otimes a_1 \otimes \cdots \otimes a_n) := \sum_{j=0}^{n-1}(-1)^j a_0 \otimes \cdots \otimes a_j a_{j+1} \otimes \cdots \otimes a_n$$

$$+ (-1)^n a_n a_0 \otimes a_1 \otimes \cdots \otimes a_{n-1} . \quad (2)$$

It is easy to prove that $b^2 = 0$. The *Hochschild homology* $HH_*(\mathcal{A})$ of the algebra $\mathcal{A}$ is the homology of this complex,

$$HH_n(\mathcal{A}) := H_n(C_*(\mathcal{A}), b) = Z_n/B_n , \quad (3)$$

with the cycles given by $Z_n := \ker(b : C_n(\mathcal{A}) \to C_{n-1}(\mathcal{A}))$ and the boundary $B_n := \mathrm{im}(b : C_{n+1}(\mathcal{A}) \to C_n(\mathcal{A}))$. We have another operator which increases the degree

$$B : C_n(\mathcal{A}) \to C_{n+1}(\mathcal{A}) , \qquad B = B_0 A , \quad (4)$$

where

$$B_0(a_0 \otimes a_1 \otimes \cdots \otimes a_n) := \mathbb{I} \otimes a_0 \otimes a_1 \otimes \cdots \otimes a_n \quad (5)$$

$$A(a_0 \otimes a_1 \otimes \cdots \otimes a_n) := \frac{1}{n+1} \sum_{j=0}^{n}(-1)^{nj} a_j \otimes a_{j+1} \otimes \cdots \otimes a_{j-1} , \quad (6)$$

with the obvious cyclic identification $n + 1 = 0$. Again it is straightforward to check that $B^2 = 0$ and that $bB + Bb = 0$.

By putting together these two operators, one gets a bi-complex $(C_*(\mathcal{A}), b, B)$ with $C_{p-q}(\mathcal{A})$ in bi-degree p, q. The *cyclic homology* $HC_*(\mathcal{A})$ of the algebra $\mathcal{A}$ is the homology of the total complex $(CC(\mathcal{A}), b + B)$, whose n-th term is given by $CC_n(\mathcal{A}) := \bigoplus_{p+q=n} C_{p-q}(\mathcal{A}) = \bigoplus_{0 \le q \le [n/2]} C_{2n-q}(\mathcal{A})$. This bi-complex may be best organized in a plane diagram whose vertical arrows are associated with the operator b and whose horizontal ones are associated with the operator B,

$$\vdots \qquad\qquad \vdots$$

$$b\downarrow \qquad\qquad b\downarrow \qquad\qquad b\downarrow$$

$$C_2(\mathcal{A}) \xleftarrow{B} C_1(\mathcal{A}) \xleftarrow{B} C_0(\mathcal{A})$$

$$b\downarrow \qquad\qquad b\downarrow \qquad\qquad\qquad (7)$$

$$C_1(\mathcal{A}) \xleftarrow{B} C_0(\mathcal{A})$$

$$b\downarrow$$

$$C_0(\mathcal{A})$$

The n-th term $CC_n(\mathcal{A})$ of the total complex is just the n-th (NW – SE) diagonal in the diagram (7). Then,

$$HC_n(\mathcal{A}) := H_n(CC(\mathcal{A}), b + B) = Z_n^\lambda / B_n^\lambda , \qquad (8)$$

with the *cyclic cycles* given by $Z_n^\lambda := \ker(b + B : CC_n(\mathcal{A}) \to CC_{n-1}(\mathcal{A}))$ and the *cyclic boundaries* given by $B_n^\lambda := \operatorname{im}(b + B : CC_{n+1}(\mathcal{A}) \to CC_n(\mathcal{A}))$.

Example 1. If M is a compact manifold, the Hochschild homology of the algebra of smooth functions $C^\infty(M)$ gives the de Rham complex (Hochschild-Konstant-Rosenberg theorem),

$$\Omega_{dR}^k(M) \simeq HH_k(C^\infty(M)) , \qquad (9)$$

with $\Omega_{dR}^k(M)$ the space of de Rham forms of order k on M. If d denotes the de Rham exterior differential, this isomorphisms is implemented by

$$a_0 da_1 \wedge \cdots \wedge da_k \mapsto \varepsilon_k(a_0 \otimes a_1 \otimes \cdots \otimes da_k) \qquad (10)$$

where ε_k is the *antisymmetrization map*

$$\varepsilon_k(a_0 \otimes a_1 \otimes \cdots \otimes da_k) := \sum_{\sigma \in S_k} sign(\sigma)(a_0 \otimes a_{\sigma(1)} \otimes \cdots \otimes da_{\sigma(k)}) \qquad (11)$$

and S_k is the symmetric group of order k. In particular one checks that $b \circ \varepsilon_k = 0$. The de Rham differential d corresponds to the operator B_* (the lift of B to homology) in the sense that

$$\varepsilon_{k+1} \circ d = (k+1)B_* \circ \varepsilon_k . \qquad (12)$$

On the other hand, the cyclic homology gives [9, 45]

$$HC_k(C^\infty(M)) = \Omega^k_{dR}(M)/d\Omega^{k-1}_{dR}(M) \oplus H^{k-2}_{dR}(M) \oplus H^{k-4}_{dR}(M) \oplus \cdots , \quad (13)$$

where $H^j_{dR}(M)$ is the j-th de Rham cohomology group. The last term in the sum is $H^0_{dR}(M)$ or $H^1_{dR}(M)$ according to wether k is even or odd.

From the fact that $C^\infty(M)$ is commutative it follows that there is a natural decomposition (the λ-decomposition) of cyclic homology in smaller pieces,

$$HC_0(C^\infty(M)) = HC^{(0)}_0(C^\infty(M)) ,$$
$$HC_k(C^\infty(M)) = HC^{(k)}_k(C^\infty(M)) \cdots \oplus HC^{(1)}_k(C^\infty(M)) , \quad (14)$$

which is obtained by suitable idempotents $e^{(i)}_k$ which commute with the operator B: $Be^{(i)}_k = e^{(i+1)}_{k+1}B$. The previous decomposition corresponds to the decomposition in (13) and give a way to extract the de Rham cohomology

$$HC^{(k)}_k(C^\infty(M)) = \Omega^k_{dR}(M)/d\Omega^{k-1}_{dR}(M) ,$$
$$HC^{(i)}_k(C^\infty(M)) = H^{2i-k}_{dR}(M) , \qquad \text{for} \quad [n/2] \le i < n , \quad (15)$$
$$HC^{(i)}_k(C^\infty(M)) = 0 , \qquad \text{for} \quad i < [n/2] ,$$

Looking at this example, one may think of cyclic homology as a generalization of de Rham cohomology to the noncommutative setting.

A Hochschild k cochain on the algebra $\mathcal{A}$ is an $(n+1)$-linear functional on $\mathcal{A}$ or a linear form on $\mathcal{A}^{\otimes(n+1)}$. Let $C^n(\mathcal{A}) = \text{Hom}(\mathcal{A}^{\otimes(n+1)}, \mathbb{C})$ be the collection of such cochains. We have a cochain complex $(C^*(\mathcal{A}) = \bigoplus_n C^n(\mathcal{A}), b)$ with a coboundary map, again denoted with the symbol b, defined by

$$b : C^n(\mathcal{A}) \to C^{n+1}(\mathcal{A}) ,$$
$$b\varphi(a_0, a_1, \cdots , a_{n+1}) := \sum_{j=0}^{n} (-1)^j \varphi(a_0, \cdots , a_j a_{j+1}, \cdots , a_{n+1})$$
$$+ (-1)^{n+1}\varphi(a_{n+1}a_0 a_1, \cdots , a_n) . \quad (16)$$

Clearly $b^2 = 0$ and the *Hochschild cohomology* $HH^*(\mathcal{A})$ of the algebra $\mathcal{A}$ is the cohomology of this complex,

$$HH^n(\mathcal{A}) := H^n(C^*(\mathcal{A}), b) = Z^n/B^n , \quad (17)$$

with the cocycles given by $Z^n := \ker(b : C^n(\mathcal{A}) \to C^{n+1}(\mathcal{A}))$ and the coboundaries given by $B^n := \text{im}(b : C^{n-1}(\mathcal{A}) \to C^n(\mathcal{A}))$.

A Hochschild 0-cocycle τ on the algebra $\mathcal{A}$ is a *trace*, since $\tau \in \text{Hom}(\mathcal{A}, \mathbb{C})$ and the cocycle condition is

$$\tau(a_0 a_1) - \tau(a_1 a_0) = b\tau(a_0, a_1) = 0 . \quad (18)$$

The trace property is extended to higher orders by saying that an n-cochain φ is *cyclic* if $\lambda\varphi = \varphi$, with

$$\lambda\varphi(a_0, a_1, \cdots, a_n) = (-1)^n \varphi(a_n, a_0, \cdots, a_{n-1}) \ . \tag{19}$$

A *cyclic cocycle* is a cyclic cochain φ for which $b\varphi = 0$.

A straightforward computation shows that the sets of cyclic n-cochains $C_\lambda^n(\mathcal{A}) = \{\varphi \in C^n(\mathcal{A}) \ , \ \lambda\varphi = \varphi\}$ are preserved by the Hochschild boundary operator: $(1 - \lambda)\varphi = 0$ implies that $(1 - \lambda)b\varphi = 0$. Thus we get a subcomplex $(C_\lambda^*(\mathcal{A}) = \bigoplus_n C_\lambda^n(\mathcal{A}), b)$ of the complex $(C^*(\mathcal{A}) = \bigoplus_n C^n(\mathcal{A}), b)$. The *cyclic cohomology* $HC^*(\mathcal{A})$ of the algebra $\mathcal{A}$ is the cohomology of this subcomplex,

$$HC^n(\mathcal{A}) := H^n(C_\lambda^*(\mathcal{A}), b) = Z_\lambda^n / B_\lambda^n \ , \tag{20}$$

with the cyclic cocycles given by $Z_\lambda^n := \ker(b : C_\lambda^n(\mathcal{A}) \to C_\lambda^{n+1}(\mathcal{A}))$ and the cyclic coboundaries given by $B^n := \operatorname{im}(b : C_\lambda^{n-1}(\mathcal{A}) \to C_\lambda^n(\mathcal{A}))$.

One can also define an operator B which is dual to the one in (4) for the homology and give a bicomplex description of cyclic cohomology by giving a diagram dual to the one in (7) with all arrows inverted and all indices "up". Since we shall not need this description later on, we only refer to [10] for all details. We mention an additional important operator, the *periodicity operator* S which is a map of degree 2 between cyclic cocycle,

$$S : Z_\lambda^{n-1} \longrightarrow Z_\lambda^{n+1} \ , \tag{21}$$

$$S\varphi(a_0, a_1, \cdots, a_{n+1}) := -\frac{1}{n(n+1)} \sum_{j=1}^{n} \varphi(a_0, \cdots, a_{j-1}a_j a_{j+1}, \cdots, a_{n+1})$$

$$- \frac{1}{n(n+1)} \sum_{1 \le i < j \le n} (-1)^{i+j} \varphi(a_0, \cdots, a_{i-1}a_i, \cdots, a_j a_{j+1}, \cdots, a_{n+1}) \ .$$

One shows that $S(Z_\lambda^{n-1}) \subseteq Z_\lambda^{n+1}$. In fact $S(Z_\lambda^{n-1}) \subseteq B^{n+1}$, the latter being the Hochschild coboundary; and cyclicity is easy to show.

The induced morphisms in cohomology $S : HC^n \to HC^{n+2}$ define two directed systems of abelian groups. Their inductive limits

$$HP^0(\mathcal{A}) := \varinjlim HC^{2n}(\mathcal{A}) \ , \quad HP^1(\mathcal{A}) := \varinjlim HC^{2n+1}(\mathcal{A}) \ , \tag{22}$$

form a $\mathbb{Z}_2$-graded group which is called the *periodic cyclic cohomology* $HP^*(\mathcal{A})$ of the algebra $\mathcal{A}$.

"Il va sans dire": there is also a *periodic cyclic homology* [10, 45].

2.2 Noncommutative Algebras from Idempotents

Let $\mathcal{A}$ be an algebra (over $\mathbb{C}$) and let $e \in \operatorname{Mat}_r(\mathcal{A})$, $e^2 = e$, be an idempotent. Its *even* (reduced) Chern character is a formal sum of chains

$$\operatorname{ch}_*(e) = \sum_k \operatorname{ch}_k(e) \ , \tag{23}$$

with the component $\mathrm{ch}_k(e)$ an element of $\mathcal{A} \otimes \overline{\mathcal{A}}^{\otimes 2k}$, where $\overline{\mathcal{A}} = \mathcal{A}/\mathbb{C}1$ is the quotient of $\mathcal{A}$ by the scalar multiples of the unit 1. The formula for $\mathrm{ch}_k(e)$ is (with λ_k a normalization constant),

$$\mathrm{ch}_k(e) = \left\langle \left(e - \frac{1}{2}\mathbb{I}_r\right) \otimes e \otimes e \cdots \otimes e \right\rangle$$

$$= \lambda_k \sum \left(e_{i_0 i_1} - \frac{1}{2}\delta_{i_0 i_1}\right) \otimes \widetilde{e}_{i_1 i_2} \otimes \widetilde{e}_{i_2 i_3} \cdots \otimes \widetilde{e}_{i_{2k} i_0} \qquad (24)$$

where δ_{ij} is the usual Kronecker symbol and only the class $\widetilde{e}_{i_j i_{j+1}} \in \overline{\mathcal{A}}$ is used in the formula. The crucial property of the character $\mathrm{ch}_*(e)$ is that it defines a cycle [9, 10, 13, 45] in the reduced (b, B)-bicomplex of cyclic homology described above,

$$(b + B)\,\mathrm{ch}_*(e) = 0 , \qquad B\,\mathrm{ch}_k(e) = b\,\mathrm{ch}_{k+1}(e) . \qquad (25)$$

It turns out that the map $e \mapsto \mathrm{ch}_*(e)$ leads to a well defined map from the K theory group $K_0(\mathcal{A})$ to cyclic homology of $\mathcal{A}$ (in fact the correct receptacle is period cyclic homology [45]). In Sect. 6 below, we shall construct some interesting examples of this Chern character on quantum spheres. For the remaining part of this Section we shall use it to define some "even" dimensional noncommutative algebras (including spheres).

For any pair of integers m, r we shall construct a universal algebra $\mathcal{A}_{m,r}$ as follows. We let $\mathcal{A}_{m,r}$ be generated by the r^2 elements e_{ij}, $i, j \in \{1, \ldots, r\}$, $e = [e_{ij}]$ on which we first impose the relations stating that e is an idempotent

$$e^2 = e . \qquad (26)$$

We impose additional relations by requiring the vanishing of all "lower degree" components of the Chern character of e,

$$\mathrm{ch}_k(e) = 0 , \qquad \forall k < m . \qquad (27)$$

Then, an admissible morphism from $\mathcal{A}_{m,r}$ to an arbitrary algebra $\mathcal{B}$,

$$\rho : \mathcal{A}_{m,r} \to \mathcal{B} , \qquad (28)$$

is given by the $\rho(e_{ij}) \in \mathcal{B}$ which fulfill $\rho(e)^2 = \rho(e)$, and

$$\mathrm{ch}_k(\rho(e)) = 0 , \qquad \forall k < m . \qquad (29)$$

We define the algebra $\mathcal{A}_{m,r}$ as the quotient of the algebra defined by (26) by the intersection of kernels of the admissible morphisms ρ. Elements of the algebra $\mathcal{A}_{m,r}$ can be represented as polynomials in the generators e_{ij} and to prove that such a polynomial $P(e_{ij})$ is non zero in $\mathcal{A}_{m,r}$ one must construct a solution to the above equations for which $P(e_{ij}) \neq 0$.

To get a C^*-algebra we endow $\mathcal{A}_{m,r}$ with the involution given by,

$$(e_{ij})^* = e_{ji} \tag{30}$$

which means that $e = e^*$ in $\mathrm{Mat}_r(\mathcal{A})$, i.e. e is a projection in $\mathrm{Mat}_r(\mathcal{A})$ (or equivalently, a self-adjoint idempotent). We define a norm by,

$$\|P\| = \mathrm{Sup} \, \|(\pi(P))\| \tag{31}$$

where π ranges through all representations of the above equations on Hilbert spaces. Such a π is given by a Hilbert space $\mathcal{H}$ and a self-adjoint idempotent,

$$E \in \mathrm{Mat}_r(\mathcal{L}(\mathcal{H})), \;\; E^2 = E, \;\; E = E^* \tag{32}$$

such that (29) holds for $\mathcal{B} = \mathcal{L}(\mathcal{H})$. For any polynomial $P(e_{ij})$ the quantity (31), i.e. the supremum of the norms, $\|P(E_{ij})\|$ is finite.

We let $A_{m,r}$ be the universal C*-algebra obtained as the completion of $\mathcal{A}_{m,r}$ for the above norm.

2.3 Noncommutative Algebras from Unitaries

In the odd case, more than projections one rather needs unitary elements and the formulæ for the *odd* (reduced) Chern character in cyclic homology are similar to those above. The Chern character of a unitary $u \in \mathrm{Mat}_r(\mathcal{A})$ is a formal sum of chains

$$\mathrm{ch}_*(u) = \sum_k \mathrm{ch}_k(u) \,, \tag{33}$$

with the component $\mathrm{ch}_{n+\frac{1}{2}}(u)$ as element of $\mathcal{A} \otimes \overline{\mathcal{A}}^{\otimes(2n-1)}$ given by

$$\mathrm{ch}_{k+\frac{1}{2}}(u) = \lambda_k \left(u_{i_1}^{i_0} \otimes (u^*)_{i_2}^{i_1} \otimes u_{i_3}^{i_2} \otimes \cdots \otimes (u^*)_{i_0}^{i_{2k+1}} \right.$$
$$\left. - (u^*)_{i_1}^{i_0} \otimes u_{i_2}^{i_1} \otimes (u^*)_{i_3}^{i_2} \otimes \cdots \otimes u_{i_0}^{i_{2k+1}} \right) \,, \tag{34}$$

and λ_k suitable normalization constants. Again $\mathrm{ch}_*(u)$ defines a cycle in the reduced (b, B)-bicomplex of cyclic homology [9, 10, 13, 45],

$$(b+B)\,\mathrm{ch}_*(u) = 0 \,, \;\;\;\; B\,\mathrm{ch}_{k+\frac{1}{2}}(e) = b\,\mathrm{ch}_{k+\frac{1}{2}+1}(e) \,, \tag{35}$$

and the map $u \mapsto \mathrm{ch}_*(u)$ leads to a well defined map from the K theory group $K_1(\mathcal{A})$ to (in fact periodic) cyclic homology.

For any pair of integers m, r we can define $\mathcal{B}_{m,r}$ to be the universal algebra generated by the r^2 elements u_{ij}, $i, j \in \{1, \ldots, r\}$, $u = [u_{ij}]$ and we impose as above the relations

$$\mathrm{ch}_{k+\frac{1}{2}}(\rho(u)) = 0 \;\;\;\; \forall k < m \,. \tag{36}$$

To get a C*-algebra we endow $\mathcal{B}_{m,r}$ with the involution given by,

$$u\,u^* = u^*\,u = 1\,, \tag{37}$$

which means that u is a unitary in $\mathrm{Mat}_r(\mathcal{A})$. As before, we define a norm by,

$$\|P\| = \mathrm{Sup}\,\|(\pi(P))\| \tag{38}$$

where π ranges through all representations of the above equations in Hilbert space.

We let $B_{m,r}$ be the universal C*-algebra obtained as the completion of $\mathcal{B}_{m,r}$ for the above norm.

3 Fredholm Modules and Spectral Triples

As we have mentioned in Sect. 2, the Chern characters $\mathrm{ch}_*(x)$ leads to well defined maps from the K theory groups $K_*(\mathcal{A})$ to (period) cyclic homology. The dual Chern characters, ch^*, of even and odd Fredholm modules provides similar maps to (period) cyclic cohomology.

3.1 Fredholm Modules and Index Theorems

A Fredholm module can be thought of as an abstract elliptic operator. The full fledged theory started with Atiyah and culminated in the KK-theory of Kasparov and the cyclic cohomology of Connes. Here we shall only mention the few facts that we shall need later on.

Let $\mathcal{A}$ be an algebra with involution. An odd Fredholm module [9] over $\mathcal{A}$ consists of

(1) a representation ψ of the algebra $\mathcal{A}$ on an Hilbert space $\mathcal{H}$;
(2) an operator F on $\mathcal{H}$ such that

$$F^2 = \mathbb{I}\,, \quad F^* = F\,,$$
$$[F, \psi(a)] \in \mathcal{K}\,, \quad \forall\,a \in \mathcal{A}\,, \tag{39}$$

where $\mathcal{K}$ are the compact operators on $\mathcal{H}$.

An even Fredholm module has also a $\mathbb{Z}_2$-grading γ of $\mathcal{H}$, $\gamma^* = \gamma$, $\gamma^2 = \mathbb{I}$, such that

$$F\gamma + \gamma F = 0\,,$$
$$\psi(a)\gamma - \gamma\psi(a) = 0\,, \quad \forall\,a \in \mathcal{A}\,. \tag{40}$$

In fact, often the first of conditions (39) needs to be weakened somehow to $F^2 - \mathbb{I} \in \mathcal{K}$.

With an even module we shall indicate with $\mathcal{H}^\pm$ and $\psi^\pm$ the component of the Hilbert space and of the representation with respect to the grading.

Given any positive integer r, one can extend the previous modules to a Fredholm module $(\mathcal{H}_r, F_r)$ over the algebra $\mathrm{Mat}_r(\mathcal{A}) = \mathcal{A} \otimes \mathrm{Mat}_r(\mathbb{C})$ by a simple procedure

$$\mathcal{H}_r = \mathcal{H} \otimes \mathbb{C}^r , \quad \psi_r = \psi \otimes \mathrm{id} , \quad \mathbb{F}_r = F \otimes \mathbb{I}_r , \tag{41}$$

and $\gamma_r = \gamma \otimes \mathbb{I}_r$, for an even module.

The importance of Fredholm modules is testified by the following theorem which can be associated with the names of Atiyah and Kasparov [2, 38],

Theorem 1.
a) *Let $(\mathcal{H}, F, \gamma)$ be an even Fredholm module over the algebra $\mathcal{A}$. And let $e \in \mathrm{Mat}_r(\mathcal{A})$ be a projection $e^2 = e = e^*$. Then we have a Fredholm operator*

$$\psi_r^-(e) F_r \psi_r^+(e) : \psi_r^+(e)\mathcal{H}_r \to \psi_r^-(e)\mathcal{H}_r , \tag{42}$$

whose index depends only on the class of the projection e in the K-theory of $\mathcal{A}$. Thus we get an additive map

$$\varphi : K_0(\mathcal{A}) \to \mathbb{Z} ,$$
$$\varphi([e]) = \mathrm{Index}\left(\psi_r^-(e) F_r \psi_r^+(e)\right) . \tag{43}$$

b) *Let $(\mathcal{H}, F)$ be an odd Fredholm module over the algebra $\mathcal{A}$, and take the projection $E = \frac{1}{2}(\mathbb{I} + F)$. Let $u \in \mathrm{Mat}_r(\mathcal{A})$ be unitary $uu^* = u^*u = \mathbb{I}$. Then we have a Fredholm operator*

$$E_r \psi_r(u) E_r : E_r \mathcal{H}_r \to E_r \mathcal{H}_r , \tag{44}$$

whose index depends only on the class of the unitary u in the K-theory of $\mathcal{A}$. Thus we get an additive map

$$\varphi : K_1(\mathcal{A}) \to \mathbb{Z} ,$$
$$\varphi([u]) = \mathrm{Index}\left(E_r \psi_r(u) E_r\right) . \tag{45}$$

If $\mathcal{A}$ is a C^*-algebra, both in the even and the odd cases, the index map φ only depends on the K-*homology* class

$$[(\mathcal{H}, F)] \in KK(\mathcal{A}, \mathbb{C}) , \tag{46}$$

of the Fredholm module in the Kasparov KK group, $K^*(\mathcal{A}) = KK(\mathcal{A}, \mathbb{C})$, which is the abelian group of stable homotopy classes of Fredholm modules over $\mathcal{A}$ [38].

Both in the even and odd cases, the index pairings (43) and (45) can be given as [10]

$$\varphi(x) = \langle \mathrm{ch}^*(\mathcal{H}, F), \mathrm{ch}_*(x) \rangle , \quad x \in K_*(\mathcal{A}) , \tag{47}$$

via the Chern characters

$$\mathrm{ch}^*(\mathcal{H}, F) \in HC^*(\mathcal{A}), \quad \mathrm{ch}_*(x) \in HC_*(\mathcal{A}), \tag{48}$$

and the pairing between cyclic cohomology $HC^*(\mathcal{A})$ and cyclic homology $HC_*(\mathcal{A})$ of the algebra $\mathcal{A}$.

The Chern character $\mathrm{ch}_*(x)$ in homology is given by (23) and (33) in the even and odd case respectively. As for the Chern character $\mathrm{ch}^*(x)$ in cohomology we shall give some fundamentals in the next Section.

3.2 The Chern Characters of Fredholm Modules

For the general theory we refer to [10]. In Sect. 6 we shall construct some interesting examples of these Chern characters on quantum spheres. Additional examples have been constructed in [36].

We recall [58] that on a Hilbert space $\mathcal{H}$ and with $\mathcal{K}$ denoting the compact operators one defines, for $p \in [1, \infty[$, the Schatten p-class, $\mathcal{L}^p$, as the ideal of compact operators for which $\mathrm{Tr}\, T$ is finite: $\mathcal{L}^p = \{T \in \mathcal{K} \ : \ \mathrm{Tr}\, T < \infty\}$. Then, the Hölder inequality states that $\mathcal{L}^{p_1} \mathcal{L}^{p_2} \cdots \mathcal{L}^{p_k} \subset \mathcal{L}^p$, with $p^{-1} = \sum_{j=1}^{k} p_j^{-1}$.

Let now $(\mathcal{H}, F)$ be Fredholm module (even or odd) over the algebra $\mathcal{A}$. We say that $(\mathcal{H}, F)$ is p-summable if

$$[F, \psi(a)] \in \mathcal{L}^p, \quad \forall\, a \in \mathcal{A}. \tag{49}$$

For simplicity, in the rest of this section, we shall drop the symbol ψ which indicates the representation on $\mathcal{A}$ on $\mathcal{H}$. The idea is then to construct "quantized differential forms" and integrate (via a trace) forms of degree higher enough so that they belong to $\mathcal{L}^1$. In fact, one need to introduce a conditional trace. Given an operator T on $\mathcal{H}$ such that $FT + TF \in \mathcal{L}^1$, one defines

$$\mathrm{Tr}'\, T := \frac{1}{2} \mathrm{Tr}\, F(FT + TF); \tag{50}$$

note that, if $T \in \mathcal{L}^1$ then $\mathrm{Tr}\, T = \mathrm{Tr}'\, T$ by cyclicity of the trace.

Let now n be a nonnegative integer and let $(\mathcal{H}, F)$ be Fredholm module over the algebra $\mathcal{A}$. We take this module to be *even* or *odd* according to whether n is even or odd; and we shall also take it to be $(n + 1)$-summable. We shall construct a so called n-dimensional *cycle* $(\Omega^* = \oplus_k \Omega^k, d, \int)$ over the algebra $\mathcal{A}$. Elements of Ω^k are *quantized differential forms*: $\Omega^0 = \mathcal{A}$ and for $k > 0$, Ω^k is the linear span of operators of the form

$$\omega = a_0[F, a_1] \cdots [F, a_n], \quad a_j \in \mathcal{A}. \tag{51}$$

By the assumption of summability, Hölder inequality gives that $\Omega^k \subset \mathcal{L}^{\frac{n+1}{k}}$. The product in Ω^* is just the product of operators $\omega\omega' \in \Omega^{k+k'}$ for any $\omega \in \Omega^k$ and $\omega' \in \Omega^{k'}$. The differential $d : \Omega^k \to \Omega^{k+1}$ is defined by

$$d\omega = F\omega - (-1)^k \omega F, \quad \omega \in \Omega^k, \tag{52}$$

and $F^2 = 1$ implies both $d^2 = 0$ and the fact that d is a graded derivation

$$d(\omega\omega') = (d\omega)\omega' + (-1)^k \omega d\omega' , \quad \omega \in \Omega^k , \quad \omega' \in \Omega^{k'} . \tag{53}$$

Finally, one defines a trace in degree n by,

$$\int : \Omega^n \to \mathbb{C} , \tag{54}$$

which is both closed ($\int d\omega = 0$) and graded ($\int \omega\omega' = (-1)^{kk'} \int \omega'\omega$).

Let us first consider the case n is odd. With $\omega \in \Omega^n$ one defines

$$\int \omega := \mathrm{Tr}' \, \omega = \frac{1}{2} \, \mathrm{Tr} \, F(F\omega + \omega F)) = \frac{1}{2} \, \mathrm{Tr} \, F d\omega , \tag{55}$$

which is well defined since $F d\omega \in \mathcal{L}^1$.

If n is even and γ is the grading, with $\omega \in \Omega^n$ one defines

$$\int \omega := \mathrm{Tr}' \, \gamma\omega = \frac{1}{2} \, \mathrm{Tr} \, F(F\gamma\omega + \gamma\omega F)) = \frac{1}{2} \, \mathrm{Tr} \, \gamma F d\omega , \tag{56}$$

(remember that $F\gamma = -\gamma F$); this is again well defined since $\gamma F d\omega \in \mathcal{L}^1$. One straightforwardly proves closeness and graded cyclicity of both the integrals (55) and (56).

The *character* of the Fredholm module is the cyclic cocycle $\tau^n \in Z_\lambda^n(\mathcal{A})$ given by,

$$\tau^n(a_0, a_1, \ldots, a_n) := \int a_0 da_1 \cdots da_n , \quad a_j \in \mathcal{A} ; \tag{57}$$

explicitly,

$$\tau^n(a_0, a_1, \ldots, a_n) = \mathrm{Tr}' \, a_0[F, a_1], \ldots, [F, a_n] , \quad n \text{ odd} , \tag{58}$$

$$\tau^n(a_0, a_1, \ldots, a_n) = \mathrm{Tr}' \, \gamma \, a_0[F, a_1], \ldots, [F, a_n] , \quad n \text{ even} . \tag{59}$$

In both cases one checks closure, $b\tau^n = 0$, and cyclicity, $\lambda\tau^n = (-1)^n \tau^n$.

We see that there is ambiguity in the choice of the integer n. Given a Fredholm module $(\mathcal{H}, F)$ over $\mathcal{A}$, the parity of n is fixed by for its precise value there is only a lower bound determined by the $(n+1)$-summability. Indeed, since $\mathcal{L}^{p_1} \subset \mathcal{L}^{p_2}$ if $p_1 \leq p_2$, one could replace n by $n+2k$ with k any integer. Thus one gets a sequence of cyclic cocycle $\tau^{n+2k} \in Z_\lambda^{n+2k}(\mathcal{A}), k \geq 0$, with the same parity. The crucial fact is that the cyclic cohomology classes of these cocycles are related by the periodicity operator S in (21). The characters τ^{n+2k} satisfy

$$S[\tau^m]_\lambda = c_m[\tau^{m+2}]_\lambda , \quad \text{in } HC^{m+2}(\mathcal{A}) , \quad m = n + 2k , \ k \geq 0 , \tag{60}$$

with c_m a constant depending on m (one could get rid of these constants by suitably normalizing the characters in (58) and (59)). Therefore, the sequence $\{\tau^{n+2k}\}_{k \geq 0}$ determine a well defined class $[\tau^F]$ in the periodic cyclic cohomology $HP^0(\mathcal{A})$ or $HP^1(\mathcal{A})$ according to whether n is even or odd. The class $[\tau^F]$ is the Chern character of the Fredholm module $(\mathcal{A}, \mathcal{H}, F)$ in periodic cyclic cohomology.

3.3 Spectral Triples and Index Theorems

As already mentioned, a noncommutative geometry is described by a spectral triple [10]

$$(\mathcal{A}, \mathcal{H}, D) . \tag{61}$$

Here $\mathcal{A}$ is an algebra with involution, together with a representation ψ of $\mathcal{A}$ as bounded operators on a Hilbert space $\mathcal{H}$ as bounded operators, and D is a self-adjoint operator with compact resolvent and such that,

$$[D, \psi(a)] \text{ is bounded } \forall\, a \in \mathcal{A} . \tag{62}$$

An even spectral triple has also a $\mathbb{Z}_2$-grading γ of $\mathcal{H}$, $\gamma^* = \gamma$, $\gamma^2 = \mathbb{I}$, with the additional properties,

$$D\gamma + \gamma D = 0 ,$$
$$\psi(a)\gamma - \gamma\psi(a) = 0 , \quad \forall\, a \in \mathcal{A} . \tag{63}$$

Given a spectral triple there is associated a fredholm module with the operator F just given by the sign of D, $F = D|D|^{-1}$ (if the kernel of D in not trivial one can still adjust things and define such an F).

The operator D plays in general the role of the Dirac operator [44] in ordinary Riemannian geometry. It specifies both the K-homology fundamental class (cf. [10]), as well as the metric on the state space of $\mathcal{A}$ by

$$d(\varphi, \psi) = \operatorname{Sup} \{|\varphi(a) - \psi(a)|; \|[D, a]\| \le 1\} . \tag{64}$$

What holds things together in this spectral point of view on noncommutative geometry is the nontriviality of the pairing between the K-theory of the algebra $\mathcal{A}$ and the K-homology class of D. There are index maps as with Fredholm modules above,

$$\varphi : K_*(\mathcal{A}) \to \mathbb{Z} \tag{65}$$

and the maps φ given by expressions like (43) and (45) with the operator D replacing the operator F there.

An operator theoretic index formula [10], [18], [34] expresses the above index pairing (65) by explicit *local* cyclic cocycles on the algebra $\mathcal{A}$. These local formulas become extremely simple in the special case where only the top component of the Chern character $\mathrm{ch}_*(e)$ in cyclic homology fails to vanish. This is easy to understand in the analogous simpler case of ordinary manifolds since the Atiyah-Singer index formula gives the integral of the product of the Chern character $\mathrm{ch}(E)$, of the bundle E over the manifold M, by the index class; if the only component of $\mathrm{ch}(E)$ is ch_n, $n = \frac{1}{2} \dim M$ only the 0-dimensional component of the index class is involved in the index formula.

For instance, in the even case, provided the components $\mathrm{ch}_k(e)$ all vanish for $k < n$ the index formula reduces to the following,

$$\varphi(e) = (-1)^n \int \gamma \left(e - \frac{1}{2} \right) [D, e]^{2n} D^{-2n} . \tag{66}$$

Here, e is a projection $e^2 = e = e^*$, γ is the $\mathbb{Z}/2$ grading of $\mathcal{H}$ as above, the resolvent of D is of order $\frac{1}{2n}$ (i.e. its characteristic values μ_k are $0(k^{-\frac{1}{2n}})$) and f is the coefficient of the logarithmic divergency in the ordinary operator trace [27] [65]. There is a similar formula for the odd case.

Example 2. The Canonical Triple over a Manifold
The basic example of spectral triple is the *canonical triple* on a closed n-dimensional Riemannian spin manifold (M, g). A spin manifold is a manifold on which it is possible to construct principal bundles having the groups $Spin(n)$ as structure groups. A manifold admits a spin structure if and only if its second Stiefel-Whitney class vanishes [44].

The canonical spectral triple $(\mathcal{A}, \mathcal{H}, D)$ over the manifold M is as follows:

(1) $\mathcal{A} = C^\infty(M)$ is the algebra of complex valued smooth functions on M.
(2) $\mathcal{H} = L^2(M, \mathcal{S})$ is the Hilbert space of square integrable sections of the irreducible, rank $2^{[n/2]}$, spinor bundle over M; its elements are spinor fields over M. The scalar product in $L^2(M, \mathcal{S})$ is the usual one of the measure $d\mu(g)$ of the metric g, $(\psi, \phi) = \int d\mu(g) \bar{\psi}(x) \cdot \phi(x)$, with the pointwise scalar product in the spinor space being the natural one in $\mathbb{C}^{2^{[n/2]}}$.
(3) D is the Dirac operator of the Levi-Civita connection of the metric g. It can be written locally as

$$D = \gamma^\mu(x)(\partial_\mu + \omega_\mu^\mathcal{S}) , \tag{67}$$

where $\omega_\mu^\mathcal{S}$ is the lift of the Levi-Civita connection to the bundle of spinors.
The curved gamma matrices $\{\gamma^\mu(x)\}$ are Hermitian and satisfy

$$\gamma^\mu(x)\gamma^\nu(x) + \gamma^\nu(x)\gamma^\mu(x) = 2g(dx^\mu, dx^n) = 2g^{\mu\nu} , \quad \mu, \nu = 1, \ldots, n . \tag{68}$$

The elements of the algebra $\mathcal{A}$ act as multiplicative operators on $\mathcal{H}$,

$$(f\psi)(x) =: f(x)\psi(x) , \quad \forall f \in \mathcal{A} , \psi \in \mathcal{H} . \tag{69}$$

For this triple, the distance in (64) is the geodesic distance on the manifold M of the metric g.

An additional important ingredient is provided by a *real structure*. In the context of the canonical triple, this is given by J, the charge conjugation operator, which is an antilinear isometry of $\mathcal{H}$. We refer to [10] for all details; for a friendly introduction see [40].

4 Examples of Isospectral Deformations

We shall now construct some examples of (a priori noncommutative) spaces $\mathrm{Gr}_{m,r}$ such that

$$A_{m,r} = C(\mathrm{Gr}_{m,r}) \quad \text{or} \quad B_{m,r} = C(\mathrm{Gr}_{m,r}) , \tag{70}$$

according to even or odd dimensions, with the C*-algebras $A_{m,r}$ and $B_{m,r}$ defined at the end of Sect. 2.2 and Sect. 2.3, and associated with the vanishing of the "lower degree" components of the Chern character of an idempotent and of a unitary respectively.

4.1 Spheres in Dimension 2

The simplest case is $m = 1$, $r = 2$. We have then

$$e = \begin{pmatrix} e_{11} & e_{12} \\ e_{21} & e_{22} \end{pmatrix} \tag{71}$$

and the condition (29) just means that

$$e_{11} + e_{22} = 1 \tag{72}$$

while (26) means that

$$e_{11}^2 + e_{12}\, e_{21} = e_{11} , \quad e_{11}\, e_{12} + e_{12}\, e_{22} = e_{12} , \tag{73}$$
$$e_{21}\, e_{11} + e_{22}\, e_{21} = e_{21} , \quad e_{21}\, e_{12} + e_{22}^2 = e_{22} .$$

By (72) we get $e_{11} - e_{11}^2 = e_{22} - e_{22}^2$, so that (73) shows that $e_{12}\, e_{21} = e_{21}\, e_{12}$. We also see that e_{12} and e_{21} both commute with e_{11}. This shows that $A_{1,2}$ is commutative and allows to check that $\mathrm{Gr}_{1,2} = S^2$ is the 2-sphere. Thus $\mathrm{Gr}_{1,2}$ is an ordinary commutative space.

4.2 Spheres in Dimension 4

Next, we move on to the case $m = 2$, $r = 4$.

Note first that the notion of admissible morphism is a non trivial piece of structure on $\mathrm{Gr}_{2,4}$ since, for instance, the identity map is not admissible [15].

Commutative solutions were found in [13] with the commutative algebra $A = C(S^4)$ and an admissible surjection $A_{2,4} \to C(S^4)$, where the sphere S^4 appears naturally as quaternionic projective space, $S^4 = P_1(\mathbb{H})$.

In [17] we found noncommutative solutions, showing that the algebra $A_{2,4}$ is noncommutative, and we constructed explicit admissible surjections,

$$A_{2,4} \to C(S_\theta^4) \tag{74}$$

where S_θ^4 is the noncommutative 4-sphere we are about to describe and whose form is dictated by natural deformations of the ordinary 4-sphere, similar in spirit to the standard deformation of the torus $\mathbb{T}^2$ to the noncommutative torus $\mathbb{T}_\theta^2$. In fact, as will become evident later on, noncommutative tori in arbitrary dimensions play a central role in the deformations.

We first determine the algebra generated by the usual matrices $\mathrm{Mat}_4(\mathbb{C})$ and a projection $e = e^* = e^2$ such that $\mathrm{ch}_0(e) = 0$ as above and whose matrix expression is of the form,

$$[e^{ij}] = \frac{1}{2} \begin{pmatrix} q_{11} & q_{12} \\ q_{21} & q_{22} \end{pmatrix} \tag{75}$$

where each q_{ij} is a 2×2 matrix of the form,

$$q = \begin{pmatrix} \alpha & \beta \\ -\lambda\beta^* & \alpha^* \end{pmatrix}, \tag{76}$$

and $\lambda = \exp(2\pi i\theta)$ is a complex number of modulus one (different from -1 for convenience). Since $e = e^*$, both q_{11} and q_{22} are self-adjoint, moreover since $\mathrm{ch}_0(e) = 0$, we can find $z = z^*$ such that,

$$q_{11} = \begin{pmatrix} 1+z & 0 \\ 0 & 1+z \end{pmatrix}, \quad q_{22} = \begin{pmatrix} 1-z & 0 \\ 0 & 1-z \end{pmatrix}. \tag{77}$$

We let $q_{12} = \begin{pmatrix} \alpha & \beta \\ -\lambda\beta^* & \alpha^* \end{pmatrix}$, we then get from $e = e^*$,

$$q_{21} = \begin{pmatrix} \alpha^* & -\bar{\lambda}\beta \\ \beta^* & \alpha \end{pmatrix}. \tag{78}$$

We thus see that the commutant $\mathcal{A}_\theta$ of $\mathrm{Mat}_4(\mathbb{C})$ is generated by t, α, β and we first need to find the relations imposed by the equality $e^2 = e$. In terms of the matrix

$$e = \frac{1}{2} \begin{pmatrix} 1+z & q \\ q^* & 1-z \end{pmatrix}, \tag{79}$$

the equation $e^2 = e$ means that $z^2 + qq^* = 1$, $z^2 + q^*q = 1$ and $[z, q] = 0$. This shows that z commutes with α, β, α^* and β^* and since $qq^* = q^*q$ is a diagonal matrix

$$\alpha\alpha^* = \alpha^*\alpha, \quad \alpha\beta = \lambda\beta\alpha, \quad \alpha^*\beta = \bar{\lambda}\beta\alpha^*, \quad \beta\beta^* = \beta^*\beta \tag{80}$$

so that the generated algebra $\mathcal{A}_\theta$ is not commutative for λ different from 1. The only further relation, besides $z = z^*$, is a sphere relation

$$\alpha\alpha^* + \beta\beta^* + z^2 = 1. \tag{81}$$

We denote by S_θ^4 the corresponding noncommutative space defined by "duality", so that its algebra of polynomial functions is $\mathcal{A}(S_\theta^4) = \mathcal{A}_\theta$. This algebra is a deformation of the commutative *-algebra $\mathcal{A}(S^4)$ of complex polynomial functions on the usual sphere S^4 to which it reduces for $\theta = 0$.

The projection e given in (79) is clearly an element in the matrix algebra $\mathrm{Mat}_4(\mathcal{A}_\theta) \simeq \mathrm{Mat}_4(\mathbb{C}) \otimes \mathcal{A}_\theta$. Then, it naturally acts on the free $\mathcal{A}_\theta$-module $\mathcal{A}_\theta^4 \simeq \mathbb{C}^4 \otimes \mathcal{A}_\theta$ and one gets as its range a finite projective module which can be thought of as the module of "section of a vector bundle" over S_θ^4. The module $e\mathcal{A}_\theta^4$ is a deformation of the usual [3] complex rank 2 instanton bundle over S^4 to which it reduces for $\theta = 0$ [41].

For the sphere S_θ^4 the deformed instanton has correct characteristic classes. The fact that $\mathrm{ch}_0(e)$ has been imposed from the very beginning and could be interpreted as stating the fact that the projection and the corresponding module (the "vector bundle") has complex rank equal to 2. Next, we shall check that the two dimensional component $\mathrm{ch}_1(e)$ of the Chern character, automatically vanishes as an element of the (reduced) (b, B)-bicomplex.

With $q = \begin{pmatrix} \alpha & \beta \\ -\lambda\beta^* & \alpha^* \end{pmatrix}$, we get,

$$\mathrm{ch}_1(e) = \frac{1}{2^3} \left\langle z \left(dq\, dq^* - dq^*\, dq \right) + q \left(dq^*\, dz - dz\, dq^* \right) + q^* \left(dz\, dq - dq\, dz \right) \right\rangle$$

where the expectation in the right hand side is relative to $\mathrm{Mat}_2(\mathbb{C})$ (it is a partial trace) and we use the notation d instead of the tensor notation. The diagonal elements of $\omega = dq\, dq^*$ are

$$\omega_{11} = d\alpha\, d\alpha^* + d\beta\, d\beta^* , \quad \omega_{22} = d\beta^*\, d\beta + d\alpha^*\, d\alpha$$

while for $\omega' = dq^*\, dq$ we get,

$$\omega'_{11} = d\alpha^*\, d\alpha + d\beta\, d\beta^* , \quad \omega'_{22} = d\beta^*\, d\beta + d\alpha\, d\alpha^* .$$

It follows that, since z is diagonal,

$$\left\langle z \left(dq\, dq^* - dq^*\, dq \right) \right\rangle = 0 . \tag{82}$$

The diagonal elements of $q\, dq^*\, dz = \rho$ are

$$\rho_{11} = \alpha\, d\alpha^*\, dz + \beta\, d\beta^*\, dz , \quad \rho_{22} = \beta^*\, d\beta\, dz + \alpha^*\, d\alpha\, dz$$

while for $\rho' = q^*\, dq\, dz$ they are

$$\rho'_{11} = \alpha^*\, d\alpha\, dz + \beta\, d\beta^*\, dz , \quad \rho'_{22} = \beta^*\, d\beta\, dz + \alpha\, d\alpha^*\, dz .$$

Similarly for $\sigma = q\, dz\, dq^*$ and $\sigma' = q^*\, dz\, dq$ one gets the required cancellations so that,

$$\mathrm{ch}_1(e) = 0 , \tag{83}$$

Summing up we thus get that the element $e \in C^\infty(S_\theta^4, \mathrm{Mat}_4(\mathbb{C}))$ given in (79) is a self-adjoint idempotent, $e = e^2 = e^*$, and satisfies $\mathrm{ch}_k(e) = 0 \; \forall k < 2$. Moreover, $\mathrm{Gr}_{2,4}$ is a noncommutative space and $S_\theta^4 \subset \mathrm{Gr}_{2,4}$.

Since $\mathrm{ch}_1(e) = 0$, it follows that $\mathrm{ch}_2(e)$ is a Hochschild cycle which will play the role of the round volume form on S_θ^4 and that we shall now compute. With the above notations one has,

$$\mathrm{ch}_2(e) = \frac{1}{2^5} \left\langle \begin{pmatrix} z & q \\ q^* & -z \end{pmatrix} \begin{pmatrix} dz & dq \\ dq^* & -dz \end{pmatrix}^4 \right\rangle . \tag{84}$$

The direct computation gives the Hochschild cycle $\mathrm{ch}_2(e)$ as a sum of five components

$$\mathrm{ch}_2(e) = z\, c_z + \alpha\, c_\alpha + \alpha^*\, c_{\alpha^*} + \beta\, c_\beta + \beta^*\, c_{\beta^*} ; \tag{85}$$

where the components $c_z, c_\alpha, c_{\alpha^*}, c_\beta, c_{\beta^*}$, which are elements in the tensor product $\mathcal{A}_\theta \otimes \overline{\mathcal{A}_\theta} \otimes \mathcal{A}_\theta \otimes \overline{\mathcal{A}_\theta}$, are explicitly given in [17]. The vanishing of $b\,\mathrm{ch}_2(e)$, which has six hundred terms, can be checked directly from the commutation relations (80). The cycle $\mathrm{ch}_2(e)$ is totally "λ-antisymmetric".

Our sphere S_θ^4 is by construction the suspension of the noncommutative 3-sphere S_θ^3 whose coordinate algebra is generated by α and β as above and say the special value $z = 0$. This 3-sphere is part of a family of spheres that we shall describe in the next Section.

Had we taken the deformation parameter to be real, $\lambda = q \in \mathbb{R}$, the corresponding 3-sphere S_q^3 would coincide with the quantum group $SU(2)_q$. Similarly, had we taken the deformation parameter in S_θ^4 to be real like in [24] we would have obtained a different deformation S_q^4 of the commutative sphere S^4, whose algebra is different from the above one. More important, the component $\mathrm{ch}_1(e)$ of the Chern character would not vanish [23].

4.3 Spheres in Dimension 3

Odd dimensional spaces, in particular spheres, are constructed out of unitaries rather than projections [15, 16, 17].

Let us consider the lowest dimensional case for which $m = 2, r = 2$. We shall use the convention that repeated indices are summed on. Greek indices like $\mu, \nu, \ldots$, are taken to be valued in $\{0, 1, 2, 3\}$ while latin indices like $j, k, \ldots$, are taken to be valued in $\{1, 2, 3\}$.

We are then looking for an algebra $\mathcal{B}$ such that

(1) $\mathcal{B}$ is generated as a unital $*$-algebra by the entries of a unitary matrix

$$u \in \mathrm{Mat}_2(\mathcal{B}) \simeq \mathrm{Mat}_2(\mathbb{C}) \otimes \mathcal{B}, \quad uu^* = u^*u = 1 , \tag{86}$$

(2) the unitary u satisfies the additional condition

$$\mathrm{ch}_{\frac{1}{2}}(u) := \sum u_i^j \otimes (u^*)_i^j - (u^*)_i^j \otimes u_i^j = 0 . \tag{87}$$

Let us take as "generators" of $\mathcal{B}$ elements z^μ, $z^{\mu*}$, $\mu \in \{0, 1, 2, 3\}$. Then using ordinary Pauli matrices σ_k, $k \in \{1, 2, 3\}$, an element in $u \in \mathrm{Mat}_2(\mathcal{B})$ can be written as

$$u = \mathbb{I}_2 z^0 + \sigma_k z^k \ . \tag{88}$$

The requirement that u be unitaries give the following conditions on the generators

$$
\begin{aligned}
z^k z^{0*} - z^0 z^{k*} + \varepsilon_{klm} z^l z^{m*} &= 0 \ , \\
z^{0*} z^k - z^{k*} z^0 + \varepsilon_{klm} z^{l*} z^m &= 0 \ , \\
\sum_{\mu=0}^{3} (z^\mu z^{\mu*} - z^{\mu*} z^\mu) &= 0 \ ,
\end{aligned}
\tag{89}
$$

together with the condition that

$$\sum_{\mu=0}^{3} z^{\mu*} z^\nu = 1 \ . \tag{90}$$

Notice that the "sphere" relation (90) is consistent with the relations (89) since the latter imply that $\sum_{\mu=0}^{3} z^{\mu*} z^\nu$ is in the center of $\mathcal{B}$.

Then, one imposes condition (87) which reads

$$\sum_{\mu=0}^{3} (z^{\mu*} \otimes z^\mu - z^\mu \otimes z^{\mu*}) = 0 \ , \tag{91}$$

and which is satisfied [15, 16] if and only if there exists a symmetric unitary matrix $\Lambda \in \mathrm{Mat}_4(\mathbb{C})$ such that

$$z^{\mu*} = \Lambda^\mu_\nu z^\nu \ . \tag{92}$$

Now, there is some freedom in the definition of the algebra $\mathcal{B}$ which is stated by the fact that the defining conditions (1) and (2) above do not change if we transform

$$z^\mu \mapsto \rho S^\mu_\nu z^\nu \ , \tag{93}$$

with $\rho \in \mathrm{U}(1)$ and $S \in \mathrm{SO}(4)$. Under this transformation, the matrix Λ in (92) transforms as

$$\Lambda \mapsto \rho^2 S^t \Lambda S \ . \tag{94}$$

One can then diagonalize the symmetric unitary Λ by a real rotation S and fix its first eigenvalue to be 1 by an appropriate choice of $\rho \in \mathrm{U}(1)$. So, we can take

$$\Lambda = \mathrm{diag}(1, e^{-i\varphi_1}, e^{-i\varphi_2}, e^{-i\varphi_3}) \ , \tag{95}$$

that is, we can put

$$z^0 = x^0, \quad z^k = e^{i\varphi_k} x^k, \quad k \in \{1, 2, 3\} \ , \tag{96}$$

with $e^{-i\varphi_k} \in U(1)$ and $(x^\mu)^* = x^\mu$. Conditions (89) translate to

$$[x^0, x^k]_- \cos\varphi_k = i\,[x^l, x^m]_+ \sin(\varphi_l - \varphi_m) ,$$
$$[x^0, x^k]_+ \sin\varphi_k = i\,[x^l, x^m]_- \cos(\varphi_l - \varphi_m) , \tag{97}$$

with (k, l, m) the cyclic permutation of $(1, 2, 3)$ starting with $k = 1, 2, 3$ and $[x, y]_\pm = xy - yx$. There is also the sphere relation (90),

$$\sum_{\mu=0}^{3} (x^{\mu*})^2 = 1 . \tag{98}$$

We have therefore a three parameters family of algebras $\mathcal{B}_\varphi$ which are labelled by an element $\varphi = (e^{-i\varphi_1}, e^{-i\varphi_2}, e^{-i\varphi_3}) \in \mathbb{T}^3$. The algebras $\mathcal{B}_\varphi$ are deformations of the algebra $A(S^3)$ of polynomial functions on an ordinary 3-sphere S^3 which is obtained for the special value $\varphi = (1, 1, 1)$. We denote by S^3_φ the corresponding noncommutative space, so that $A(S^3_\varphi) = \mathcal{B}_\varphi$. Next, one computes $\mathrm{ch}_{\frac{3}{2}}(u_\varphi)$) and shows that is a non trivial cycle ($b\,\mathrm{ch}_{\frac{3}{2}}(u_\varphi)) = 0$) on $\mathcal{B}_\varphi$ [15].

A special value of the parameter φ gives the 3-sphere S^3_θ described at the end of previous Section. Indeed, put $\varphi_1 = \varphi_2 = -\pi\theta$ and $\varphi_3 = 0$ and define

$$\alpha = x^0 + i\,x^3 , \quad \alpha^* = x^0 - i\,x^3 ,$$
$$\beta = x^1 + i\,x^2 , \quad \beta^* = x^1 - i\,x^2 . \tag{99}$$

then $\alpha, \alpha^*, \beta, \beta^*$ satisfies conditions (80), with $\lambda = \exp(2\pi\,i\,\theta)$, together with $\alpha\alpha^* + \beta\beta^* = 1$, thus defining the sphere S^3_θ of Sect. 4.2.

In Sect. 4.6 we shall describe some higher dimensional examples.

4.4 The Noncommutative Geometry of S^4_θ

Next we will analyze the metric structure, via a Dirac operator D, on our noncommutative 4-spheres S^4_θ. The operator D will give a solution to the following quartic equation,

$$\left\langle \left(e - \frac{1}{2} \right) [D, e]^4 \right\rangle = \gamma \tag{100}$$

where $\langle\ \rangle$ is the projection on the commutant of $4\,4$ $\mathbb{C}$-matrices (in fact, it is a partial trace on the matrix entries) and $\gamma = \gamma_5$, in the present four dimensional case, is the grading operator.

Let $C^\infty(S^4_\theta)$ be the algebra of smooth functions on the noncommutative sphere S^4_θ. We shall construct a spectral triple $(C^\infty(S^4_\theta), \mathcal{H}, D)$ which describes the geometry on S^4_θ corresponding to the round metric.

In order to do that we first need to find good coordinates on S^4_θ in terms of which the operator D will be easily expressed. We choose to parametrize α, β and z as follows,

$$\alpha = u \cos \varphi \cos \psi , \quad \beta = v \sin \varphi \cos \psi , \quad z = \sin \psi . \qquad (101)$$

Here φ and ψ are ordinary angles with domain $0 \le \varphi \le \frac{\pi}{2}$, $-\frac{\pi}{2} \le \psi \le \frac{\pi}{2}$, while u and v are the usual unitary generators of the algebra $C^\infty(\mathbb{T}_\theta^2)$ of smooth functions on the noncommutative 2-torus. Thus the presentation of their relations is

$$uv = \lambda vu , \quad uu^* = u^*u = 1 , \quad vv^* = v^*v = 1 . \qquad (102)$$

One checks that α, β, z given by (2) satisfy the basic presentation of the generators of $C^\infty(S_\theta^4)$ which thus appears as a *subalgebra* of the algebra generated (and then closed under smooth calculus) by $e^{i\varphi}$, $e^{i\psi}$, u and v.

For $\theta = 0$ the round metric is given as,

$$G = d\alpha \, d\bar\alpha + d\beta \, d\bar\beta + dz^2 \qquad (103)$$

and in terms of the coordinates, φ, ψ, u, v one gets,

$$G = \cos^2 \varphi \cos^2 \psi \, du \, d\bar{u} + \sin^2 \varphi \cos^2 \psi \, dv \, d\bar{v} + \cos^2 \psi \, d\varphi^2 + d\psi^2 . \qquad (104)$$

Its volume form is given by

$$\omega = \frac{1}{2} \sin \varphi \cos \varphi \, (\cos \psi)^3 \, \bar{u} \, du \wedge \bar{v} \, dv \wedge d\varphi \wedge d\psi . \qquad (105)$$

In terms of these rectangular coordinates we get the following simple expression for the Dirac operator,

$$D = (\cos \varphi \cos \psi)^{-1} \, u \, \frac{\partial}{\partial u} \, \gamma_1 + (\sin \varphi \cos \psi)^{-1} \, v \, \frac{\partial}{\partial v} \, \gamma_2 + \qquad (106)$$
$$+ \frac{i}{\cos \psi} \left(\frac{\partial}{\partial \varphi} + \frac{1}{2} \cot g \, \varphi - \frac{1}{2} \operatorname{tg} \varphi \right) \gamma_3 + i \left(\frac{\partial}{\partial \psi} - \frac{3}{2} \operatorname{tg} \psi \right) \gamma_4 .$$

Here γ_μ are the usual Dirac 4×4 matrices with

$$\{\gamma_\mu, \gamma_\nu\} = 2 \, \delta_{\mu\nu} , \quad \gamma_\mu^* = \gamma_\mu . \qquad (107)$$

It is now easy to move on to the noncommutative case, the only tricky point is that there are nontrivial boundary conditions for the operator D, which are in particular antiperiodic in the arguments of both u and v. We shall just leave them unchanged in the noncommutative case, the only thing which changes is the algebra and the way it acts in the Hilbert space as we shall explain in more detail in the next section. The formula for the operator D is now,

$$D = (\cos \varphi \cos \psi)^{-1} \, \delta_1 \, \gamma_1 + (\sin \varphi \cos \psi)^{-1} \, \delta_2 \, \gamma_2 + \qquad (108)$$
$$+ \frac{i}{\cos \psi} \left(\frac{\partial}{\partial \varphi} + \frac{1}{2} \cot g \, \varphi - \frac{1}{2} \operatorname{tg} \varphi \right) \gamma_3 + i \left(\frac{\partial}{\partial \psi} - \frac{3}{2} \operatorname{tg} \psi \right) \gamma_4 .$$

where the γ_μ are the usual Dirac matrices and where δ_1 and δ_2 are the derivations of the noncommutative torus so that

$$\delta_1(u) = u, \quad \delta_1(v) = 0,$$
$$\delta_2(u) = 0, \quad \delta_2(v) = v; \tag{109}$$

One can then check that the corresponding metric is the round one.

In order to compute the operator $\langle (e - \frac{1}{2}) [D, e]^4 \rangle$ (in the tensor product by $\mathrm{Mat}_4(\mathbb{C})$) we need the commutators of D with the generators of $C^\infty(S_\theta^4)$. They are given by the following simple expressions,

$$[D, \alpha] = u \{ \gamma_1 - \mathrm{i}\, \sin(\phi)\, \gamma_3 - \mathrm{i}\, \cos\phi)\, \sin(\psi)\, \gamma_4 \}, \tag{110}$$
$$[D, \alpha^*] = -u^* \{ \gamma_1 + \mathrm{i}\, \sin(\phi)\, \gamma_3 + \mathrm{i}\, \cos(\phi)\, \sin(\psi)\, \gamma_4 \},$$
$$[D, \beta] = v \{ \gamma_2 + \mathrm{i}\, \cos(\phi)\, \gamma_3 - \mathrm{i}\, \sin(\phi)\, \sin(\psi)\, \gamma_4 \},$$
$$[D, \beta^*] = -v^* \{ \gamma_2 - \mathrm{i}\, \cos(\phi)\, \gamma_3 + \mathrm{i}\, \sin(\phi)\, \sin(\psi)\, \gamma_4 \},$$
$$[D, z] = \mathrm{i}\, \cos(\psi)\, \gamma_4 .$$

We check in particular that they are all bounded operators and hence that for any $f \in C^\infty(S_\theta^4)$ the commutator $[D, f]$ is bounded. Then, a long but straightforward calculation shows that equation (100) is valid: the operator $\langle (e - \frac{1}{2}) [D, e]^4 \rangle$ is a multiple of $\gamma = \gamma_5 := \gamma_1\gamma_2\gamma_3\gamma_4$. One first checks that it is equal to $\pi(\mathrm{ch}_2(e))$ where $\mathrm{ch}_2(e)$ is the Hochschild cycle in (85) and π is the canonical map from the Hochschild chains to operators given by

$$\pi(a_0 \otimes a_1 \otimes ... \otimes a_n) = a_0[D, a_1]...[D, a_n] . \tag{111}$$

4.5 Isospectral Noncommutative Geometries

We shall describe fully a noncommutative geometry for S_θ^4 with the couple $(\mathcal{H}, D)$ just the "commutative" ones associated with the commutative sphere S^4; hence realizing an isospectral deformation. We shall in fact describe a very general construction of isospectral deformations of noncommutative geometries which implies in particular that any compact spin Riemannian manifold M whose isometry group has rank ≥ 2 admits a natural one-parameter isospectral deformation to noncommutative geometries M_θ. The deformation of the algebra will be performed along the lines of [54] (see also [62] and [59]).

Let us start with the canonical spectral triple $(\mathcal{A} = C^\infty(S^4), \mathcal{H}, D)$ associated with the sphere S^4. We recall that $\mathcal{H} = L^2(S^4, \mathcal{S})$ is the Hilbert space of spinors and D is the Dirac operator. Also, there is a real structure provided by J, the charge conjugation operator, which is an antilinear isometry of $\mathcal{H}$.

Recall that on the sphere S^4 there is an isometric action of the 2-torus, $\mathbb{T}^2 \subset \mathrm{Isom}(S^4)$ with $\mathbb{T} = \mathbb{R}/2\pi\mathbb{Z}$ the usual torus. We let $U(s), s \in \mathbb{T}^2$, be the corresponding (projective) unitary representation in $\mathcal{H} = L^2(S^4, \mathcal{S})$ so that by construction

$$U(s)\, D = D\, U(s)\,, \quad U(s)\, J = J\, U(s)\,. \tag{112}$$

Also,

$$U(s)\, a\, U(s)^{-1} = \alpha_s(a)\,, \quad \forall\, a \in \mathcal{A}\,, \tag{113}$$

where $\alpha_s \in \mathrm{Aut}(\mathcal{A})$ is the action by isometries on functions on S^4.

We let $p = (p_1, p_2)$ be the generator of the two-parameter group $U(s)$ so that

$$U(s) = \exp(2\pi i (s_1 p_1 + s_2 p_2))\,. \tag{114}$$

The operators p_1 and p_2 commute with D but anticommute with J (due to the antilinearity of the latter). Both p_1 and p_2 have half-integral spectrum,

$$\mathrm{Spec}(2\, p_j) \subset \mathbb{Z}\,, \ j = 1, 2\,. \tag{115}$$

Next, we define a bigrading of the algebra of bounded operators in $\mathcal{H}$ with the operator T declared to be of bidegree (n_1, n_2) when,

$$\alpha_s(T) := U(s)\, T\, U(s)^{-1} = \exp(2\pi i (s_1 n_1 + s_2 n_2))\, T\,, \ \forall\, s \in \mathbb{T}^2\,. \tag{116}$$

Any operator T of class C^∞ relative to α_s (i.e. such that the map $s \to \alpha_s(T)$ is of class C^∞ for the norm topology) can be uniquely written as a doubly infinite norm convergent sum of homogeneous elements, .

$$T = \sum_{n_1, n_2} \widehat{T}_{n_1, n_2}\,, \tag{117}$$

with $\widehat{T}_{n_1, n_2}$ of bidegree (n_1, n_2) and where the sequence of norms $\|\widehat{T}_{n_1, n_2}\|$ is of rapid decay in (n_1, n_2).

Let now $\lambda = \exp(2\pi i\, \theta)$. For any operator T in $\mathcal{H}$ of class C^∞ relative to the action of $\mathbb{T}^2$ we define its left twist $l(T)$ by

$$l(T) = \sum_{n_1, n_2} \widehat{T}_{n_1, n_2}\, \lambda^{n_2 p_1}\,, \tag{118}$$

and its right twist $r(T)$ by

$$r(T) = \sum_{n_1, n_2} \lambda^{n_1 p_2}\, \widehat{T}_{n_1, n_2}\,, \tag{119}$$

Since $|\lambda| = 1$ and p_1, p_2 are self-adjoint, both series converge in norm. The construction involves in the case of half-integral spin the choice of a square root of λ.

One has the following,

Lemma 1.
a) *Let x be a homogeneous operator of bidegree (n_1, n_2) and y be a homogeneous operator of bidegree (n_1', n_2'). Define*

$$x * y = \lambda^{n_1' n_2} xy;; \tag{120}$$

then $l(x)l(y) = l(x * y)$.

b) *Let x and y be homogeneous operators as before. Then,*

$$l(x)\, r(y) - r(y)\, l(x) = (x\, y - y\, x)\, \lambda^{n_1'(n_2 + n_2')} \lambda^{n_2 p_1 + n_1' p_2}\ . \tag{121}$$

In particular, $[l(x), r(y)] = 0$ *if* $[x, y] = 0$.

To check a) and b) one simply uses the following commutation rule which follows from (116) and it is fulfilled for any homogeneous operator T of bidegree (m, n),

$$\lambda^{a p_1 + b p_2}\, T = \lambda^{am + bn}\, T\, \lambda^{a p_1 + b p_2}\ , \quad \forall a, b \in \mathbb{Z}\ . \tag{122}$$

The product $*$ defined in equation (120) extends by linearity to an associative $*$-product on the linear space of smooth operators.

One could also define a deformed "right product". If x is homogeneous of bidegree (n_1, n_2) and y is homogeneous of bidegree (n_1', n_2') the product is defined by

$$x *_r y = \lambda^{-n_1' n_2}\, xy\ . \tag{123}$$

Then, as with the previous lemma one shows that $r(x)r(y) = r(x *_r y)$.

By Lemma 1 a) one has that $l(C^\infty(S^4))$ is still an algebra and we shall identify it with (the image on the Hilbert space $\mathcal{H}$ of) the algebra $C^\infty(S_\theta^4)$ of smooth functions on the deformed sphere S_θ^4).

We can then define a new spectral triple $\big(l(C^\infty(S^4)) \simeq C^\infty(S_\theta^4), \mathcal{H}, D\big)$ where both the Hilbert space $\mathcal{H}$ and the operator D are unchanged while the algebra $C^\infty(S^4)$ is modified to $l(C^\infty(S^4)) \simeq C^\infty(S_\theta^4)$. Since D is of bidegree $(0, 0)$ one has that

$$[D,\, l(a)] = l([D,\, a]) \tag{124}$$

which is enough to check that $[D, x]$ is bounded for any $x \in l(\mathcal{A})$.

Next, we also deform the real structure by twisting the charge conjugation isometry J by

$$\tilde{J} = J \lambda^{-p_1 p_2}\ . \tag{125}$$

Due to the antilinearity of J one has that $\tilde{J} = \lambda^{p_1 p_2} J$ and hence

$$\tilde{J}^2 = J^2\ . \tag{126}$$

Lemma 2.

For x homogeneous of bidegree (n_1, n_2) one has that

$$\tilde{J}\, l(x)\, \tilde{J}^{-1} = r(J\, x\, J^{-1})\ . \tag{127}$$

For the proof one needs to check that $\tilde{J}l(x) = r(J x J^{-1})\tilde{J}$. One has

$$\lambda^{-p_1 p_2} x = x \lambda^{-(p_1+n_1)(p_2+n_2)} = x \lambda^{-n_1 n_2} \lambda^{-(p_1 n_2 + n_1 p_2)} \lambda^{-p_1 p_2} . \qquad (128)$$

Then

$$\tilde{J}l(x) = J \lambda^{-p_1 p_2} x \lambda^{n_2 p_1} = J x \lambda^{-n_1 n_2} \lambda^{-n_1 p_2} \lambda^{-p_1 p_2} , \qquad (129)$$

while

$$r(J x J^{-1})\tilde{J} = \lambda^{-n_1 p_2} J x J^{-1} J \lambda^{-p_1 p_2} = J x \lambda^{-n_1(p_2+n_2)} \lambda^{-p_1 p_2} . \qquad (130)$$

Thus one gets the required equality of Lemma 2.

For $x, y \in l(\mathcal{A})$ one checks that

$$[x, y^0] = 0, \quad y^0 = \tilde{J} y^* \tilde{J}^{-1} . \qquad (131)$$

Indeed, one can assume that x and y are homogeneous and use Lemma 2 together with Lemma 1 a). Combining equation (131) with equation (124) one then checks the order one condition

$$[[D, x], y^0] = 0, \quad \forall x, y \in l(\mathcal{A}). \qquad (132)$$

Summing up, we have the following

Theorem 2.
a) *The spectral triple* $(C^\infty(S_\theta^4), \mathcal{H}, D)$ *fulfills all axioms of noncommutative manifolds.*
b) *Let* $e \in C^\infty(S_\theta^4, \mathrm{Mat}_4(\mathbb{C}))$ *be the canonical idempotent given in (79). The Dirac operator* D *fulfills*

$$\left\langle \left(e - \frac{1}{2}\right)[D, e]^4 \right\rangle = \gamma$$

where $\langle\ \rangle$ *is the projection on the commutant of* $\mathrm{Mat}_4(\mathbb{C})$ *(i.e. a partial trace) and* γ *is the grading operator.*

Moreover, the real structure is given by the twisted involution $\tilde{J}$ defined in (125). One checks using the results of [55] and [12] that Poincaré duality continues to hold for the deformed spectral triple.

Theorem 2 can be extended to all metrics on the sphere S^4 which are invariant under rotation of u and v and have the same volume form as the round metric. In fact, by paralleling the construction for the sphere described above, one can extend it quite generally [17]:

Theorem 3.
Let M *be a compact spin Riemannian manifold whose isometry group has rank* ≥ 2 *(so that one has an inclusion* $\mathbb{T}^2 \subset \mathrm{Isom}(M)$*). Then* M *admits a natural one-parameter isospectral deformation to noncommutative (spin) geometries* M_θ.

Let $(\mathcal{A}, \mathcal{H}, D)$ be the canonical spectral triple associated with a compact Riemannian spin manifold M as described in Ex. 2. Here $\mathcal{A} = C^\infty(M)$ is the algebra of smooth functions on M; $\mathcal{H} = L^2(M, \mathcal{S})$ is the Hilbert space of spinors and D is the Dirac operator. Finally, there is the charge conjugation operator J, an antilinear isometry of $\mathcal{H}$ which gives the real structure.

The deformed spectral triple is given by $(l(\mathcal{A}), \mathcal{H}, D)$ with $\mathcal{H} = L^2(M, \mathcal{S})$ the Hilbert space of spinors, D the Dirac operator and $l(\mathcal{A})$ is really the algebra of smooth functions on M with product deformed to a $*$-product defined in a way exactly similar to (120). The real structure is given by the twisted involution $\tilde{J}$ defined as in (125). And again, by the results of [55] and [12], Poincaré duality continues to hold for the deformed spectral triple.

4.6 Noncommutative Spherical Manifolds

As we have seen, on the described deformations one changes the algebra and the way it acts on the Hilbert space while keeping the latter and the Dirac operator unchanged, thus getting isospectral deformations. From the decomposition (116) and the deformed product (120) one sees that a central role is played by tori and their noncommutative generalizations. We are now going to describe in more details this use of the noncommutative tori.

Let $\theta = (\theta_{jk} = -\theta_{kj})$ be a real antisymmetric $n \times n$ matrix. The non-commutative torus $\mathbb{T}_\theta^n$ of "dimension" n and twist θ is the "quantum space" whose algebra of polynomial functions $A(\mathbb{T}_\theta^n)$ is generated by n independent unitaries $u_1, \ldots, u_n$, subject to the commutation relations [8, 53]

$$u_j u_k = e^{2\pi i \theta_{jk}} u_k u_j. \tag{133}$$

The corresponding C*-algebra of continuous functions is the universal C*-algebra $C(\mathbb{T}_\theta^n)$ with the same generators and relations. There is an action τ of $\mathbb{T}^n$ on this C*-algebra. If $\alpha = (\alpha_1, \ldots, \alpha_n)$, this action is given by

$$\tau(e^{2\pi i \alpha}) : u_j \mapsto e^{2\pi i \alpha_j} u_j .$$

The smooth subalgebra $C^\infty(\mathbb{T}_\theta^n)$ of $C(\mathbb{T}_\theta^n)$ under this action consists of rapidly convergent Fourier series of the form $\sum_{r \in \mathbb{Z}^n} a_r u^r$, with $a_r \in \mathbb{C}$, where

$$u^r := e^{-\pi i r_j \theta_{jk} r_k} u_1^{r_1} u_2^{r_2} \ldots u_n^{r_n} .$$

The unitary elements $\{u^r : r \in \mathbb{Z}^n\}$ form a Weyl system [34], since

$$u^r u^s = e^{\pi i r_j \theta_{jk} s_k} u^{r+s} .$$

The phase factors

$$\rho_\theta(r, s) := \exp\{\pi i r_j \theta_{jk} s_k\} \tag{134}$$

form a 2-cocycle for the group $\mathbb{Z}^n$, which is skew (i.e., $\rho_\theta(r, r) = 1$) since θ is skew-symmetric. This also means that $C(\mathbb{T}_\theta^n)$ may be defined as the twisted group C*-algebra $C(\mathbb{Z}^n, \rho_\theta)$.

Let now M be a compact manifold (with no boundary) carrying a smooth action σ of a torus $\mathbb{T}^n$ of dimension $n \geq 2$. By averaging the translates of a given Riemannian metric on M over this torus, we may assume that M has a $\mathbb{T}^n$-invariant metric g, so that $\mathbb{T}^n$ acts by isometries.

The general θ-deformation of M can be accomplished in two equivalent ways. Firstly, in [17] the deformation is given by a star product of ordinary functions along the lines of [54] (see also [62]). Indeed, the algebra $C^\infty(M)$ may be decomposed into spectral subspaces which are indexed by the dual group $\mathbb{Z}^n = \widehat{\mathbb{T}^n}$. Now, each $r \in \mathbb{Z}^n$ labels a character $e^{2\pi i\alpha} \mapsto e^{2\pi i r \cdot \alpha}$ of $\mathbb{T}^n$, with the scalar product $r \cdot \alpha := r_1\alpha_1 + \cdots + r_n\alpha_n$. The r-th spectral subspace for the action σ of $\mathbb{T}^n$ on $C^\infty(M)$ consists of those smooth functions f_r for which

$$\sigma(e^{2\pi i\alpha})f_r = e^{2\pi i r \cdot \alpha} f_r ,$$

and each $f \in C^\infty(M)$ is the sum of a unique (rapidly convergent) series $f = \sum_{r \in \mathbb{Z}^n} f_r$.

The θ-deformation of $C^\infty(M)$ may be defined by replacing the ordinary product by a *Moyal product*, defined on spectral subspaces by

$$f_r \star_\theta g_s := \rho_\theta(r, s) f_r g_s , \tag{135}$$

with $\rho_\theta(r, s)$ the phase factor in (134). Thus the deformed product is also taken to respects the $\mathbb{Z}^n$-grading of functions.

In particular, when $M = \mathbb{T}^n$ with the obvious translation action, the algebras $(C^\infty(\mathbb{T}^n), \star_\theta)$ and $C^\infty(\mathbb{T}^n_\theta)$ are isomorphic.

In the general case, we write $C^\infty(M_\theta) := (C^\infty(M), \star_\theta)$. Thus, at the level of smooth algebras the deformation is given explicitly by the star product of ordinary smooth functions. It is shown in [54] that there is a natural completion of the algebra $C^\infty(M_\theta)$ to a C*-algebra $C(M_\theta)$ whose smooth subalgebra (under the extended action of $\mathbb{T}^n$) is precisely $C^\infty(M_\theta)$.

An equivalent approach [15], is to define $C(M_\theta)$ as the fixed-point C*-subalgebra of $C(M) \otimes C(\mathbb{T}^n_\theta)$ under the action $\sigma \times \tau^{-1}$ of $\mathbb{T}^n$ defined by

$$e^{2\pi i\alpha} \cdot (f \otimes a) := \sigma(e^{2\pi i\alpha}) f \otimes \tau(e^{-2\pi i\alpha}) a ;$$

that is,

$$C(M_\theta) := \left(C(M) \otimes C(\mathbb{T}^n_\theta)\right)^{\sigma \times \tau^{-1}} .$$

The smooth subalgebra is then given by

$$C^\infty(M_\theta) := \left(C^\infty(M) \widehat{\otimes} C^\infty(\mathbb{T}^n_\theta)\right)^{\sigma \times \tau^{-1}} , \tag{136}$$

with $\widehat{\otimes}$ denoting the appropriate (projective) tensor product of Fréchet algebras. This approach has the advantage that $C^\infty(M_\theta)$ may be determined by generators and relations with the algebra structure specified by the basic commutation relations (133) [15].

4.7 The θ-Deformed Planes and Spheres in Any Dimensions

We shall briefly describe these classes of spaces while referring to [15] for more details.

Let $\theta = (\theta_{jk} = -\theta_{kj})$ be a real antisymmetric $n \times n$ matrix. And denote $\lambda^{jk} = e^{2\pi i \theta_{jk}}$; then we have that $\lambda^{kj} = (\lambda^{jk})^{-1}$ and $\lambda^{jj} = 1$.

Let $\mathcal{A}(\mathbb{R}^{2n}_\theta)$ be the complex unital $*$-algebra generated by $2n$ elements $(z^j, z^{k*}, j, k = 1, \ldots, n)$ with relations

$$z^j z^k = \lambda^{jk} z^k z^j , \quad z^{j*} z^{k*} = \lambda^{jk} z^{k*} z^{j*}, \quad z^{j*} z^k = \lambda^{kj} z^k z^{j*} , \tag{137}$$

with $j, k = 1, \ldots, n$. The $*$-algebra $\mathcal{A}(\mathbb{R}^{2n}_\theta)$ can be thought of as the algebra of complex polynomials on the noncommutative $2n$-plane $\mathbb{R}^{2n}_\theta$ since it is a deformation of the commutative $*$-algebra $\mathcal{A}(\mathbb{R}^{2n})$ of complex polynomial functions on $\mathbb{R}^{2n}$ to which it reduces for $\theta = 0$. From relations (137), it follows that the elements $z^{j*} z^j = z^j z^{j*}$, $j = 1, \ldots, n$, are in the center of $\mathcal{A}(\mathbb{R}^{2n}_\theta)$. Since $\sum_{j=1}^n z^j z^{j*}$ is central as well, it makes sense to define $\mathcal{A}(S^{2n-1}_\theta)$ to be the quotient of the $*$-algebra $\mathcal{A}(\mathbb{R}^{2n}_\theta)$ by the ideal generated by $\sum_{j=1}^n z^j z^{j*} - 1$. The $*$-algebra $\mathcal{A}(S^{2n-1}_\theta)$ can be thought of as the algebra of complex polynomials on the noncommutative $(2n-1)$-sphere S^{2n-1}_θ since it is a deformation of the commutative $*$-algebra $\mathcal{A}(S^{2n-1})$ of complex polynomial functions on the usual sphere S^{2n-1}.

Next, one defines $\mathcal{A}(\mathbb{R}^{2n+1}_\theta)$ to be the complex unital $*$-algebra generated by $2n+1$ elements made of $(z^j, z^{j*}, j = 1, \ldots, n)$ and of an addition hermitian element $x = x^*$ with relations like (137) and in addition

$$z^j x = x z^j , \quad j = 1, \ldots, n . \tag{138}$$

The $*$-algebra $\mathcal{A}(\mathbb{R}^{2n+1}_\theta)$ is the algebra of complex polynomials on the noncommutative $(2n+1)$-plane $\mathbb{R}^{2n+1}_\theta$.

By the very definition, the elements $z^{j*} z^j = z^j z^{j*}$, $j = 1, \ldots, n$, and x are in the center of $\mathcal{A}(\mathbb{R}^{2n+1}_\theta)$ and so is the element $\sum_{j=1}^n z^j z^{j*} + x^2$. Then one defines $\mathcal{A}(S^{2n}_\theta)$ to be the quotient of the $*$-algebra $\mathcal{A}(\mathbb{R}^{2n+1}_\theta)$ by the ideal generated by $\sum_{j=1}^n z^j z^{j*} + x^2 - 1$. The $*$-algebra $\mathcal{A}(S^{2n}_\theta)$ is the algebra of complex polynomials on the noncommutative $2n$-sphere S^{2n}_θ and is a deformation of the commutative $*$-algebra $\mathcal{A}(S^{2n})$ of complex polynomial functions on a usual sphere S^{2n}. By construction the sphere S^{2n}_θ is a suspension of the sphere S^{2n-1}_θ.

Next, let $\mathrm{Cliff}(\mathbb{R}^{2n}_\theta)$ be the unital associative $*$-algebra over $\mathbb{C}$ generated by $2n$ elements Γ^j, Γ^{k*}, $j, j = 1, \ldots, n$, with relations

$$\Gamma^j \Gamma^k + \lambda^{kj} \Gamma^k \Gamma^j = 0 ,$$
$$\Gamma^{j*} \Gamma^{k*} + \lambda^{kj} \Gamma^{k*} \Gamma^{j*} = 0 ,$$
$$\Gamma^j \Gamma^{k*} + \lambda^{jk} \Gamma^k \Gamma^{j*} = \delta^{jk} \mathbb{I} , \tag{139}$$

where $\mathbb{I}$ is the unit of the algebra and δ^{jk} is the usual flat metric. For $\theta = 0$ one gets the usual Clifford algebra $\mathrm{Cliff}(\mathbb{R}^{2n}$ of $\mathbb{R}^{2n}$. The element $\gamma \in \mathrm{Cliff}(\mathbb{R}^{2n}_\theta)$ defined by

$$\gamma = [\Gamma^{1*}, \Gamma^1] \cdot \ldots \cdot [\Gamma^{n*}, \Gamma^n] \tag{140}$$

is hermitian, $\gamma = \gamma^*$, satisfies

$$\gamma^2 = \mathbb{I}, \quad \gamma\Gamma^j + \Gamma^j\gamma = 0, \quad \gamma\Gamma^{j*} + \Gamma^{j*}\gamma = 0, \tag{141}$$

and determines a $\mathbb{Z}_2$-grading of $\mathrm{Cliff}(\mathbb{R}^{2n}_\theta)$, $\Lambda \mapsto \gamma\Lambda\gamma$. In fact, one shows [15] that $\mathrm{Cliff}(\mathbb{R}^{2n}_\theta)$ is isomorphic to the usual Clifford algebra $\mathrm{Cliff}(\mathbb{R}^{2n}$ as a $*$-algebra and as a $\mathbb{Z}_2$-graded algebra. Furthermore, there is a representation of $\mathrm{Cliff}(\mathbb{R}^{2n}_\theta)$ for which $\gamma = \begin{pmatrix} \mathbb{I} & 0 \\ 0 & \mathbb{I} \end{pmatrix}$ and $\Gamma^j \in \mathrm{Mat}_{2^n}(\mathbb{C})$ of the form

$$\Gamma^j = \begin{pmatrix} 0 & \sigma^j \\ \bar{\sigma}^{j*} & 0 \end{pmatrix}, \quad \Gamma^{j*} = \begin{pmatrix} 0 & \bar{\sigma}^{j*} \\ \sigma^j & 0 \end{pmatrix}, \tag{142}$$

with σ^j and $\bar{\sigma}^{j*}$ in $\mathrm{Mat}_{2^{n-1}}(\mathbb{C})$.

Theorem 4.
a) *There is a canonical projection $e \in \mathrm{Mat}_{2^n}(\mathcal{A}(S^{2n}_\theta))$ given by*

$$e = \frac{1}{2}\left(\mathbb{I} + \sum_{j=1}^{n}(\Gamma^{j*}z^j + \Gamma^j z^{j*} + \gamma x)\right), \tag{143}$$

where (z^j, z^{k*}, x) are the generators of $\mathcal{A}(S^{2n}_\theta)$. Moreover, one has that

$$\mathrm{ch}_k(e) = 0, \quad 0 \le k \le n-1. \tag{144}$$

a) *There is a canonical unitary $u \in \mathrm{Mat}_{2^{n-1}}(\mathcal{A}(S^{2n-1}_\theta))$ given by*

$$u = \sum_{j+1}^{n}(\bar{\sigma}^j z^j + \sigma^j z^{j*}), \tag{145}$$

where (z^j, z^{k*}) are the generators of $\mathcal{A}(S^{2n-1}_\theta)$. Moreover, one has that

$$\mathrm{ch}_{k+\frac{1}{2}}(e) = 0, \quad 0 \le k \le n-1. \tag{146}$$

For a proof we refer to [15].

The projection e in (143) and the unitary u in (145) provide non-commutative solutions, via constraints (144) and (146), for the algebras $A_{m,r} = C(\mathrm{Gr}_{m,r})$ and $B_{m,r} = C(\mathrm{Gr}_{m,r})$ defined in (70). Thus, there are admissible surjections

$$A_{2n,2^n} \to C(S^{2n}_\theta), \quad B_{2n-1,2^{n-1}} \to C(S^{2n-1}_\theta) \tag{147}$$

The projection (143) generalizes to higher dimensions the projection constructed in (79) for the four dimensional sphere S_θ.

4.8 Gauge Theories

From Theorem 3 we know that the deformed spheres, and in particular the even ones, S_θ^{2n}, can be endowed with the structure of a noncommutative spin manifold, via a spectral triple $(C^\infty(S_\theta^{2n}), \mathcal{H}, D)$, which is isospectral since both the Hilbert space $\mathcal{H} = L^2(S^{2n}, \mathcal{S})$ and the Dirac operator D are the usual one on the commutative sphere S^{2n} whereas only the algebra and its representation on $\mathcal{H}$ are changed. In particular one could take the Dirac operator of the usual "round" metric. Then out of this one can define a suitable Hodge operator $*_H$ on S_θ^{2n}. It turns out that the canonical projection (143) satisfies self-duality equations

$$*_H \, e(de)^n = \mathrm{i}^n e(de)^n \ . \tag{148}$$

These equations were somehow "in the air". For the four dimensional case I mentioned them during a talk in Ancona in February 2001 [42]. For the general case they were derived in [15] and also in [1]. Their "commutative" counterparts were proposed in [28, 29] together with a description of gauge theories in terms of projectors.

In particular on the four dimensional sphere S_θ, one can develop Yang-Mills theory, since there are all the required structures, namely the algebra, the calculus (by means of the Dirac operator) and the "vector bundle" e.

The Yang-Mills action is given by,

$$YM(\nabla) = \int \theta^2 \, ds^4 \ , \tag{149}$$

where $\theta = \nabla^2$ is the curvature, and $ds = D^{-1}$. This action has a strictly positive lower bound [10] given by a topological invariant which is just the index (66),

$$\varphi(e) = \int \gamma \left(e - \frac{1}{2} \right) [D, e]^4 \, ds^4 \ . \tag{150}$$

For the canonical projection (79), owing to (100), this topological invariant turns out to be just 1,

$$\varphi(e) = 1 \ . \tag{151}$$

An important problem, which is still open, is the construction and the classification of Yang-Mills connections in the noncommutative situation along the line of the ADHM construction [3]. For the noncommutative torus this was done in [19] and for a noncommutative $\mathbb{R}^4$ in [50].

5 Euclidean and Unitary Quantum Spheres

The contents of this Section is essentially a subset of the paper of Eli Hawkins and myself [35].

The quantum Euclidean spheres in any dimensions, S_q^{N-1}, are (quantum) homogeneous spaces of quantum orthogonal groups, $SO_q(N)$ [30]. The natural coaction of $SO_q(N)$ on $\mathbb{R}_q^N$,

$$\delta : A(\mathbb{R}_q^N) \to A(SO_q(N)) \otimes A(\mathbb{R}_q^N) \,, \qquad (152)$$

preserves the "radius of the sphere" and yields a coaction of the quantum group $SO_q(N)$ on S_q^{N-1}.

Similarly, "odd dimensional" quantum spheres S_q^{2n-1} can be constructed as noncommutative homogeneous spaces of quantum unitary groups $SU_q(n)$ [61] (see also [64]). Then, analogously to (152), there is also a coaction of the quantum group $SU_q(n)$ on S_q^{2n-1}

$$\delta : A(S_q^{2n-1}) \to A(SU_q(n)) \otimes A(S_q^{2n-1}) \,. \qquad (153)$$

In fact, it was realized in [35] that odd quantum Euclidean spheres are the same as unitary ones. This fact extends the classical result that odd dimensional spheres are simultaneously homogeneous spaces of orthogonal and of unitary groups.

The *-algebra $A(S_q^{N-1})$ of polynomial functions on each of the spheres S_q^{N-1} is given by generators and relations which were expressed in terms of a self-adjoint, unipotent matrix (a matrix of functions whose square is the identity) which is defined recursively. Instead in [43] the algebra was described by means of a suitable self-adjoint idempotent (a matrix of functions whose square is itself). Let us then describe the algebra $A(S_q^{N-1})$. It is generated by elements $\{x_0 = x_0*, x_i, x_i^*, i = 1, \ldots, n\}$ for $N = 2n + 1$ while for $N = 2n$ there is no x_0. These generators obey the following commutation relations,

$$\begin{aligned} x_i x_j &= q x_j x_i, \quad 0 \le i < j \le n, \\ x_i^* x_j &= q x_j x_i^*, \quad i \ne j, \end{aligned} \qquad (154a)$$

$$[x_i, x_i^*] = (1 - q^{-2})s_{i-1}, \qquad (154b)$$

with the understanding that $x_0 = 0$ if $N = 2n$, so that in this case the generator x_1 is normal,

$$x_1 x_1^* = x_1^* x_1 \quad \text{in} \quad A(S_q^{2n-1}) \,. \qquad (155)$$

The "partial radii" $s_i \in A(S_q^{2n})$, are given recursively by

$$\begin{aligned} s_i &:= s_{i-1} + x_i^* x_i = q^{-2} s_{i-1} + x_i x_i^* \,, \\ s_0 &:= x_0^2 \,, \end{aligned} \qquad (156)$$

and the last one s_n which can be shown to be central, is normalized to

$$s_n = 1 \qquad (157)$$

We see that the equality of the two formulæ for the elements s_i in (156) is equivalent to the commutation relation (154b). These s_i are self-adjoint and related as

$$0 \leq s_0 \leq \cdots \leq s_{n-1} \leq s_n = 1 . \tag{158}$$

From the commutation relations (154a) it follows for $i < j$ that $x_i^* x_i x_j = q^2 x_j x_i^* x_i$; on the other hand $x_j^* x_j x_i = x_i x_j^* x_j$. By induction, we deduce that

$$s_i x_j = \begin{cases} q^2 x_j s_i & : i < j \\ x_j s_i & : i \geq j, \end{cases} \qquad s_i x_j^* = \begin{cases} q^{-2} x_j^* s_i & : i < j \\ x_j^* s_i & : i \geq j \end{cases}$$

and that the s_i's are mutually commuting. They can be used to construct representations of the algebra as we shall show later on.

As we have mentioned, in [43] it was shown that the defining relations of the algebra $A(S_q^{N-1})$ are equivalent to the condition that a certain matrix over $A(S_q^{N-1})$ be idempotent. In [35] it was proven that this is also equivalent to the condition that another matrix be unipotent, as we shall explain presently. First consider the even spheres S_q^{2n} for any integer $n > 0$. The algebra $A(S_q^{2n})$ is generated by elements $\{x_0, x_i, x_i^*, i = 1, \ldots, n\}$. Let us first consider the free unital $*$-algebra $F := \mathbb{C}\langle 1, x_0, x_i, x_i^*, i = 1, \ldots, n \rangle$ on $2n+1$ generators. We recursively define self-adjoint matrices $u_{(2n)} \in \mathrm{Mat}_{2^n}(F)$ for all n by,

$$u_{(2n)} := \begin{pmatrix} q^{-1} u_{(2n-2)} & x_n \\ x_n^* & -u_{(2n-2)} \end{pmatrix}, \tag{159}$$

with $u_{(0)} = x_0$. The $*$-algebra $A(S_q^{2n})$ is then defined by the relations that $u_{(2n)}$ is unipotent, $u_{(2n)}^2 = 1$, and self-adjoint, $u_{(2n)}^* = u_{(2n)}$. That is, the algebra is the quotient of the free algebra F by these relations.

The self-adjointness relations merely give that x_i^* is the adjoint of x_i and x_0 is self-adjoint. Unipotency gives a matrix of 2^{2n} relations, although many of these are vacuous or redundant. These can be deduced inductively from (159) which gives,

$$u_{(2n)}^2 = \begin{pmatrix} q^{-2} u_{(2n-2)}^2 + x_n x_n^* & q^{-1} u_{(2n-2)} x_n - x_n u_{(2n-2)} \\ q^{-1} x_n^* u_{(2n-2)} - u_{(2n-2)} x_n^* & u_{(2n-2)}^2 + x_n^* x_n \end{pmatrix} . \tag{160}$$

The condition that $u_{(2n)}^2 = 1$ means in particular that $u_{(2n)}^2$ is diagonal with all the diagonal entries equal. Looking at (160), we see that the same must be true of $u_{2n-2}^2 \in \mathrm{Mat}_{2^{n-1}}(A(S_q^{2n}))$, and so on. Thus, the diagonal relations require that all the diagonal entries of (each) $u_{(2j)}^2$ are equal. If this is true for $u_{(2j-2)}^2$, then the relation for $u_{(2j)}^2$ is that the same element (the diagonal entry) can be written in two different ways. This element is simply s_j and the two ways of writing it are those given in (156). Finally, $u_{(2n)}^2 = 1$ gives the relation $s_n = 1$. The off-diagonal relations are $q^{-1} u_{(2j-2)} x_j = x_j u_{(2j-2)}$ and $q^{-1} x_j^* u_{(2j-2)} = u_{(2j-2)} x_j^*$ for every $j = 1, \ldots, n$. Because the matrix

$u_{(2j-2)}$ is constructed linearly from all of the generators x_i and x_i^* for $i <$ j, this conditions are equivalent to the commutation relations (154a). This presentation of the relations by the unipotency of $u_{(2n)}$ is also the easiest way to see that there is an isomorphism $A(S_{1/q}^{2n}) \cong A(S_q^{2n})$ which is obtained by the substitutions $q \leftrightarrow q^{-1}$, $x_0 \to (-q)^n x_0$, and $x_i \to (-q)^{n-i} x_i^*$. This transforms the matrix $u_{(2n)} \to \widetilde{u}_{(2n)}$ and the latter is unipotent and self-adjoint if and only if $u_{(2n)}$ is. Thus there is an isomorphism $A(S_{1/q}^{2n}) \cong A(S_q^{2n})$, and we can assume that $|q| > 1$ without loss of generality.

Next, consider the odd spheres S_q^{2n-1} for any integer $n > 0$. We can construct a unipotent $u_{(2n-1)} \in \mathrm{Mat}_{2^n}[A(S_q^{2n-1})]$, simply by setting $x_0 = 0$ in $u_{(2n)}$. Once again, the unipotency condition, $u_{(2n-1)}^2 = 1$, is equivalent to the relations defining the algebra $A(S_q^{2n-1})$ of polynomial functions on S_q^{2n-1}. Again, one defines self-adjoint elements $s_i \in A(S_q^{2n-1})$ such that $s_i = s_{i-1} + x_i^* x_i = q^{-2} s_{i-1} + x_i x_i^*$ with now $s_0 = x_0^2 = 0$. The commutation relations are again given by (154) but now (154b) gives in particular that the generator x_1 is normal, $x_1 x_1^* = x_1^* x_1$. The previous argument also shows that $A(S_q^{2n-1})$ is the quotient of $A(S_q^{2n})$ by the ideal generated by x_0. Geometrically, we may think of S_q^{2n-1} as a noncommutative subspace of S_q^{2n}. Because of the isomorphism $A(S_{1/q}^{2n}) \cong A(S_q^{2n})$, we have another isomorphism $A(S_{1/q}^{2n-1}) \cong A(S_q^{2n-1})$, and again we can assume that $|q| > 1$ without any loss of generality.

Remark 1. The algebras of our spheres, both in even and odd "dimensions", are generated by the entries of a projections. This is the same as the condition of *full projection* used by S. Waldmann in his analysis of Morita equivalence of star products [63].

There is also a way of realizing even spheres as noncommutative subspaces of odd ones. Consider S_q^{2n+1}, set $x_1 = x_1^* = x_0$ and relabel x_2 as x_1, *et cetera*; let $u'_{(2n+1)}$ be the matrix obtained from $u_{(2n+1)}$ with these substitutions. The matrix $u'_{(2n+1)}$ is the same as $u_{(2n)}$ in which we substitute

$$x_0 \to \begin{pmatrix} 0 & x_0 \\ x_0 & 0 \end{pmatrix}, \qquad x_j \to \begin{pmatrix} x_j & 0 \\ 0 & x_j \end{pmatrix}, \qquad j \neq 0.$$

Then the unipotency of $u'_{(2n+1)}$ yields precisely the same relations coming from the unipotency of $u_{(2n)}$. This shows that $A(S_q^{2n})$ is the quotient of $A(S_q^{2n+1})$ by the $*$-ideal generated by $x_1 - x_1^*$. Geometrically, we may think of S_q^{2n} as a noncommutative subspace of S_q^{2n+1}.

Summing up, every sphere contains a smaller sphere of dimension one less; by following this tower of inclusions to its base, we see that every sphere contains a classical S^1, because the circle does not deform. From this, it is easy to see that the spheres S_q^{N-1} have a S^1 worth of classical points. Indeed, with $\lambda \in \mathbb{C}$ such that $|\lambda|^2 = 1$, there is a family of 1-dimensional representations (characters) of the algebra $A(S_q^{N-1})$ given by

$$\psi_\lambda(1) = 1, \quad \psi_\lambda(x_n) = \lambda, \quad \psi_\lambda((x_n)^*) = \bar{\lambda},$$
$$\psi_\lambda(x_i) = \psi_\lambda((x_i)^*) = 0, \tag{161}$$

for $i = 0, 1, \ldots, n-1$ or $i = 1, \ldots, n-1$ according to whether $N = 2n+1$ or $N = 2n$, respectively.

Each even sphere algebra has an involutive automorphism

$$\sigma : A(S_q^{2n}) \to A(S_q^{2n})$$
$$x_0 \mapsto -x_0; \quad x_j \mapsto x_j, \quad j \neq 0, \tag{162}$$

which corresponds to flipping (reflecting) the classical S^{2n} across the hyperplane $x_0 = 0$. The coinvariant algebra of σ is the quotient of $A(S_q^{2n})$ by the ideal generated by x_0, which, as we have noted, is simply $A(S_q^{2n-1})$.

Geometrically this means that S_q^{2n-1} is the "equator" of S_q^{2n}, the subspace fixed by the flip.

As for odd spheres, they have an action $\rho : \mathbb{T} \to \mathrm{Aut}[A(S^{2n})]$ of the torus group $\mathbb{T}$, defined by multiplying x_1 by a phase and leaving the other generators unchanged,

$$\rho(\lambda) : A(S_q^{2n+1}) \to A(S_q^{2n+1})$$
$$x_1 \mapsto \lambda x_1; \quad x_j \mapsto x_j, \quad j \neq 1. \tag{163}$$

The coinvariant algebra is given by setting $x_1 = 0$. Now, let $u''_{(2n+1)}$ be the matrix obtained by setting $x_1 = 0$ and relabeling x_2 as x_1, *et cetera*, in the matrix $u_{(2n+1)}$. Then, $u''_{(2n+1)}$ is equivalent to tensoring $u_{(2n-1)}$ with $\left(\begin{smallmatrix} 1 & 0 \\ 0 & 1 \end{smallmatrix}\right)$,

$$u''_{(2n+1)} = u_{(2n-1)} \otimes \left(\begin{smallmatrix} 1 & 0 \\ 0 & 1 \end{smallmatrix}\right)$$

and the result is unipotent if and only if $u_{(2n-1)}$ is; that is the unipotency of $u''_{(2n+1)}$ yields all and only the same relations coming from the unipotency of $u_{(2n-1)}$. This shows that $A(S_q^{2n-1})$ is the quotient of $A(S_q^{2n+1})$ by the $*$-ideal generated by x_1 and S_q^{2n-1} is the noncommutative subspace of S_q^{2n+1} fixed by the $\mathbb{T}$-action in (163).

5.1 The Structure of the Deformations

For each deformed sphere S_q^{N-1}, we have a one parameter family of algebras $A(S_q^{N-1})$ which, at $q = 1$, gives $A(S_1^{N-1}) = A(S^{N-1})$, the algebra of polynomial functions on a classical sphere S^{N-1}. It is possible to identify this one-parameter family of algebras to a fixed vector space and view the product as varying with the parameter: let us indicate this product with the symbol $*_q$. We can then construct a Poisson bracket on $A(S^{N-1})$ from the first derivative of the product at the classical parameter value, $q = 1$,

$$\{f, g\} := -i \frac{d}{dq}\bigg|_{q=1} (f *_q g - g *_q f). \tag{164}$$

The usual properties of a Poisson bracket (Leibniz and Jacobi identities) are simple consequences of associativity.

In general, given such a one-parameter deformation from a commutative manifold $\mathcal{M}$ into noncommutative algebras, we can construct a Poisson bracket on functions. This Poisson algebra, $A(\mathcal{M})$ with the commutative product and the Poisson bracket, describes the deformation to first order. A deformation is essentially a path through an enormous space of possible algebras, and the Poisson algebra is just a tangent. Nevertheless, if the deformation is well behaved the Poisson algebra does indicate where it is heading.

It has been proved by Gromov that on the manifold $\mathbb{R}^{2n}$ there exists exotic symplectic structures [33], that is symplectic structures that are essentially different from the standard one. This means that any such an exotic structure cannot be obtained as the pull back of the standard one via an embedding of $\mathbb{R}^{2n}$ into itself (see [46] for additional information). To our knowledge, deformations of $\mathbb{R}^{2n}$ with an exotic structure have not been constructed yet.

On the other hand, the standard symplectic structure yields a well known deformation. If we complete to a C*-algebra, then the deformation of $\mathcal{C}_0(\mathbb{R}^{2n})$ (continuous functions vanishing at infinity) will be the algebra, $\mathcal{K}$, of compact operators on a countably infinite-dimensional Hilbert space. We shall use only this deformation of $\mathbb{R}^{2n}$ in what follows.

Let us go back to the spheres S_q^{N-1} and look more closely at them. We have seen that the S_q^{2n-1} noncommutative subspace of S_q^{2n} corresponds to the equator, $S^{2n-1} \subset S^{2n}$, where $x_0 = 0$ and the Poisson structure on S^{2n} is degenerate. On the remaining $S^{2n} \setminus S^{2n-1}$, the Poisson structure is nondegenerate. So, topologically, we have a union of two copies of symplectic $\mathbb{R}^{2n}$. Then, the kernel of the quotient map $A(S_q^{2n}) \to A(S_q^{2n-1})$ should be a deformation of the subalgebra of functions on S^{2n} which vanish at the equator. If we complete to C*-algebras, this should give us the direct sum of two copies of $\mathcal{K}$, one for each hemisphere. Thus we expect that the C*-algebra $\mathcal{C}(S_q^{2n})$ will be an extension:

$$0 \to \mathcal{K} \oplus \mathcal{K} \to \mathcal{C}(S_q^{2n}) \to \mathcal{C}(S_q^{2n-1}) \to 0 \,. \tag{165}$$

In odd dimensions, the Poisson structure is necessarily degenerate. However, the S_q^{2n-1} noncommutative subspace of S_q^{2n+1} corresponds classically to the Poisson structure being more degenerate on $S^{2n-1} \subset S^{2n+1}$. It is of rank $2n$ at most points, but of rank $2n-2$ (or less) along S^{2n-1}. The complement $S^{2n+1} \setminus S^{2n-1}$ has a symplectic foliation by $2n$ dimensional leaves which is invariant under the $\mathbb{T}$ action; the simplest possibility is that this corresponds to the product in the identification

$$S^{2n+1} \setminus S^{2n-1} \cong S^1 \times \mathbb{R}^{2n} \,.$$

If we complete to C*-algebras, then the deformation of this should give the algebra $\mathcal{C}(S^1) \otimes \mathcal{K}$. The kernel of the quotient map $A(S_q^{2n+1}) \to A(S_q^{2n-1})$ should be this deformation, so we expect another extension,

$$0 \to \mathcal{C}(S^1) \otimes \mathcal{K} \to \mathcal{C}(S_q^{2n+1}) \to \mathcal{C}(S_q^{2n-1}) \to 0 . \qquad (166)$$

The extensions (165) and (166) turn out to be correct. As we have mentioned, the odd dimensional spheres we are considering are equivalent to the "unitary" odd quantum spheres of Vaksman and Soibelman [61]. In [37] Hong and Szymański obtained the C*-algebras $\mathcal{C}(S_q^{2n+1})$ as Cuntz-Krieger algebras of suitable graphs. From this construction they derived the extension (166). They also considered even spheres, defined as quotients of odd ones by the ideal generated by $x_1 - x_1^*$. These are thus isomorphic to the even spheres we are considering here. They also obtained these as Cuntz-Krieger algebras and derived the extension (165). However, as explicitly stated in the introduction to [37], they were unable to realize even spheres as quantum homogeneous spaces of quantum orthogonal groups, thus also failing to realize that "unitary" and "orthogonal" odd quantum spheres are the same.

5.2 Representations

We shall now exhibit all representations of the algebra $A(S_q^{N-1})$ which in turn extend to the C*-algebra $\mathcal{C}(S_q^{N-1})$.

Representations of the odd dimensional spheres were constructed in [61]. The primitive spectra of all these spheres were compute in [37], which amounts to a classification of representations. The representations for quantum Euclidean spheres have also been constructed in [31] by thinking of them as quotient algebras of quantum Euclidean planes.

The structure of the representations can be anticipated from the construction of S_q^{N-1} via the extensions (165) and (166) and by remembering that an irreducible representation ψ can be partially characterized by its kernel. Moreover, an irreducible representation of a C*-algebra restricts either to an irreducible or a trivial representation of any ideal; and conversely, an irreducible representation of an ideal extends to an irreducible representation of the C*-algebra (see for instance [26]).

For an even sphere S_q^{2n}, the kernel of an irreducible representation ψ will contain one or both of the copies of $\mathcal{K} \subset \mathcal{C}(S_q^{2n})$. If $\mathcal{K} \oplus \mathcal{K} \subseteq \ker \psi$, then ψ factors through $\mathcal{C}(S_q^{2n-1})$ and is given by a representation of that algebra. If one copy of $\mathcal{K}$ is not in $\ker \psi$, then ψ restricts to a representation of this $\mathcal{K}$. However, $\mathcal{K}$ has only one irreducible representation. Since $\mathcal{K}$ is an ideal in $\mathcal{C}(S_q^{2n})$, the unique irreducible representation of $\mathcal{K}$ uniquely extends to a representation of $\mathcal{C}(S_q^{2n})$ (with the other copy of $\mathcal{K}$ in its kernel).

Thus, up to isomorphism the irreducible representations of S_q^{2n} should be:

1. all irreducible representations of S_q^{2n-1},
2. a unique representation with kernel the second copy of $\mathcal{K}$,
3. a unique representation with kernel the first copy of $\mathcal{K}$.

From the extension (165) we expect that the generator x_0 is a self-adjoint element of $\mathcal{K} \oplus \mathcal{K} \subset \mathcal{C}(S_q^{2n})$ and it should have almost discrete, real spectrum: it will therefore be used to decompose the Hilbert space in a representation.

Similarly, from the construction of S_q^{2n+1} by the extension (166), one can anticipate the structure of its representations. Firstly, if $\mathcal{C}(S^1) \otimes \mathcal{K} \subseteq \ker \psi$, then ψ factors through $\mathcal{C}(S_q^{2n-1})$ and is really a representation of S_q^{2n-1}. Otherwise, ψ restricts to an irreducible representation of $\mathcal{C}(S^1) \otimes \mathcal{K}$. This factorizes as the tensor product of an irreducible representation of $\mathcal{C}(S^1)$ with one of $\mathcal{K}$. The irreducible representations of $\mathcal{C}(S^1)$ are simply given by the points of S^1, and as we have mentioned, $\mathcal{K}$ has a unique irreducible representation. The representations of $\mathcal{C}(S^1) \otimes \mathcal{K}$ are thus classified by the points of S^1. These representations extend uniquely from the ideal $\mathcal{C}(S^1) \otimes \mathcal{K}$ to the whole algebra $\mathcal{C}(S_q^{2n+1})$.

Thus, up to isomorphism, the irreducible representations of S_q^{2n+1} should be:

1. all irreducible representations of S_q^{2n-1},
2. a family of representations parameterized by S^1.

In the construction of the representations, a simple identity regarding the spectra of operators will be especially useful (see, for instance [56]). If x is an element of any C*-algebra, then

$$\{0\} \cup \operatorname{Spec} x^* x = \{0\} \cup \operatorname{Spec} x x^* \ . \tag{167}$$

5.3 Even Sphere Representations

To illustrate the general structure we shall start by describing the lowest dimensional case, namely S_q^2. This is isomorphic to the so-called equator sphere of Podleś [51]. For this sphere, the representations were also constructed in [48] in a way close to the one presented here.

Let us then consider the sphere S_q^2. As we have discussed, we expect that, in some faithful representation, x_0 is a compact operator and thus has an almost discrete, real spectrum. However, we cannot assume a priori that x_0 has eigenvalues, let alone that its eigenvectors form a complete basis of the Hilbert space. The sphere relation $1 = x_0^2 + x_1^* x_1 = q^{-2} x_0^2 + x_1 x_1^*$ shows that $x_0^2 \le 1$ and thus $\|x_0\| \le 1$. As x_0 is self-adjoint, this shows that $\operatorname{Spec} x_0 \subseteq [-1, 1]$. By (167) we have also,

$$\{0\} \cup \operatorname{Spec} x_1^* x_1 = \{0\} \cup \operatorname{Spec} x_1 x_1^*$$
$$\{0\} \cup \operatorname{Spec}(1 - x_0^2) = \{0\} \cup \operatorname{Spec}(1 - q^{-2} x_0^2)$$
$$\{1\} \cup \operatorname{Spec} x_0^2 = \{1\} \cup q^{-2} \operatorname{Spec} x_0^2 \ .$$

Because we have assumed that $|q| \ge 1$, the only subsets of $[0, 1]$ that satisfy this condition are $\{0\}$ and $\{0, q^{-2k} \mid k = 0, 1, \dots\}$.

If $x_0 \ne 0 \in \mathcal{C}(S_q^2)$ then $\operatorname{Spec} x_0^2$ is the latter set. We cannot simply assume that $x_0 \ne 0$, since not every *-algebra is a subalgebra of a C*-algebra; however, our explicit representations will show that that is the case here.

Now let $\mathcal{H}$ be a separable Hilbert space and suppose that we have an irreducible $*$-representation, $\psi : A(S_q^2) \to \mathcal{L}(\mathcal{H})$.

If $\psi(x_0) = 0$ then $1 = \psi(x_1)\psi(x_1)^* = \psi(x_1)^*\psi(x_1)$. Thus $\psi(x_1)$ is unitary, and by the assumption of irreducibility, it is a number $\lambda \in \mathbb{C}$, $|\lambda| = 1$. So, $\mathcal{H} = \mathbb{C}$ and the representation is $\psi_\lambda^{(1)}$ defined by,

$$\psi_\lambda^{(1)}(x_0) = 0; \qquad \psi_\lambda^{(1)}(x_1) = \lambda, \quad \lambda \in S^1 . \tag{168}$$

Thus we have an S^1 worth of representations with x_0 in the kernel.

If $\psi(x_0) \neq 0$, then $1 \in \operatorname{Spec} x_0^2$; it is an isolated point in the spectrum and therefore an eigenvalue. For some sign $\pm$ there exists a unit vector $|0\rangle \in \mathcal{H}$ such that $\psi(x_0)|0\rangle = \pm |0\rangle$. The relation $x_0 x_1 = q x_1 x_0$ suggests that x_1 and x_1^* shift the eigenvalues of x_0. Indeed, for $k = 0, 1, \ldots$, the vector $\psi(x_1^*)^k |0\rangle$ is an eigenvector as well, because

$$\psi(x_0)\psi(x_1^*)^k |0\rangle = q^{-k}\psi(x_1^{*k} x_0)|0\rangle = \pm q^{-k}\psi(x_1^*)^k |0\rangle .$$

By normalizing, we obtain a sequence of unit eigenvectors, defined by

$$|k\rangle := (1 - q^{-2k})^{-1/2}\psi(x_1^*)|k-1\rangle .$$

We have thus two representations $\psi_+^{(2)}$ and $\psi_-^{(2)}$, and direct computation shows that

$$\begin{aligned}
\psi_\pm^{(2)}(x_0)|k\rangle &= \pm q^{-k}|k\rangle, \\
\psi_\pm^{(2)}(x_1)|k\rangle &= (1 - q^{-2k})^{1/2}|k-1\rangle, \\
\psi_\pm^{(2)}(x_1^*)|k\rangle &= (1 - q^{-2(k+1)})^{1/2}|k+1\rangle.
\end{aligned} \tag{169}$$

The eigenvectors $\{|k\rangle \mid k = 0, 1, \ldots\}$ are mutually orthogonal because they have distinct eigenvalues, and by the assumption of irreducibility they form a basis for the Hilbert space $\mathcal{H}$.

Notice that any power of $\psi_\pm^{(2)}(x_0)$ is a trace class operator, while this is not the case for the operators $\psi_\pm^{(2)}(x_1)$ and $\psi_\pm^{(2)}(x_1^*)$ nor for any of their powers.

Note also that the representations (169) are related by the automorphism σ in (162), as

$$\psi_\pm^{(2)} \circ \sigma = \psi_\mp^{(2)} . \tag{170}$$

If we set a value of q with $|q| < 1$ in (169), the operators would be unbounded. This is the reason for assuming that $|q| > 1$. The assumption was used in computing $\operatorname{Spec} x_0$. Not only is $\|x_0\| \leq 1$, but by a similar calculation $\|x_0\| \leq |q|$. Which bound is more relevant obviously depends on whether q is greater or less than 1. For $|q| < 1$ the appropriate formulæ for the representations can be obtained from (169) by replacing the index k with $-k - 1$. As a consequence, the role of x_1 and x_1^* as lowering and raising operators is exchanged.

For the general even spheres S_q^{2n} the structure of the representations is similar to that for S_q^2, but more complicated. The element x_0 is no longer sufficient to completely decompose the Hilbert space of the representation and we need to use all the commuting self-adjoint elements $s_i \in A(S_q^{2n})$ defined in (156).

Suppose that $\psi : A(S_q^{2n}) \to \mathcal{L}(\mathcal{H})$ is an irreducible $*$-representation. If $\psi(x_0) = 0$, then ψ factors through $A(S_q^{2n-1})$. Thus ψ is an irreducible representation of $A(S_q^{2n-1})$; these will be discussed later.

If $\psi(x_0) \neq 0$, then $\psi(s_0) \neq 0$, and by the relations (158), all of the $\psi(s_i)$'s are nonzero. Proceeding recursively, we find that there is a simultaneous eigenspace with eigenvalue 1 for all the $\psi(s_i)$'s. That is, there must exist a unit vector $|0,\ldots,0\rangle \in \mathcal{H}$ such that $\psi(s_i)|0,\ldots,0\rangle = |0,\ldots,0\rangle$ for all i and $\psi(x_0)|0,\ldots,0\rangle = \pm|0,\ldots,0\rangle$. More unit vectors are defined by

$$|k_0,\ldots,k_{n-1}\rangle \sim \psi(x_1^*)^{k_0}\ldots\psi(x_n^*)^{k_{n-1}}|0,\ldots,0\rangle$$

modulo a positive normalizing factor. Working out the correct normalizing factors we get two representations $\psi_\pm^{(2n)}$ defined by,

$$\psi_\pm^{(2n)}(x_0)|k_0,\ldots,k_{n-1}\rangle = \pm q^{-(k_0+\cdots+k_{n-1})}|k_0,\ldots,k_{n-1}\rangle \qquad (171)$$

$$\psi_\pm^{(2n)}(x_i)|k_0,\ldots,k_{n-1}\rangle = (1-q^{-2k_{i-1}})^{1/2}q^{-(\sum_{j=i}^{n-1}k_j)}|\ldots,k_{i-1}-1,\ldots\rangle$$

$$\psi_\pm^{(2n)}(x_i^*)|k_0,\ldots,k_{n-1}\rangle = (1-q^{-2(k_{i-1}+1)})^{1/2}q^{-(\sum_{j=i}^{n-1}k_j)}|\ldots,k_{i-1}+1,\ldots\rangle$$

with $i = 1,\ldots,n$. With this values for the index i, we see that x_i lowers k_{i-1}, whereas x_i^* raises k_{i-1}. From irreducibility the collection of vectors $\{|k_0,\ldots,k_{n-1}\rangle, k_i \geq 0\}$ constitute a complete basis for the Hilbert space $\mathcal{H}$.

As before, the two representations (171) are related by the automorphism σ,

$$\psi_\pm^{(2n)} \circ \sigma = \psi_\mp^{(2n)} . \qquad (172)$$

Again the formulæ (171) for the representations are corrected for $|q| > 1$; and again the representations for $|q| < 1$ can be obtained by replacing all indices k_i with $-k_i - 1$ in (171).

In all of the irreducible representations of $A(S_q^{2n})$, the representative of x_0 is compact; in fact it is trace class. We can deduce from this that the C*-ideal generated by $\psi_\pm^{(2n)}(x_0)$ in $\mathcal{C}(S_q^{2n})$ is isomorphic to $\mathcal{K}(\mathcal{H})$, the ideal of all compact operators on $\mathcal{H}$. By using the continuous functional calculus, we can apply any function $f \in \mathcal{C}[-1,1]$ to x_0. If f is supported on $[0,1]$, then $f(x_0) \in \ker \psi_-^{(2n)}$. Likewise if f is supported in $[-1,0]$, then $f(x_0) \in \ker \psi_+^{(2n)}$. From this we deduce that the C*-ideal generated by x_0 in $\mathcal{C}(S_q^{2n})$ is $\mathcal{K} \oplus \mathcal{K}$. One copy of $\mathcal{K}$ is $\ker \psi_+^{(2n)}$; the other is $\ker \psi_-^{(2n)}$. Thus we get exactly the extension (165).

5.4 Odd Sphere Representations

Again, to illustrate the general strategy we shall work out in detail the simplest case, that of the sphere S_q^3. This can be identified with the underlying noncommutative space of the quantum group $SU_q(2)$ and as such the representations of the algebra are well known [66].

The generators $\{x_i, x_i^* \mid i = 1, 2\}$ of the algebra $A(S_q^3)$ satisfy the commutation relations $x_1 x_2 = q x_2 x_1$, $x_i^* x_j = q x_j x_i^*, i \neq j$, $[x_1, x_1^*] = 0$, and $[x_2, x_2^*] = (1 - q^{-2}) x_1 x_1^*$. Furthermore, there is also the sphere relation $1 = x_2^* x_2 + x_1^* x_1 = x_2 x_2^* + q^{-2} x_1 x_1^*$.

The normal generator x_1 plays much the same role for the representations of S_q^3 that x_0 does for those of S_q^2. The sphere relation shows that $\|x_1\| \leq 1$ and

$$\{0\} \cup \operatorname{Spec} x_2^* x_2 = \{0\} \cup \operatorname{Spec} x_2 x_2^*$$
$$\{0\} \cup \operatorname{Spec}(1 - x_1^* x_1) = \{0\} \cup \operatorname{Spec}(1 - q^{-2} x_1 x_1^*)$$
$$\{1\} \cup \operatorname{Spec} x_1^* x_1 = \{1\} \cup q^{-2} \operatorname{Spec} x_1^* x_1 ,$$

which shows that either $x_1 = 0$ or $\operatorname{Spec} x_1^* x_1 = \{0, q^{-2k} \mid k = 0, 1, \ldots\}$. Let $\psi : A(S_q^3) \to \mathcal{L}(\mathcal{H})$ be an irreducible $*$-representation.

If $\psi(x_1) = 0$ then the relations reduce to $1 = \psi(x_2)\psi(x_2)^* = \psi(x_2)^*\psi(x_2)$. Thus $\psi(x_2)$ is unitary and by the assumption of irreducibility, it is a scalar, $\psi(x_2) = \lambda \in \mathbb{C}$ with $|\lambda| = 1$. Thus, as before, we have an S^1 of representations of this kind. If $\psi(x_1) \neq 0$, then $1 \in \operatorname{Spec} \psi(x_1^* x_1)$ and is an isolated point in the spectrum. Thus, there exists a unit vector $|0\rangle \in \mathcal{H}$ such that $\psi(x_1^* x_1)|0\rangle = |0\rangle$, and by the assumption of irreducibility, there is some $\lambda \in \mathbb{C}$ with $|\lambda| = 1$ such that $\psi(x_1)|0\rangle = \lambda|0\rangle$. We see then that $\psi(x_2^*)^k |0\rangle$ is an eigenvector

$$\psi(x_1)\psi(x_2^*)^k |0\rangle = q^{-k}\psi(x_2^{*k} x_1)|0\rangle = \lambda q^{-k}\psi(x_2^*)^k |0\rangle .$$

By normalizing, we get a sequence of unit eigenvectors recursively defined by

$$|k\rangle := (1 - q^{-2k})^{-1/2}\psi(x_2^*)|k - 1\rangle .$$

A family of representations $\psi_\lambda^{(3)}$, $\lambda \in S^1$, is then defined by

$$\psi_\lambda^{(3)}(x_1)|k\rangle = \lambda q^{-k}|k\rangle ,$$
$$\psi_\lambda^{(3)}(x_1^*)|k\rangle = \bar\lambda q^{-k}|k\rangle ,$$
$$\psi_\lambda^{(3)}(x_2)|k\rangle = (1 - q^{-2k})^{1/2}|k - 1\rangle ,$$
$$\psi_\lambda^{(3)}(x_2^*)|k\rangle = (1 - q^{-2(k+1)})^{1/2}|k + 1\rangle . \tag{173}$$

We notice that any power of $\psi_\lambda^{(3)}(x_1)$ or $\psi_\lambda^{(3)}(x_1^*)$ is a trace class operator, while this is not the case for the operators $\psi_\lambda^{(3)}(x_2)$ and $\psi_\lambda^{(3)}(x_2^*)$ nor for any of their powers.

Next, for the general odd spheres S_q^{2n+1} , let $\psi : A(S_q^{2n+1}) \to \mathcal{L}(\mathcal{H})$ be an irreducible representation.

If $\psi(x_1) = 0$ then ψ factors through $A(S_q^{2n-1})$ and is an irreducible representation of that algebra.

If $\psi(x_1) \neq 0$ then $\psi(s_1) \neq 0, \psi(s_2) \neq 0$, et cetera. By the same arguments as for S_q^{2n}, there must exist a simultaneous eigenspace with eigenvalue 1 for all of $s_1, \ldots s_n$. By the assumption of irreducibility, this eigenspace is 1-dimensional. Let $|0, \ldots, 0\rangle \in \mathcal{H}$ be a unit vector in this eigenspace. Then $s_i |0, \ldots, 0\rangle = |0, \ldots, 0\rangle$ for $i = 1, \ldots, n$. The restriction of $\psi(x_1)$ to this subspace is unitary and thus for some $\lambda \in \mathbb{C}$ with $|\lambda| = 1$, one has that $\psi(x_1) |0, \ldots, 0\rangle = \lambda |0, \ldots, 0\rangle$. We can construct more simultaneous eigenvectors of the s_i's by defining

$$|k_1, \ldots, k_n\rangle \sim \psi(x_2)^{k_1} \ldots \psi(x_{n+1})^{k_n} |0, \ldots, 0\rangle$$

modulo a positive normalizing constant. Working out the normalization, one has a family of representations $\psi_\lambda^{(2n+1)}$,

$$\psi_\lambda^{(2n+1)}(x_1) |k_1, \ldots, k_n\rangle = \lambda q^{-(k_1 + \cdots + k_n)} |k_1, \ldots, k_n\rangle ,$$
$$\psi_\lambda^{(2n+1)}(x_1^*) |k_1, \ldots, k_n\rangle = \bar{\lambda} q^{-(k_1 + \cdots + k_n)} |k_1, \ldots, k_n\rangle ,$$
$$\psi_\lambda^{(2n+1)}(x_i) |k_1, \ldots, k_n\rangle = (1 - q^{-2k_{i-1}})^{1/2} q^{-(\sum_{j=i}^n k_j)} |\ldots, k_{i-1} - 1, \ldots\rangle , \tag{174}$$
$$\psi_\lambda^{(2n+1)}(x_i^*) |k_1, \ldots, k_n\rangle = (1 - q^{-2(k_{i-1}+1)})^{1/2} q^{-(\sum_{j=i}^n k_j)} |\ldots, k_{i-1} + 1, \ldots\rangle ,$$

for $i = 2, \ldots, n+1$. With this values for the index i, x_i lowers k_{i-1}, whereas x_i^* raises k_{i-1}. From irreducibility the vectors $\{|k_1, \ldots, k_n\rangle, k_i \geq 0\}$ form an orthonormal basis of the Hilbert space $\mathcal{H}$.

As for the even case, the formulæ (174) give bounded operators only for $|q| > 1$; and as before, the representations for $|q| < 1$ can be obtained by replacing all indices k_i with $-k_i - 1$.

Again, as in the even case, we can verify that $\psi_\lambda^{(2n+1)}(x_1)$ is compact (indeed, trace class) and that the ideal generated by $\psi_\lambda^{(2n+1)}(x_1)$ is $\mathcal{K}(\mathcal{H})$, in the C*-algebra completion of the image $\psi_\lambda^{(2n+1)}(A(S_q^{2n+1}))$. The representations $\psi_\lambda^{(2n+1)}$ can be assembled into a single representation by adjointable operators on a Hilbert $\mathcal{C}(S^1)$-module. With this we can verify that the ideal generated by x_1 in $\mathcal{C}(S_q^{2n+1})$ is $\mathcal{C}(S^1) \otimes \mathcal{K}$ and this verifies the extension (166).

Summing up, we get a complete picture of the set of irreducible representations of all these spheres S_q^N; or equivalently, of the primitive spectrum of the C*-algebra $\mathcal{C}(S_q^N)$ of continuous functions on S_q^N.

For the odd spheres S_q^{2n+1}, the set of irreducible representations is indexed by the union of $n+1$ copies of S^1. These run from the representations $\psi_\lambda^{(2n+1)}$ of S_q^{2n+1} given in (174) down to the one dimensional representations $\psi_\lambda^{(1)}$ that factor through the undeformed S^1.

For the even spheres S_q^{2n}, the set of irreducible representations is indexed by the union of n copies of S^1 and 2 points. The isolated points correspond to the 2 representations $\psi_\pm^{(2n)}$ specific to S_q^{2n} and given in (171); the circles correspond to representations $\psi_\lambda^{(2m+1)}$ coming from lower odd dimensional spheres, down to the undeformed S^1.

6 K-Homology and K-Theory for Quantum Spheres

We explicitly construct complete sets of generators for the K-theory (by non-trivial self-adjoint idempotents and unitaries) and the K-homology (by non-trivial Fredholm modules) of the spheres S_q^{N-1}. We also construct the corresponding Chern characters in cyclic homology and cohomology and compute the pairing of K-theory with K-homology.

We shall study generators of the K-homology and K-theory of the spheres S_q^{N-1}. The K-theory classes will be given by means of self-adjoint idempotents (naturally associated with the aforementioned unipotents) and of unitaries in algebras of matrices over $A(S_q^{N-1})$. The K-homology classes will be given as (homotopy classes of) suitable 1-summable Fredholm modules using the representations constructed previously.

For odd spheres (i.e. for N even) the odd K-homology generators are first given in terms of unbounded Fredholm modules. These are given by means of a natural unbounded operator D which, while failing to have compact resolvent, has bounded commutators with all elements in the algebra $A(S_q^{2n-1})$.

In fact, in order to compute the pairing of K-theory with K-homology, it is more convenient to first compute the Chern characters and then use the pairing between cyclic homology and cohomology [10]. Thus, together with the generators of K-theory and K-homology we shall also construct the associated Chern characters in the cyclic homology $HC_*[A(S_q^{N-1})]$ and cyclic cohomology $HC^*[A(S_q^{N-1})]$ respectively.

Needless to say, the pairing is integral (it comes from a noncommutative index theorem). The non-vanishing of the pairing will testify to the non-triviality of the elements that we construct in both K-homology K-theory.

It is worth recalling the K-theory and homology of the classical spheres. For an even dimensional sphere S^{2n}, the groups are

$$K^0(S^{2n}) \cong \mathbb{Z}^2, \qquad K^1(S^{2n}) = 0\,,$$
$$K_0(S^{2n}) \cong \mathbb{Z}^2, \qquad K_1(S^{2n}) = 0\,.$$

One generator of the K-theory $[1] \in K^0(S^{2n})$ is given by the trivial 1-dimensional bundle. The other generator of $K^0(S^{2n})$ is the left handed spinor bundle. One K-homology generator $[\varepsilon] \in K_0(S^{2n})$ is "trivial" and is the push-forward of the generator of $K_0(*) \cong \mathbb{Z}$ by the inclusion $\iota : * \hookrightarrow S^{2n}$ of a point (any point) into the sphere. The other generator, $[\mu] \in K_0(S^{2n})$, is the K-orientation of S^{2n} given by its structure as a spin manifold [10].

For an odd dimensional sphere, the groups are

$$K^0(S^{2n+1}) \cong \mathbb{Z}, \qquad K^1(S^{2n+1}) \cong \mathbb{Z},$$
$$K_0(S^{2n+1}) \cong \mathbb{Z}, \qquad K_1(S^{2n+1}) \cong \mathbb{Z}.$$

The generator $[1] \in K^0(S^{2n+1})$ is the trivial 1-dimensional bundle. The generator of $K^1(S^{2n+1})$ is a nontrivial unitary matrix-valued function on S^{2n+1}; for instance, it may be takes as the matrix (187) in the limit $q = 1$. The trivial generator $[\varepsilon] \in K_0(S^{2n+1})$ is again given by the inclusion of a point. The generator $[\mu] \in K_1(S^{2n+1})$ is the K-orientation of S^{2n+1} given by its structure as a spin manifold [10].

There is a natural pairing between K-homology and K-theory. If we pair $[\varepsilon]$ with a vector bundle we get the rank of the vector bundle, i.e. the dimension of its fibers. If we pair $[\mu]$ with a vector bundle it gives the "degree" of the bundle, a measure of its nontriviality. Similarly, pairing with $[\mu]$ measures the nontriviality of a unitary.

The K-theory and K-homology of the quantum Euclidean spheres are isomorphic to that of the classical spheres; that is, for any N and q, one has that $K_*[\mathcal{C}(S_q^{N-1})] \cong K^*(S^{N-1})$ and $K^*[\mathcal{C}(S_q^{N-1})] \cong K_*(S^{N-1})$. In the case of K-theory, this was proven by Hong and Szymański in [37] using their construction of the C*-algebras as Cuntz-Krieger algebras of graphs. The groups K_0 and K_1 were given as the cokernel and the kernel respectively, of a matrix canonically associated with the graph. The result for K-homology can be proven using the same techniques [21, 52]: the groups K^0 and K^1 are now given as the kernel and the cokernel respectively, of the transposed matrix. The K-theory and the K-homology for the particular case of S_q^2 (in fact for all Podleś spheres S_{qc}^2) was worked out in [48]) while for $S^3 \cong SU_q(2)$ it was spelled out in [47].

6.1 K-Homology

Because the K-homology of these deformed spheres is isomorphic to the K-homology of the ordinary spheres, we need to construct two independent generators. First consider the "trivial" generator of $K^0[\mathcal{C}(S_q^{N-1})]$. This can be constructed in a manner closely analogous to the undeformed case.

As we have already mentioned, the trivial generator of $K_0(S^{N-1})$ is the image of the generator of the K-homology of a point by the functorial map $K_*(\iota) : K_0(*) \to K_0(S^{N-1})$, where $\iota : * \hookrightarrow S^{N-1}$ is the inclusion of a point into the sphere. The quantum Euclidean spheres do not have as many points, but they do have some. We have seen that the relations among the various spheres always include a homomorphism $A(S_q^{N-1}) \to A(S^1)$. Equivalently, every S_q^{N-1} has a circle S^1 as a classical subspace; thus for every $\lambda \in S^1$ there is a point, i.e., the homomorphism $\psi_\lambda^{(1)} : \mathcal{C}(S_q^{N-1}) \to \mathbb{C}$.

We can construct an element $[\varepsilon_\lambda] \in K^0[\mathcal{C}(S_q^{N-1})]$ by pulling back the generator of $K^0(\mathbb{C})$ by $\psi_\lambda^{(1)}$. This construction factors through $K_0(S^1)$. Because S^1 is path connected, the points of S^1 all define homotopic (and hence K-homologous) Fredholm modules. Thus there is a single K-homology class $[\varepsilon_\lambda] \in K^0[\mathcal{C}(S_q^{N-1})]$, independent of $\lambda \in S^1$.

The canonical generator of $K^0(\mathbb{C})$ is given by the following Fredholm module: The Hilbert space is $\mathbb{C}$; the grading operator is $\gamma = 1$; the representation is the obvious representation of $\mathbb{C}$ on $\mathbb{C}$; the Fredholm operator is 0. If we pull this back to $K^0[\mathcal{C}(S_q^{N-1})]$ using $\psi_\lambda^{(1)}$, then the Fredholm module ε_λ is given in the same way but with $\psi_\lambda^{(1)}$ for the representation.

Given this construction of ε_λ, it is straightforward to compute its Chern character $\mathrm{ch}^*(\varepsilon_\lambda) \in HC^*[A(S_q^{N-1})]$: It is the pull back of the Chern character of the canonical generator of $K^0(\mathbb{C})$. An element of the cyclic cohomology HC^0 is a trace. The degree 0 part of the Chern character of the canonical generator of $K^0(\mathbb{C})$ is given by the identity map $\mathbb{C} \to \mathbb{C}$, which is trivially a trace. Pulling this back we find $\mathrm{ch}^0(\varepsilon_\lambda) = \psi_\lambda^{(1)} : A(S_q^{N-1}) \to \mathbb{C}$ which is also a trace because it is a homomorphism to a commutative algebra. These are distinct elements of $HC^0[A(S_q^{N-1})]$ for different values of λ. However, because the Fredholm modules ε_λ all lie in the same K-homology class, their Chern characters are all equivalent in *periodic* cyclic cohomology defined in (22). Indeed, applying the periodicity operator (21) once one gets that the cohomology classes $S(\psi_\lambda^{(1)}) \in HC^2[A(S_q^{N-1})]$ are all the same. For the computation of the pairing between K-theory and K-homology, any trace determining the same periodic cyclic cohomology class can be used. The most symmetric choice of trace is given by averaging $\psi_\lambda^{(1)}$ over $\lambda \in S^1 \subset \mathbb{C}$:

$$\tau^0(a) := \oint_{S^1} \psi_\lambda^{(1)}(a) \frac{d\lambda}{2\pi i \lambda} \, .$$

The result is normalized, $\tau^0(1) = 1$, and vanishes on all the generators. The higher degree parts of $\mathrm{ch}^*(\varepsilon_\lambda)$ depend only on the K-homology class $[\varepsilon_\lambda]$ and can be constructed from τ^0 by the periodicity operator (21).

6.2 Fredholm Modules for Even Spheres

We will now construct an element $[\mu_{\mathrm{ev}}] \in K^0[\mathcal{C}(S_q^{2n})]$ by giving a suitable even Fredholm module $\mu := (\mathcal{H}, F, \gamma)$.

Identify the Hilbert spaces for the representations $\psi_\pm^{(2n)}$ given in (171) by identifying their bases, and call this $\mathcal{H}$. The representation for the Fredholm module is

$$\psi := \psi_+^{(2n)} \oplus \psi_-^{(2n)}$$

acting on $\mathcal{H} \oplus \mathcal{H}$. The grading operator and the Fredholm operator are respectively,

$$\gamma = \begin{pmatrix} 1 & 0 \\ 0 & -1 \end{pmatrix}, \quad F = \begin{pmatrix} 0 & 1 \\ 1 & 0 \end{pmatrix}.$$

It is obvious that F is odd (since it anticommutes with γ) and Fredholm (since it is invertible). The remaining property to check is that for any $a \in A(S_q^{2n})$, the commutator $[F, \psi(a)]_-$ is compact. Indeed,

$$[F, \psi(a)]_- = \begin{pmatrix} 0 & -\psi_+^{(2n)}(a) + \psi_-^{(2n)}(a) \\ \psi_+^{(2n)}(a) - \psi_-^{(2n)}(a) & 0 \end{pmatrix}.$$

However, $\psi_+^{(2n)}(a) - \psi_-^{(2n)}(a) = \psi_+^{(2n)}[a - \sigma(a)]$ and $a - \sigma(a)$ is always proportional to a power of x_0. Thus this is not only compact, it is trace class. This also shows that we have (at least) a 1-summable Fredholm module. This is in contrast to the fact that the analogous element of $K_0(S^{2n})$ for the undeformed sphere is given by a $2n$-summable Fredholm module.

The corresponding Chern character [10] $\mathrm{ch}^*(\mu_{\mathrm{ev}})$ has a component in degree 0, $\mathrm{ch}^0(\mu_{\mathrm{ev}}) \in HC^0[A(S_q^{2n})]$. From the general construction (59), the element $\mathrm{ch}^0(\mu_{\mathrm{ev}})$ is the trace

$$\tau^1(a) := \tfrac{1}{2} \mathrm{Tr}\left(\gamma F([F, \psi(a)])\right) = \mathrm{Tr}\left[\psi_+^{(2n)}(a) - \psi_-^{(2n)}(a)\right]. \tag{175}$$

As we have mentioned, $\psi_+^{(2n)}(a) - \psi_-^{(2n)}(a) = \psi_+^{(2n)}[a - \sigma(a)]$ is trace class since $a - \sigma(a)$ is always proportional to a power of x_0. The higher degree parts of $\mathrm{ch}^*(\mu_{\mathrm{ev}})$ can be obtained via the periodicity operator (21).

For S_q^2 our Fredholm module coincides with the one constructed in [48].

6.3 Fredholm Modules for Odd Spheres

The element $[\mu_{\mathrm{odd}}] \in K^1[\mathcal{C}(S_q^{2n+1})]$ is most easily given by an unbounded Fredholm module. The corresponding unbounded operator D which, while failing to have compact resolvent, has bounded commutators with all elements in the algebra $A(S_q^{N-1})$.

Let the representation ψ be the direct integral (over $\lambda \in S^1$) of the representations $\psi_\lambda^{(2n+1)}$ given in (174). The operator is the unbounded "Dirac" operator $D := \lambda^{-1}\frac{d}{d\lambda}$.

From (174), we see that the representative of x_1 is proportional to λ and as a consequence,

$$[D, \psi(x_1)]_- = \psi(x_1) \tag{176a}$$

whereas for $i > 1$ the representative of x_i does not involve λ and therefore

$$[D, \psi(x_i)]_- = 0, \quad i > 0. \tag{176b}$$

Since $a \mapsto [D, \psi(a)]_-$ is a derivation, this shows that $[D, \psi(a)]_-$ is bounded for any $a \in A(S_q^{2n+1})$. Note however that for $n > 0$ (i.e., except for S^1)

all eigenvalues of D have infinite degeneracy and therefore D does not have compact resolvent.

This triple can be converted in to a bounded Fredholm module by applying a cutoff function to D. A convenient choice is $F = \chi(D)$ where

$$\chi(m) := \begin{cases} 1 & : m > 0 \\ -1 & : m \leq 0. \end{cases}$$

To be more explicit, use a Fourier series basis for the Hilbert space,

$$|k_0, k_1, \ldots, k_n\rangle := \lambda^{k_0} |k_1, \ldots, k_n\rangle,$$

in which the representation is given by,

$$\psi(x_1)|k_0, \ldots, k_n\rangle = q^{-(k_1 + \cdots + k_n)} |k_0 + 1, \ldots, k_n\rangle,$$

$$\psi(x_1^*)|k_0, \ldots, k_n\rangle = q^{-(k_1 + \cdots + k_n)} |k_0 - 1, \ldots, k_n\rangle,$$

$$\psi(x_i)|k_0, \ldots, k_n\rangle = (1 - q^{-2k_i - 1})^{1/2} q^{-(k_i + \cdots + k_n)} |\ldots, k_{i-1} - 1, \ldots\rangle,$$

$$\psi(x_i^*)|k_0, \ldots, k_n\rangle = (1 - q^{-2(k_{i-1}+1)})^{1/2} q^{-(\sum_{j=i}^{n} k_j)} |\ldots, k_{i-1} + 1, \ldots\rangle,$$

for $i = 1, \ldots, n$. The Fredholm operator is then given by

$$F|k_0, \ldots, k_n\rangle = \chi(k_0) |k_0, \ldots, k_n\rangle .$$

The only condition to check is that the commutator $[F, \psi(a)]_-$ is compact for any $a \in \mathcal{C}(S_q^{2n+1})$. Since $a \mapsto [F, \psi(a)]_-$ is a derivation, it is sufficient to check this on generators. One finds

$$[F, \psi(x_i)]_- = 0 , \quad i > 1 ,$$

$$[F, \psi(x_1)]_- |k_0, \ldots, k_n\rangle = \begin{cases} 2q^{-(k_1 + \cdots + k_n)} |1, k_1, \ldots, k_n\rangle & : k_0 = 0 \\ 0 & : k_0 \neq 0, \end{cases} \quad (177)$$

which is indeed compact, and in fact trace class.

Thus, this is a 1-summable Fredholm module. Again this is in contrast to the fact that the analogous element of $K_1(S^{2n+1})$ for the undeformed sphere is given by a $(2n + 1)$-summable Fredholm module.

Its Chern character [10] begins with $\mathrm{ch}^{\frac{1}{2}}(\mu_{\mathrm{odd}}) \in HC^1[A(S_q^{2n+1})]$. From the general construction (58), the element $\mathrm{ch}^{\frac{1}{2}}(\mu_{\mathrm{odd}})$ is given by the cyclic 1-cocycle φ defined by

$$\varphi(a, b) := \tfrac{1}{2} \mathrm{Tr}\left(\psi(a)[F, \psi(b)]_-\right). \quad (178)$$

One checks directly cyclicity, i.e. $\varphi(a, b) = -\varphi(b, a)$, and closure under b, i.e. $\varphi(ab, c) - \varphi(a, bc) + \varphi(ca, b) = b\varphi(a, b, c)$.

The higher degree parts of $\mathrm{ch}^*(\mu_{\mathrm{odd}})$ can be obtained via the periodicity operator (21).

For $S_q^3 \cong \mathrm{SU}_q(2)$ our Fredholm module coincides with the one in [47].

6.4 Singular Integrals

We could interpret the classes $[\mu_{ev}] \in K^0[\mathcal{C}(S_q^{2n})]$ and $[\mu_{odd}] \in K^1[\mathcal{C}(S_q^{2n+1})]$ as giving "singular" integrals over the corresponding quantum spheres. With the associated Chern characters given in (175) and (178) respectively, and from the general expression (57), these integrals are given by,

$$\int_{S_q^{2n}} a = \tau^1(a) , \quad \forall a \in \mathcal{A}(S_q^{2n}) , \tag{179}$$

$$\int_{S_q^{2n+1}} a \, db = \phi(a,b) , \quad \forall a,b \in \mathcal{A}(S_q^{2n+1}) . \tag{180}$$

As a way of illustration, let us compute them on generators. We indicate with δ_{ij} the usual Kronecker delta which is equal to 1 if $i = j$ and 0 otherwise.

Firstly, for even spheres we find

$$\int_{S_q^{2n}} x_i = \tau^1(x_i) = \mathrm{Tr}\left[\psi_+^{(2n)}(x_i - \sigma(x_i))\right] = 2\,\mathrm{Tr}\left[\psi_+^{(2n)}(x_i)\right]\delta_{i0} . \tag{181}$$

Thus, we need to compute

$$\mathrm{Tr}[\psi_+^{(2n)}(x_0)] = \sum_{k_0=0}^{\infty} \cdots \sum_{k_{n-1}=0}^{\infty} q^{-(k_0+\cdots+k_{n-1})} = \left(\sum_{k=0}^{\infty} q^{-k}\right)^n$$
$$= (1 - q^{-1})^{-n},$$

and in turn,

$$\int_{S_q^{2n}} x_i = \frac{2}{(1 - q^{-1})^n}\,\delta_{i0} . \tag{182}$$

Similarly, for odd spheres we find

$$\int_{S_q^{2n+1}} x_i \, dx_j^* = \phi(x_i, x_i^*) = \frac{1}{2}\,\mathrm{Tr}\left(\psi(x_1^*)[F, \psi(x_1)]_-\right)\delta_{i1}\delta_{j1} . \tag{183}$$

We have already computed $[F, \psi(x_1)]_-$ in ((177)). From that, we get

$$\psi(x_1^*)[F, \psi(x_1)]_- |k_0, \ldots, k_n\rangle = \begin{cases} 2q^{-2(k_1+\cdots+k_n)} |0, k_1, \ldots, k_n\rangle & : k_0 = 0 \\ 0 & : k_0 \neq 0 . \end{cases}$$

Thus,

$$\mathrm{Tr}\left(\psi(x_1^*)[F, \psi(x_1)]_-\right) = \sum_{k_1=0}^{\infty} \cdots \sum_{k_n=0}^{\infty} 2q^{-2(k_1+\cdots+k_n)} = 2\left(\sum_{k=0}^{\infty} q^{-2k}\right)^n$$
$$= 2(1 - q^{-2})^{-n},$$

and in turn,

$$\int_{S_q^{2n+1}} x_i \, dx_j^* = \frac{1}{(1 - q^{-2})^n}\,\delta_{i1}\delta_{j1} . \tag{184}$$

6.5 K-Theory for Even Spheres

For S_q^{2n} we construct two classes in the K-theory group $K_0[\mathcal{C}(S_q^{2n})] \cong \mathbb{Z}^2$.

The first class is trivial. The element $[1] \in K_0[\mathcal{C}(S_q^{2n})]$ is the equivalence class of $1 \in \mathcal{C}(S_q^{2n})$ which is of course an idempotent. In order to compute the pairing with K-homology, we need the degree 0 part of its Chern character, $\mathrm{ch}_0[1]$, which is represented by the cyclic cycle 1.

The second, nontrivial, class was presented in [43]. It is given by an idempotent $e_{(2n)}$ constructed from the unipotent (159) as

$$e_{(2n)} = \tfrac{1}{2}(\mathbb{I} + u_{(2n)}) \tag{185}$$

(again, for the sphere S_q^2 the idempotent (185) was already in [48]). Its degree 0 Chern character, $\mathrm{ch}_0(e_{(2n)}) \in HC_0[A(S_q^{2n})]$, is

$$\mathrm{ch}_0(e_{(2n)}) = \mathrm{tr}(e_{(2n)} - \tfrac{1}{2}\mathbb{I}_{2n}) = \tfrac{1}{2}\,\mathrm{tr}(u_{(2n)})$$
$$= \tfrac{1}{2}(q^{-1} - 1)^n x_0 , \tag{186}$$

since the recursive definition (159) of the unipotent $u_{(2n)}$ shows that,

$$\mathrm{tr}(u_{(2n)}) = (q^{-1} - 1)\,\mathrm{tr}(u_{(2n-2)}) = (q^{-1} - 1)^n x_0 .$$

Now, we can pair these classes with the two K-homology elements which we constructed in Sect. 6.1. First,

$$\langle \varepsilon_\lambda, [1] \rangle := \tau^0(1) = 1 ,$$

which is hardly surprising. Second, the "rank" of the idempotent $e_{(2n)}$ is

$$\langle \varepsilon_\lambda, e_{(2n)} \rangle := \tau^0(\mathrm{tr}(e_{(2n)})) = 2^{n-1} .$$

Also not surprising is the "degree" of $[1]$,

$$\langle \mu_{\mathrm{ev}}, [1] \rangle := \tau^1(1) = \mathrm{Tr}\big[\psi_+^{(2n)}(1) - \psi_-^{(2n)}(1)\big] = \mathrm{Tr}(1 - 1) = 0 .$$

The more complicated pairing is,

$$\langle \mu_{\mathrm{ev}}, e_{(2n)} \rangle := \tau^1(\mathrm{ch}_0\, e_{(2n)})$$
$$= \mathrm{Tr} \circ \psi_+^{(2n)} \circ (1 - \sigma)\,\big(2^{n-1} + \tfrac{1}{2}[q^{-1} - 1]^n x_0\big)$$
$$= (q^{-1} - 1)^n\,\mathrm{Tr}[\psi_+^{(2n)}(x_0)] = (q^{-1} - 1)^n(1 - q^{-1})^{-n}$$
$$= (-1)^n .$$

The fact that the matrix of pairings,

	$[1]$	$[e_{(2n)}]$
$[\varepsilon_\lambda]$	1	2^{n-1}
$[\mu_{\mathrm{ev}}]$	0	$(-1)^n$

is invertible over the integers proves that the classes $[1], [e_{(2n)}]$, both elements of $K_0[\mathcal{C}(S_q^{2n})] \cong \mathbb{Z}^2$, and the classes $[\varepsilon_\lambda], [\mu_{ev}]$, both in $K^0[\mathcal{C}(S_q^{2n})] \cong \mathbb{Z}^2$, are nonzero and that no one of them may be a multiple of another class; thus they are generators of the respective groups.

Classically, the "degree" of the left-handed spinor bundle is -1. So, the K-homology class which correctly generalizes the classical K-orientation class $[\mu] \in K_0(S^{2n})$ is actually $(-1)^{n+1}[\mu_{ev}]$.

6.6 K-Theory for Odd Spheres

Again, define $[1] \in K_0[\mathcal{C}(S_q^{2n+1})]$ as the equivalence class of $1 \in \mathcal{C}(S_q^{2n+1})$. The pairing with our element $[\varepsilon_\lambda] \in K^0[\mathcal{C}(S_q^{2n+1})]$ is again,

$$\langle \varepsilon_\lambda, [1] \rangle := \tau^0(1) = 1 .$$

There is no other independent generator in $K_0[\mathcal{C}(S_q^{2n+1})] \cong \mathbb{Z}$.

Instead, $K_1[\mathcal{C}(S_q^{2n+1})] \cong \mathbb{Z}$ is nonzero. So we need to construct a generator there. An odd K-theory element is an equivalence class of unitary matrices over the algebra. We can construct an appropriate sequence of unitary matrices recursively, just as we constructed the unipotents and idempotents.

Let $V_{(2n+1)} \in \mathrm{Mat}_{2^n}(A(S_q^{2n+1}))$ be defined recursively by

$$V_{(2n+1)} = \begin{pmatrix} x_{n+1} & q^{-1}V_{(2n-1)} \\ -V_{(2n-1)}^* & x_{n+1}^* \end{pmatrix} , \tag{187}$$

with $V_{(1)} = x_1$. By using the defining relations (154) one directly proves that it is unitary:

$$V_{(2n+1)}V_{(2n+1)}^* = V_{(2n+1)}^*V_{(2n+1)} = 1 . \tag{188}$$

In order to pair our K-homology element $[\mu_{odd}] \in K^1[\mathcal{C}(S_q^{2n+1})]$ with the unitary $V_{(2n+1)}$, we need the lower degree part $\mathrm{ch}_{\frac{1}{2}}(V_{(2n+1)}) \in HC_1[A(S_q^{2n+1})]$ of its Chern character. It is given by the cyclic cycle,

$$\mathrm{ch}_{\frac{1}{2}}(V_{(2n+1)}) := \tfrac{1}{2}\,\mathrm{tr}\left(V_{(2n+1)} \otimes V_{(2n+1)}^* - V_{(2n+1)}^* \otimes V_{(2n+1)}\right)$$
$$= \tfrac{1}{2}(q^{-2}-1)^n(x_1 \otimes x_1^* - x_1^* \otimes x_1) . \tag{189}$$

Now, compute the pairing,

$$\langle \mu_{odd}, V_{(2n+1)} \rangle := \langle \varphi, \mathrm{ch}_{\frac{1}{2}}(V_{(2n+1)}) \rangle$$
$$= -(q^{-2}-1)^n\varphi(x_1^*, x_1)$$
$$= -\tfrac{1}{2}(q^{-2}-1)^n\,\mathrm{Tr}\,(\psi(x_1^*)[F,\psi(x_1)]_-)$$
$$= -\tfrac{1}{2}(q^{-2}-1)^n 2(1-q^{-2})^{-n}$$
$$= (-1)^{n+1} .$$

This proves that $[V_{(2n+1)}] \in K_1[\mathcal{C}(S_q^{2n+1})]$ and $[\mu_{odd}] \in K^1[\mathcal{C}(S_q^{2n+1})]$ are nonzero and that neither may be a multiple of another class. Thus $[V_{(2n+1)}]$ and $[\mu_{odd}]$ are indeed generators of these groups.

Acknowledgement

These lecture notes are based on work done with Alain Connes, Eli Hawkins, John Madore and Joe Várilly; I am most grateful to them. I thank Satoshi Watamura, Yoshiaki Maeda and Ursula Carow-Watamura for their kind invitation to Sendai and for the fantastic hospitality there. I also thank all participants of the conference for the great time we had together and Tetsuya Masuda for his help and hospitality in Sendai and Tokyo. Finally, I thank Ludwik Dąbrowski, Eli Hawkins and Denis Perrot for very useful suggestions which improved the compuscript; and Nicola Ciccoli for making me aware of the existence of exotic symplectic structures on $\mathbb{R}^{2n}$.

References

1. P. Aschieri, F. Bonechi, *On the Noncommutative Geometry of Twisted Spheres*, Lett. Math. Phys. 59 (2002) 133–156.

2. M.F. Atiyah, *Global theory of elliptic operators*, in: 'Proc. Intn'l Conf. on functional analysis and related topics, Tokyo 1969, Univ. of Tokyo Press 1970, pp. 21–30.

3. M.F. Atiyah, *Geometry of Yang-Mills Fields*, Accad. Naz. Dei Lincei, Scuola Norm. Sup. Pisa, 1979.

4. F. Bonechi, N. Ciccoli, M. Tarlini, *Quantum even spheres Σ_q^{2n} from Poisson double suspension*, Commun. Math. Phys., 243 (2003) 449–459.

5. J. Brodzki, *An Introduction to K-theory and Cyclic Cohomology*, Polish Scientific Publishers 1998.

6. P.S. Chakraborty, A. Pal, *Equivariant Spectral triple on the Quantum SU(2)-group*, math.KT/0201004.

7. P.S. Chakraborty, A. Pal, *Spectral triples and associated Connes-de Rham complex for the quantum SU(2) and the quantum sphere*, math.QA/0210049.

8. A. Connes, *C* algèbres et géométrie différentielle*, C.R. Acad. Sci. Paris, Ser. A-B, 290 (1980) 599–604.

9. A. Connes, *Noncommutative differential geometry*, Inst. Hautes Etudes Sci. Publ. Math., 62 (1985) 257–360.

10. A. Connes, *Noncommutative geometry*, Academic Press 1994.

11. A. Connes, *Noncommutative geometry and reality*, J. Math. Physics, 36 (1995) 6194–6231.

12. A. Connes, *Gravity coupled with matter and foundation of noncommutative geometry*, Commun. Math. Phys., 182 (1996) 155–176.

13. A. Connes, *A short survey of noncommutative geometry*, J. Math. Physics, 41 (2000) 3832–3866.

14. A. Connes, *Cyclic Cohomology, Quantum group Symmetries and the Local Index Formula for $SU_q(2)$*, math.QA/0209142,

15. A. Connes, M. Dubois-Violette, *Noncommutative finite-dimensional manifolds. I. Spherical manifolds and related examples*, Commun. Math. Phys. 230 (2002) 539–579.

16. A. Connes, M. Dubois-Violette, *Moduli space and structure of noncommutative 3-spheres*, math.QA/0308275.

17. A. Connes, G. Landi, *Noncommutative manifolds, the instanton algebra and isospectral deformations*, Commun. Math. Phys. 221 (2001) 141–159.
18. A. Connes, H. Moscovici, *The local index formula in noncommutative geometry*, GAFA 5 (1995) 174–243.
19. A. Connes, M. Rieffel, *Yang-Mills for noncommutative two tori*, In: *Operator algebras and mathematical physics* (Iowa City, Iowa, 1985). Contemp. Math. Oper. Algebra Math. Phys., 62, Amer. Math. Soc., Providence, 1987, 237-266.
20. A. Connes, D. Sullivan, N. Teleman, *Quasi-conformal mappings, operators on Hilbert spaces and a local formula for characteristic classes*, Topology 33 (1994) 663–681.
21. J. Cuntz, *On the homotopy groups for the space of endomorphisms of a C*-algebra (with applications to topological Markov chains)*, in *Operator algebras and group representations*, Pitman 1984, pp. 124-137
22. L. Dąbrowski, *The Garden of Quantum Spheres*, Banach Center Publications, 61 (2003) 37–48.
23. L. Dąbrowski, G. Landi, *Instanton algebras and quantum 4-spheres*, Differ. Geom. Appl. 16 (2002) 277–284.
24. L. Dąbrowski, G. Landi, T. Masuda, *Instantons on the quantum 4-spheres S_q^4*, Commun. Math. Phys. 221 (2001) 161–168.
25. L. Dąbrowski, A. Sitarz, *Dirac Operator on the Standard Podleś Quantum Sphere*, Banach Center Publications, 61 (2003) 49–58.
26. K.R. Davidson, *C*-algebras by example*, American Mathematical Society 1996.
27. J. Dixmier, *Existence de traces non normales*, C.R. Acad. Sci. Paris, Ser. A-B, 262 (1966) 1107–1108.
28. M. Dubois-Violette, *Equations de Yang et Mills, modèles σ à deux dimensions et généralisation*, in '*Mathématique et Physique*', Progress in Mathematics, vol. 37, Birkhäuser 1983, pp. 43–64.
29. M. Dubois-Violette, Y. Georgelin, *Gauge Theory in Terms of Projector Valued Fields*, Phys. Lett. 82B (1979) 251–254.
30. L.D. Faddeev, N.Y. Reshetikhin, L.A. Takhtajan, *Quantization of Lie groups and Lie algebras*, Leningrad Math. Jour. 1 (1990) 193.
31. G. Fiore, *The Euclidean Hopf algebra $U_q(e^N)$ and its fundamental Hilbert space representations*, J. Math. Phys. 36 (1995) 4363.
    G. Fiore, *The q-Euclidean algebra $U_q(e^N)$ and the corresponding q-Euclidean lattice*, Int. J. Mod. Phys. A11 (1996) 863.
32. I. Frenkel, M. Jardim *Quantum Instantons with Classical Moduli Spaces*, Commun. Math. Phys., 237 (2003) 471–505.
33. M. Gromov, *Pseudo holomorphic curves in symplectic manifolds*, Inventiones Mathematicæ, 82 (1985) 307–347.
34. J.M. Gracia-Bondía, J.C. Várilly, H. Figueroa, *Elements of Noncommutative Geometry*, Birkhäuser, Boston, 2001.
35. E. Hawkins, G. Landi, *Fredholm Modules for Quantum Euclidean Spheres*, J. Geom. Phys. 49 (2004) 272–293.
36. T. Hadfield, *Fredholm Modules over Certain Group C*-algebras*, math.OA/0101184.
    T. Hadfield, *K-homology of the Rotation Algebra A_θ*, math.OA/0112235.
    T. Hadfield, *The Noncommutative Geometry of the Discrete Heisenberg Group*, math.OA/0201093.
37. J.H. Hong, W. Szymański, *Quantum spheres and projective spaces as graph algebras*, Commun. Math. Phys. 232 (2002) 1, 157–188.

38. G. Kasparov, *Topological invariants of elliptic operators, I. K-homology*, Math. URSS Izv. 9 (1975) 751–792.

39. U. Krähmer, *Dirac Operators on Quantum Flag Manifolds*, math.QA/0305071.

40. G. Landi, *An Introduction to Noncommutative Spaces and Their Geometries*, Springer, 1997; Online corrected edition, Springer Server, September 2002. A preliminary version is available as hep-th/9701078.

41. G. Landi, *Deconstructing Monopoles and Instantons*, Rev. Math. Phys. 12 (2000) 1367–1390.

42. G. Landi, talk at the Mini-workshop on *Noncommutative Geometry Between Mathematics and Physics*, Ancona, February 23–24, 2001.

43. G. Landi, J. Madore, *Twisted Configurations over Quantum Euclidean Spheres*, J. Geom. Phys. 45 (2003) 151–163.

44. B. Lawson, M.L. Michelsohn, *Spin Geometry*, Princeton University Press, 1989.

45. J.-L. Loday, *Cyclic Homology*, Springer, Berlin, 1992.

46. D. McDuff, D. Salamon, *Introduction to Symplectic Topology*, Oxford University Press, Oxford 1995.

47. T. Masuda, Y. Nakagami, J. Watanabe, *Noncommutative differential geometry on the quantum SU(2). I: An algebraic viewpoint*, K-Theory 4 (1990) 157.

48. T. Masuda, Y. Nakagami, J. Watanabe, *Noncommutative differential geometry on the quantum two sphere of P. Podleś, I: An algebraic viewpoint*, K-Theory 5 (1991) 151.

49. T. Natsume, C.L. Olsen, *A new family of noncommutative 2-spheres*, J. Func. Anal., 202 (2003) 363-391.
T. Natsume, this proceedings.

50. N. Nekrasov, A. Schwarz, *Instantons in noncommutative $\mathbb{R}^4$ and (2,0) superconformal six dimensional theory*, Commun. Math. Phys. 198 (1998) 689–703.

51. P. Podleś, *Quantum spheres*, Lett. Math. Phys. 14 (1987) 193.

52. I. Raeburn, W. Szymański, *Cuntz-Krieger algebras of infinite graphs and matrices*, University of Newcastle Preprint, December 1999.

53. M.A. Rieffel, *C^*-algebras associated with irrational rotations*, Pac. J. Math. 93 (1981) 415–429.

54. M.A. Rieffel, *Deformation Quantization for Actions of $\mathbb{R}^d$*, Memoirs of the Amer. Math. Soc. 506, Providence, RI, 1993.

55. M.A. Rieffel, *K-groups of C^*-algebras deformed by actions of $\mathbb{R}^d$*, J. Funct. Anal. 116 (1993) no. 1, 199–214.

56. S. Sakai, *C^*-Algebras and W^*-Algebras*, Springer 1998.

57. K. Schmüdgen, E. Wagner, *Dirac operator and a twisted cyclic cocycle on the standard Podleś quantum sphere*, math.QA/0305051.

58. B. Simon, *Trace Ideals and their Applications*, Cambridge Univ. Press, 1979.

59. A. Sitarz, *Twists and spectral triples for isospectral deformations*, Lett. Math. Phys. 58 (2001) 69–79.

60. A. Sitarz, *Dynamical Noncommutative Spheres*, Commun. Math. Phys., 241 (2003) 161–175.

61. L.L. Vaksman, Y.S. Soibelman, *The algebra of functions on quantum SU(n+1) group and odd-dimensional quantum spheres*, Leningrad Math. Jour. 2 (1991) 1023.

62. J.C. Várilly, *Quantum symmetry groups of noncommutative spheres*, Commun. Math. Phys. 221 (2001) 511–523.

63. S. Waldmann, *Morita Equivalence, Picard Groupoids and Noncommutative Field Theories*, math.QA/0304011 and this proceedings.

64. M. Welk, *Differential Calculus on Quantum Spheres*, Czechoslovak J. Phys. 50 (2000) no. 11, 1379–1384.
65. M. Wodzicki, *Noncommutative residue, Part I. Fundamentals*, In K-theory, arithmetic and geometry. Lecture Notes Math., 1289, Springer, 1987, 320–399.
66. S.L. Woronowicz, *Twisted* SU(2) *group. An example of a noncommutative differential calculus*, Publ. Res. Inst. Math. Sci. 23 (1987) 117.

Some Noncommutative Spheres

T. Natsume

Division of Mathematics, Nagoya Institute of Technology, Showa-ku, Nagoya
466-8555, Japan

1 Introduction

Noncommutative geometry provides us various means to deal with noncommutative phenomena in mathematics, and C^*-algebras (and von Neumann
algebras) are primary tools there. Let us discuss an example. When a locally compact group G acts on a space X, the orbit space, equipped with
the quotient topology, is not even a T_1-space in general. Thus ordinary topological, geometric methods do no work in this situation. Noncommutative
geometry gives us an alternative approach, where highly noncommutative
croseed prossed product C^*-algebra $C_0(X) \rtimes G$ is the key player.

According to Gel'fand-Naimark theory, the correspondence $X \mapsto C_0(X)$
is an equivalence of the category of locally compact spaces and propercontin-
uous maps with the category of abelian C^*-algebras and $*$-homomorphisms.
In noncommutative geometry a noncommutative C^*-algebra A is regarded
as the C^*-algebra of continuous functions on a virtual space. In this sense
A is sometimes called a noncommutative space. Applying geomtric methods
to A, information on the underlying noncommutative phenomena will be extracted. In order to develop useful machineries in noncommutative geometry,
we need interesting examples of noncommutative spaces.

Primary examples are the most celebrated noncommutative tori and the
Moyal products, which are obtained by "deforming" the commuting product
s of functions on the manifolds $\mathbb{T}^2$ and $\mathbb{R}^{2n}$, respectively.

Definition 1 Let A be a C^*-algebra. A *deformation* of A is given by a
continuous field of C^*-algebras (A_t) over $[0, \epsilon)$ for some $\epsilon > 0$ such that (1)
$A_0 = A$, (2) A_t has the same K-theory as A_0.

We will call (A_t) a *strong* deformation of the C^*-algebra A, if (A_t) is
obtained by altering relations on the generators of A, as follows.

Let A be a C^*-algebra characterized as the universal C^*-algebra generated
by $x_1, \ldots, x_k$ subject to the algebraic relations

$$R_1(x_1, \ldots, x_k, x_1^*, \ldots, x_k^*) = 0, \ldots, R_j(x_1, \ldots, x_k, x_1^*, \ldots, x_k^*) = 0 .$$

Assume that there exist one-parameter families $R_1^t, \ldots, R_j^t$ of relations such
that $R_1^0 = R_0, \ldots, R_j^0 = R_j$, and assume that each A_t is the universal C^*-
algebra generated by k generators subject to the relations $R_0^t = 0, \ldots, R_j^t = 0$

T. Natsume: *Some Noncommutative Spheres*, Lect. Notes Phys. **662**, 57–66 (2005)
www.springerlink.com   © Springer-Verlag Berlin Heidelberg 2005

for $0 \leq t < \epsilon$ for some $\epsilon > 0$. If also (A_t) is a nontrivial deformation of A in the sense of Definition 1.1, then we say (A_t) is a *strong deformation of the C^*-algebra* A.

When there exists a sequence (t_j) with $\lim t_j = 0$ such that A_{t_j} is not isomorphic to A, we say that the deformation (A_t) is nontrivial.

The work of M.A. Rieffel in his study of continuous fields of C^*-algebras and of deformation quantization is seminal to this area of investigation. (See [6, 7]).

In the present paper we study deformations of the spheres in dimensions 2, 3 and 4. The study of eformations of low-dimensional spheres has been a major subject in noncommutative geometry. O. Bratteli, G.A. Elliott, D. Evans and A. Kishimoto defined [2] noncommutative 2-spheres as the fixed point algebras of noncommutative tori with respect to automorphisms of order two. Meanwhile, P. Podleś defined [12] noncommutative 2-spheres as also fixed point algebras of S. Woronowicz's quantum $SU_\mu(2)$ with respect to the maximal abelian subgroup. As for S^3, of course the quantum $SU_\mu(2)$'s are deformations of the $S^3 \cong SU(2)$. At this point it should be mentioned that from complex analytic viewpoint K. Matsumoto introduced noncommutative 3-spheres [8] (see also [10]). Taking the suspension of Matsumoto's noncommutative 3-spheres yields noncommutative 4-spheres on which A. Connes and G. Landi studied instanton [3]. For the reader who is interested in these noncommutative spheres, a survey paper is available [4].

2 Construction of Noncommutative 2-Spheres

Regard $C(S^2)$ as the universal C^*-algebra generated by three mutually commuting self-adjoint elements x, y, z subject to the relation: $x^2 + y^2 + z^2 = 1$. Alternatively, set $\zeta = x + iy$, so that $C(S^2)$ is the universal C^*-algebra generated by a normal element ζ and a commuting self-adjoint element z subject to the relation $\zeta^* \zeta + z^2 = 1$.

Let ι be he canonical inclusion $S^2 \subset \mathbb{R}^3$. The standard $SO(3)$-invariant volume form is

$$\omega_0 = -\iota^*(xdy \wedge dz - ydx \wedge dz + zdx \wedge dy) \,.$$

In cylindrical coordinates, $\omega_0 = dz \wedge d\theta$. Thus

$$L^2(S^2, \omega_0) \cong L^2([-1, 1], dz) \otimes L^2(S^1, d\theta) \,.$$

Via the Fourier transform, identify $L^2(S^1, d\theta)$ with $\ell^2(\mathbb{Z})$. Consider the C^*-algebra $C([-1, 1]) \otimes C^*(\mathbb{Z})$ acting on $L^2([-1, 1], dz) \otimes \ell^2(\mathbb{Z})$. Denote by U the canonical unitary in $C^*(\mathbb{Z})$ corresponding to the generator $1 \in \mathbb{Z}$. Using the isomorphisms of Hilbert spaces above, $C(S^2)$ can be identified with the C^*-subalgebra of $C([-1, 1]) \otimes C^*(\mathbb{Z})$ generated by $\zeta = \sqrt{1 - z^2} \otimes U$ and $z \otimes 1$.

We wish to deform $C(S^2)$ by using a suitable crossed product. Set $I = [-1, 1]$. For $t \geq 0$, define a map $\phi_t : I \longrightarrow I$ by

$$\phi_t(z) = tz^2 + z - t .$$

If $0 \leq t < \frac{1}{2}$, then ϕ_t is a homeomorphism. Denote the associated automorphism of $C(I)$ by α_t, i.e.,

$$\alpha_t(f) = f \circ \phi_t^{-1}, \quad f \in C(I) .$$

Consider the crossed product $C(I) \rtimes_{\alpha_t} \mathbf{Z}$, and let U_t be the canonical unitary, corresponding to $1 \in \mathbf{Z}$. Set

$$z_t = z \in C(I) \rtimes_{\alpha_t} \mathbf{Z} , \quad \text{and}$$

$$\zeta_t = U_t \sqrt{1 - z_t^2} \in C(I) \rtimes_{\alpha_t} \mathbf{Z} .$$

Let A_t be the C^*-algebra generated by ζ_t and z_t in the crossed product $C(I) \rtimes_{\alpha_t} \mathbf{Z}$. From the definition of the crossed product, we get:

$$\alpha_t(z_t) = U_t z_t U_t^* ,$$

$$\alpha_t^{-1}(z_t) = U_t^* z_t U_t = t z_t^2 + z_t - t ,$$

$$z_t = \alpha_t \alpha_t^{-1}(z_t) = t U_t z_t^2 U_t^* + U_t z_t U_t^* - t .$$

Proposition 1 The elements ζ_t and z_t satisfy the relations:

(1) $\zeta_t^* \zeta_t + z_t^2 = 1$,
(2) $\zeta_t z_t - z_t \zeta_t = t \, \zeta_t (1 - z_t^2)$,
(3) $\zeta_t \zeta_t^* + (t \, \zeta_t \zeta_t^* + z_t)^2 = 1$.

Proof. As for (1),

$$\zeta_t^* \zeta_t = \sqrt{1 - z_t^2} U_t^* U_t \sqrt{1 - z_t^2} = 1 - z_t^2 .$$

As for (2),

$$\begin{aligned}
\zeta_t z_t - z_t \zeta_t &= U_t \sqrt{1 - z_t^2} z_t - U_t (U_t^* z_t U_t) \sqrt{1 - z_t^2} \\
&= U_t \sqrt{1 - z_t^2} z_t - U_t (\alpha_t^{-1}(z_t)) \sqrt{1 - z_t^2} \\
&= U_t \sqrt{1 - z_t^2} z_t - U_t (t z_t^2 + z_t - t) \sqrt{1 - z_t^2} \\
&= t \zeta_t (1 - z_t^2) .
\end{aligned}$$

Finally, for the relation (3), first of all we have

$$\zeta_t \zeta_t^* = U_t(1 - z_t^2)U_t^* = 1 - U_t z_t^2 U_t^* .$$

Since $z_t = tU_t z_t^2 U_t^* + U_t z_t U_t^* - t$, we have $U_t z_t^2 U_t^* = \frac{1}{t}(z_t + t - U_t z_t U_t^*)$. Then $\zeta_t \zeta_t^* = 1 - \frac{1}{t}(z_t + t - U_t z_t U_t^*)$. Therefore $t\zeta_t \zeta_t^* = U_t z_t U_t^* - z_t$. Thus $U_t z_t U_t^* = t\zeta_t \zeta_t^* + z_t$. Consequently,

$$(t\zeta_t \zeta_t^* + z_t)^2 = U_t z_t^2 U_t^* = 1 - \zeta_t \zeta_t^* ,$$

hence the relation (3) holds.

The main result of this section is the following.

Theorem 1 The C^*-algebra A_t with generators ζ_t and self-adjoint z_t has the universal property for the relations (1)-(3), for $0 \le t < \frac{1}{2}$.

Moreover the collection (S_t^2) has a structure of a nontrivial strong deformation of $C(S^2)$.

Proof. Let a_t, b_t be bounded operators on a Hilbert space $\mathcal{H}$, where b_t is self-adjoint and satisfying the relations:

(1) $a_t^* a_t + b_t^2 = 1$,
(2) $a_t b_t - b_t a_t = t a_t(1 - b_t^2)$,
(3) $a_t a_t^* + (t a_t a_t^* + b_t)^2 = 1$, for $0 \le t < \frac{1}{2}$.

To establish the desired universal property for A_t we need to show the existence of a $*$-homomorphism from A_t to the C^*-algebra generated by a_t and b_t which sends ζ_t to a_t and z_t to b_t. The proof proceeds by a reduction to the universal property of the crossed product. In order to do so, spectral theory plays a crucial role. For a detailed proof, see [11].

3 Analytic and Topological Properties of S_t^2

In this section we investigtae analytic and topological properties of the noncommutative 2-spheres.

We observe that the collection (S_t^2) has the structure of a continuous field of C^*-algebras over $[0, \frac{1}{2})$ with (ζ_t) and (z_t) being continuous sections, and show that this field restricted to $(0, \frac{1}{2})$ is trivial (in particular, S_t^2 and S_s^2 are isomorphic for $t, s > 0$).

We start with the short exact sequence:

$$0 \longrightarrow C_0((-1,1)) \longrightarrow C(I) \longrightarrow \mathbb{C}^2 \longrightarrow 0 ,$$

where the map $C(I) \longrightarrow \mathbb{C}^2$ evaluates functions at $\{\pm 1\}$. The homeomorphism ϕ_t fixes these endpoints, so the $\mathbb{Z}$-action on $C(I)$ preserves the ideal $C_0((-1,1))$. Hence there exists an exact sequence :

$(*)$ $0 \longrightarrow C_0((-1,1)) \rtimes_{\alpha_t} \mathbb{Z} \longrightarrow C(I) \rtimes_{\alpha_t} \mathbb{Z} \longrightarrow \mathbb{C}^2 \otimes C^*(\mathbb{Z}) \longrightarrow 0 .$

Denote by π the surjection $C(I) \rtimes_{\alpha_t} \mathbb{Z} \longrightarrow \mathbb{C}^2 \otimes C^*(\mathbb{Z})$. Recall that $S_t^2 \subset C(I) \rtimes_{\alpha_t} \mathbb{Z}$.

Proposition 2 The restriction of π onto S_t^2 induces a short exact sequence:

$$0 \longrightarrow C_0((-1,1)) \rtimes_{\alpha_t} \mathbb{Z} \longrightarrow S_t^2 \longrightarrow \mathbb{C}^2 \longrightarrow 0 .$$

We compute the K-theory of the noncommutative spheres, and exhibit two generators.

For facts about the K-theory of C^*-algebras, see [16].

Proposition 3 The K-theory of the noncommutative spheres $S_t^2, 0 < t < \frac{1}{2}$ is given by

$$K_0(S_t^2) \cong \mathbb{Z}^2 \quad \text{and} \quad K_1(S_t^2) = 0 .$$

Proof. Applying the six-term exact sequence [16, Theorem 9.3.2] to the sequence from Proposition 3.1, we get an exact sequence of K-groups:

$$\begin{array}{ccc}
\mathbb{Z} \longrightarrow K_0(S_t^2) \longrightarrow & K_0(\mathbb{C}^2) \cong \mathbb{Z}^2 \\
\uparrow & \downarrow \\
0 \longleftarrow K_1(S_t^2) \longleftarrow & K_1(C_0((-1,1)) \rtimes_{\alpha_t} \mathbb{Z}) .
\end{array}$$

To determine $K_0(S_t^2)$ and $K_1(S_t^2)$ we analyze the connecting map

$$\delta : \mathbb{Z}^2 \cong K_0(\mathbb{C}^2) \longrightarrow K_1(C_0((-1,1)) \rtimes_{\alpha_t} \mathbb{Z}) \cong \mathbb{Z} .$$

Let p and q be the projections in $\mathbb{C}^2$:

$$p = (1,1), \quad q = (1,0) .$$

Referring to 3.1, since $\pi(1) = p$, we have that $\delta([p]) = 0$. The function $f(z) = \frac{z+1}{2}$ in $C(I) \subset S_t^2$ is self-adjoint, and $\pi(f) = (1,0)$ (where the first coordinate is evaluation at $+1$, the second coordinate at -1). Set

$$u = e^{2\pi i f} = e^{2\pi i(\frac{z+1}{2})} = e^{\pi i(z+1)} .$$

Then by definition, $\delta([q]) = [u]$. Using the Pimsner-Voiculescu exact sequence for crossed products [16, pp. 171] it is routine to check that the class $[u]$ is a generator of $\mathbb{Z} \cong K_1(C_0((-1,1)) \rtimes_{\alpha_t} \mathbb{Z})$. Hence the map $\delta : \mathbb{Z}^2 \longrightarrow \mathbb{Z}$ is surjective. From the exactness of the six-term sequence we conclude $K_0(S_t^2) \cong \mathbb{Z}^2$, and $K_1(S_t^2) = 0$, completing the proof.

4 A Strict Quantization of a Poisson S^2

The deformation theory of complex manifols is analytic in nature. As an algebraic counterpart M. Gerstenhaber introduced formal deformation on (commutative) algebras [5]. Motivated by the latter, F. Bayer et al. [1] studied deformation quantization of symplectic manifolds. Let (M, Ω) be a symplectic manifold. The space $C^\infty(M)$ of C^∞-functions has the structure of a Lie algebra under the Poisson bracket associated with the symplectic structure.

Recall that a *-*product* is an associative product on the space of formal power series $C_c^\infty(M)[[t]]$, of the form

$$f * g = \sum_{i,j,k}^{\infty} t^{i+j+k} C_i(f_j, g_k) \quad \text{for} \quad f = \sum t^k f_k, \ g = \sum t^j g_j ,$$

satisfying

1. $C_0(f, g) = fg$, $C_1(f, g) = \frac{1}{2}\{f, g\}$,
2. $C_j(\cdot, \cdot)$ is a bidifferential operator.

The space $C^\infty(M)[[t]]$ equipped with a *-product is called a *deformation quantization* of M. In other words, a deformation quantization is a formal deformation of the commutative algebra $C^\infty(M)$ in the direction of the Poisson bracket.

This *-product involves formal power series, and ignores the issue of convergence. This is an important notion for the study of symplectic manifolds in the context of mathematical physics. However the absence of convergence considerations in this context precludes completion for nonzero values of the parameter. Thus it precludes the existence of what would be the resulting noncommutative C^*-algebras. In this sense, the algebraic deformation quantization lacks interest from the viewpoint of operator algebras. M. Rieffel introduced an analytic version of deformation quantization in [14], embracing important examples such as the noncommutative tori. Other examples such as fields of Toeplitz algebras are excluded, but are included in the more general notion of *strict quantization* introduced in [9].

Definition 2 By a *strict quantization* of a Poisson manifold M, we mean a continuous field of C^*-algebras (A_t) over $[0, \epsilon)$ for some $\epsilon > 0$ equipped with linear maps $\pi_t : C_c^\infty(M) \longrightarrow A_t$, $t \in [0, \epsilon)$ satisfying the conditions

(1) $A_0 = C_0(M)$, and π_0 is the canonical inclusion of $C_c^\infty(M)$ into A_0,

(2) for every $f \in C_c^\infty(M)$, the vector field $(\pi_t(f))$ is continuous,

(3) $\|\frac{1}{it}(\pi_t(f)\pi_t(g) - \pi_t(g)\pi_t(f)) - \pi_t(\{f, g\})\| \longrightarrow 0$ as $t \longrightarrow 0$ for any fixed f and g in $C_c^\infty(M)$, and

(4) the C^*-algebra generated by the linear subspace $\pi_t(C_c^\infty(M))$ is dense in A_t for every t. If M is compact, the C^*-algebra A_t is unital.

Condition (3) of the definition above is refered to as *the Correspondence Principle*.

The standard $SO(3)$-invariant volume form ω_0 is a symplectic form. Denote by $\{,\}_0$ the associated Poisson bracket. Define a skew-symmetric bilinear form $\{,\}$ by

$$\{f,g\} = (1 - z^2)\{f,g\}_0, \quad f,g \in C^\infty(S^2).$$

It is straightforward to verify that $\{,\}$ is indeed a Poisson bracket. In cylindrical coordinates,

$$\{a,b\} = (1 - z^2)\left(\frac{\partial a}{\partial z}\frac{\partial b}{\partial \theta} - \frac{\partial a}{\partial \theta}\frac{\partial b}{\partial z}\right).$$

We compute this Poisson bracket for some basic examples:

$$\{\zeta, z\} = i(1 - z^2)\zeta,$$
$$\{\bar\zeta, z\} = -i(1 - z^2)\bar\zeta,$$
$$\{\zeta, \bar\zeta\} = -2i(1 - z^2)z.$$

Theorem 2 The Poisson manifold $(S^2, \{,\})$ has a strict quantization $\{(\pi_t, A_t)\}$ in such a way that π_t's are $*$-preserving. More precisely, there exist $*$-preserving linear maps $\pi_t : C^\infty(S^2) \longrightarrow S_t^2$ such that $\{(\pi_t, S_t^2)\}$ is a strict quantization with respect to the Poisson structure $\{,\}$.

For the proof, see [11].

Podleś's quantum 2-spheres [12] yield a strict quaintization of S^2 with a Poisson structure degenerate along the equator. Let S_+ (resp. S_-) be the northern (resp. southern) hemisphere, and let $S^1 = S_+ \cap S_-$ be the equator. Then $C(S^2)$ is a pull-back C^*-algebra characterized by the diagram :

$$
\begin{array}{ccc}
C(S^2) & \longrightarrow & C(S_+) \\
\downarrow & & \downarrow \\
C(S_-) & \longrightarrow & C(S^1).
\end{array}
$$

The hemispheres S_+, S_- are topologically isomorphic to the closed 2-disk $\overline{\mathbb{D}}$. Substituting a weighted unilateral shift for the generator of $C(\overline{\mathbb{D}})$ deforms $C(\overline{\mathbb{D}})$ to the Toeplitz algebra $\mathcal{T}_\mu$ [6]. Then the pull-back construction gives a family of C^*-algebras $C(S_\mu^2)$:

$$
\begin{array}{ccc}
C(S_\mu^2) & \longrightarrow & \mathcal{T}_\mu \\
\downarrow & & \downarrow \\
\mathcal{T}_\mu & \longrightarrow & C(S^1).
\end{array}
$$

The sequence

$$0 \longrightarrow C_0(\mathbb{D}) \longrightarrow C(\overline{\mathbb{D}}) \longrightarrow C(S^1) \longrightarrow 0$$

becomes

$$0 \longrightarrow \mathcal{K} \longrightarrow \mathcal{T}_\mu \longrightarrow C(S^1) \longrightarrow 0 \,,$$

and the pull-back induces the short exact sequence

$$0 \longrightarrow \mathcal{K} \oplus \mathcal{K} \longrightarrow C(S^2_\mu) \longrightarrow C(\mathbb{T}) \longrightarrow 0 \,.$$

The C^*-algebras $C(S^2_\mu)$ with parameter μ yield a strict quantization of $C(S^2)$ with the Poisson structure $z^3\{,\}_0$. Klimek-Lesniewski deformation ([6]) of the closed 2-disk produces a strict quantization for the Poisson structure:

$$\{f,g\} = (1 - x^2 - y^2)^{-2} \left(\frac{\partial f}{\partial x}\frac{\partial g}{\partial y} - \frac{\partial g}{\partial x}\frac{\partial f}{\partial y} \right), \quad f,g \in C^\infty(\overline{\mathbb{D}}) \,.$$

The canonical projection from the northern hemisphere of $S^2 \subset \mathbb{R}^3$ onto the closed 2-disk $\overline{\mathbb{D}} \subset \mathbb{R}^2$ pulls back the Poisson structure on the 2-disk to a Poisson structue on the hemisphere $z^3\{,\}_0$.

Applying the Weyl quantization to open northern and southern hemispheres of the 2-sphere, A.-L. Sheu [15] showed the existence of a strict quantization of the Poisson 2-sphere $(S^2, (c-z)\{,\}_0)$.

Recently Hanfeng Li showed that any Poisson manifold has a strict quantization [7].

5 Noncommutative Spheres in Dimensions 3 and 4

We apply topological methods to S^2_t to create higher dimensional noncommutative spheres.

5.1. *Unreduced suspension.* Let X be a Hausdorff space. In the space $X \times [0,1]$, identify the closed subspace $X \times 0$ to one point and $X \times 1$ to another. The quotient space space under these identifications is the (unreduced) suspension of X. When X is S^n, its unreduced suspension is S^{n+1}.

A C^*-algebraic version can be given in the following way. Suppose that A is a unital C^*-algebra. Then the unreduced suspension of A is the C^*-algebra of all continuous functions f on $[0,1]$ with values in A satisfying the condition that $f(0), f(1) \in \mathbb{C} \subset A$.

Define a noncommutative 3-sphere S^3_t as the reduced suspension of S^2_5. Define S^2_t-valued continuous functions $\tilde{\zeta}_t$ and $\tilde{z}_t$ on $[0,1]$ by

$$\tilde{\zeta}_t(s) = \sqrt{s(1-s)}\zeta_t, \tilde{z}_t = \sqrt{s(1-s)}z_t \,,$$

respectively. Then these two elements together with the canonical coordinate function s on $[0,1]$ generate S^3_t. These generators satisfy the following relations:

(1) $\tilde{\zeta}_t^* \tilde{\zeta}_t + \tilde{z}_t^2 = s - s^2$,

(2) $\sqrt{s(1-s)}\{\tilde{\zeta}_t \tilde{z}_t - \tilde{z}_t \tilde{\zeta}_t\} = t\tilde{\zeta}_t(s - s^2 - \tilde{z}_t^2)$,

(3) $s(1-s)\tilde{\zeta}_t \tilde{\zeta}_t^* + \{t\tilde{\zeta}_t \tilde{\zeta}_t^* + \sqrt{s(1-s)}\tilde{z}_t\}^2 = s^2(1-s)^2$.

Thus the generators satisfy rather lousy analytic relations involving square roots.

We now modify the unreduced suspension so that the resulting noncommutative 3-spheres are characterized by algebraic relations. Consider a continuous filed (A_s) of C^*-algebras parametrized by $s \in [0,1]$ given by $A_s = S^2_{t\sqrt{s(1-s)}}$. Define sections of the field by

$$\tilde{\zeta}_t(s) = \sqrt{s(1-s)}\zeta_{t\sqrt{s(1-s)}}, \tilde{z}_t(s) = \sqrt{s(1-s)}z_{t\sqrt{s(1-s)}} .$$

Then we have

(4) the function s commutes with both $\tilde{\zeta}_t$ and $\tilde{z}_t$,

(5) $\tilde{\zeta}_t^* \tilde{\zeta}_t + \tilde{z}_t^2 = s - s^2$,

(6) $\tilde{\zeta}_t \tilde{z}_t - \tilde{z}_t \tilde{\zeta}_t = t\tilde{\zeta}_t(s - s^2 - \tilde{z}_t^2)$,

(7) $\tilde{\zeta}_t \tilde{\zeta}_t^* + (t\tilde{\zeta}_t \tilde{\zeta}_t^* + \tilde{z}_t)^2 = s - s^2$.

Define a C^*-algebra $\mathbb{S}_t^3$ as the C^*-algebra generated by $\tilde{\zeta}_t, \tilde{z}_t, s$ in the C^*-algebras of continuous sections of the field (A_s). Following the line of the proof of Theorem 2.2 of [11], we can show that $\mathbb{S}_t^3$ is characterized as the C^*-algebra having the universality for the relations (4)–(7).

It is not hard to define quantiztion maps $C^\infty(S^3) \longrightarrow \mathbb{S}_t^3$ to construct a strict quantization of a Poisson structure degenerate along a great cirle on S^3.

5.2. *Smash products.* Let $(X, x_0), (Y, y_0)$ be pointed toplogical spaces. The subspace $X \times \{y_0\} \cup \{x_0\} \times Y$ is called the one-point union and is denoted $X \vee Y$. Identify $X \vee Y$ to a single point. The quotient space $X \wedge Y$ of $X \times Y$ under the identification is called the *smash product* of $(X, x_0), (Y, y_0)$. We get that $S^2 \wedge S^2 = S^4$.

Suppose that X and Y are compact. Define $\epsilon_X : C(X \times Y) \longrightarrow C(Y)$ by $\epsilon_X(f)(y) = f(x_0, y)$. Similarly define $\epsilon_Y : C(X \times Y) \longrightarrow C(X)$. Then $C(X \wedge Y) \cong \epsilon_X^{-1}(\mathbb{C}) \cap \epsilon_Y^{-1}(\mathbb{C}) \subset C(X \times Y)$.

Define a *noncommutative 4-sphere* $S_{t,s}^4$ as the smash product of the C^*-algebras S_t^2, S_s^2 as follows. Let τ_1 be the "evaluation" map $S_t^2 \longrightarrow \mathbb{C}$ at the north pole. Similarly define the "evaluation" map $\tau_2 : S_s^2 \longrightarrow \mathbb{C}$. Denote by J_t (resp. J_s) the kernel of the homomorphism τ_1 (resp. τ_2). Then it is not difficult to check that

$$S_{t,s}^4 = S_t^2 \wedge S_s^2 \cong (J_t \otimes J_s)^\sim .$$

References

1. F. Bayen, M. Flato, C. Fronsdal, A. Lichnerowicz, D. Sternheimer, Deformation theory and quantization, I. II, Ann. Phys. 110(1978), 62–110, 111–151.
2. O. Bratteli, G.A. Elliott, D. Evans, A. Kishimoto, Noncommutative spheres. I, Internat. J. Math. 2(1991), 139–166
3. A. Connes, G. Landi, Noncommutative manifolds, the instanton algebra and isospectral deformations, math. QA/0011194v3.
4. L. Dabrowski, The garden of quantum spheres, arXive: math. QA/0212264v1.
5. M. Gerstenhaber, On the deformation of rings and algebras, Ann. Math. 79(1964), 59–103.
6. S. Klimek, A. Lesniewski, Quantum Riemann surfaces I, the unit disc, Comm. Math. Phys. 146(1992), 105–122.
7. H. Li, arXiv: math. QA/0303078.
8. K. Matsumoto, Noncommutative three spheres, Japan. J. Math. 17(1991), 333–356.
9. T. Natsume, R. Nest and I. Peter, Strict quantizations of symplectic manifolds, Lett. Math. Phys., *in press*.
10. T. Natsume, C.L. Olsen, Toeplitz operators on noncommutative spheres and an index theorem, Indiana Univ. Math. J. 46(1999), 1055–1112.
11. T. Natsume, C.L. Olsen, A new family of noncommutative 2-spheres, Jour. of Funct. Analysis, *in press*.
12. P. Podleś, Quantum spheres, Lett. Math. Phys. 14(1987), 193–202.
13. M.A. Rieffel, Continuous fields of C^*-algebras coming from group cocycles and actions, Math. Ann. 283(1989), 631–643.
14. M.A. Rieffel, Deformation quantization for Heisenberg manifolds, Comm. Math. Phys. 122(1989), 531–562.
15. A. J.-L. Sheu, Quantization of the Poisson $SU(2)$ and its Poisson homogeneous space-the 2-sphere, Comm. Math. Phys. 135(1991), 217–232.
16. N.E. Wegge-Olsen, K-Theory and C^*-algebras, Oxford University Press, 1993.

From Quantum Tori
to Quantum Homogeneous Spaces

S. Kamimura

Department of Mathematics, Keio University, Yokohama 223-8522, Japan
kamimura@math.keio.ac.jp

Summary. We construct dual objects for quantum complex projective spaces as quantum homogeneous spaces of quantum unitary groups, in which the deformation parameters are antisymmetric matrices.

1 Introduction

We start from comparison between q-deformation of dual objects of some linear Lie groups and θ-deformation of ones with respect to the dimention of their shadows.

$$
\begin{aligned}
dim_H(C^{alg}(G_q)) &< dim(G) \qquad \text{[4, 5])} \\
dim_H(C^{alg}(G_\theta)) &= dim(G) \qquad \text{[1])}
\end{aligned}
\tag{1}
$$

Here, dim_H, dim, C^{alg} and θ stands for the Hochschild dimension of **C**-algebras, the ordinary dimention of the classical spaces, coordinate ring functor of the quantum groups and antisymmetric matrix of some size, respectively.

The above comparison says that the shadows of $C^{alg}(G_q)$)'s are degenerate, but those of $C^{alg}(G_\theta)$'s are nondegenerate.

These degeneracy phenomena in q-deformation must have some singularities, but in many cases they are not so clear and the explanations of theirs are not so successful.

So we will take the way of θ-deformation in defining quantum homogeneous spaces.

2 From Quantum Tori to Quantum Groups

Most of this section is quoted from [1] except for a little arrangement and some remarks.

2.1 Quantum Tori and Euclidian Spaces

It is well-known that the Weyl's canonical commutation relations for one-parameter unitary groups can be obtained by exponetiating the Heisenberg's canonical commutation relations for self-adjoint operators.

S. Kamimura: *From Quantum Tori to Quantum Homogeneous Spaces*, Lect. Notes Phys. **662**, 67–74 (2005)
www.springerlink.com

The most important and basic quantum torus is nothing but obtained by discretizing the parameters in the Weyl's CCR. So the quantum tori are the most fundamental objects in the concept of θ-deformation. Anyway, here is the definition of quantum tori.

Definition 1. *Let $C^{alg}(T_\theta^n)$ be the unital $*$-algebra generated by n unitary elements*

$$\bar{u}^i u^i = u^i \bar{u}^i = 1 \quad (1 \leq i \leq n)$$

with commutation relations

$$u^i u^j = \lambda^{ij} u^j u^i, \quad u^i \bar{u}^j = \bar{\lambda}^{ij} \bar{u}^j u^i$$

$$\bar{u}^i u^j = \bar{\lambda}^{ij} u^j \bar{u}^i, \quad \bar{u}^i \bar{u}^j = \lambda^{ij} \bar{u}^j \bar{u}^i .$$

Here

$$\lambda^{ij} = \exp(\sqrt{-1}\theta^{ij}), \quad \theta = (\theta^{ij}) \in A(n; \mathbf{R}) = o(n) = Lie(O(n)) ,$$

and we use $^-$ instead of $*$-operation.

Only replacing the above unitary conditions to the corresponding normal conditions leads us to a natural definition of the unital $*$-algebra $C^{alg}(\mathbf{R}_\theta^{2n})$. That is,

Definition 2. *$C^{alg}(\mathbf{R}_\theta^{2n})$ is generated by n normal elements*

$$\bar{z}^i z^i = z^i \bar{z}^i \quad (1 \leq i \leq n)$$

with the same commutation relations as above,

$$z^i z^j = \lambda^{ij} z^j z^i, \quad z^i \bar{z}^j = \bar{\lambda}^{ij} \bar{z}^j z^i$$

$$\bar{z}^i z^j = \bar{\lambda}^{ij} z^j \bar{z}^i, \quad \bar{z}^i \bar{z}^j = \lambda^{ij} \bar{z}^j \bar{z}^i .$$

In order to check the correspondance between the the above quantum formulation and the following classical formulation,

$$T^n \subset \mathbf{R}^{2n} \cong \mathbf{C}^n$$

$$u_{cl}^i = \exp(\sqrt{-1}t_{cl}^i) = \cos t_{cl}^i + \sqrt{-1}\sin t_{cl}^i$$

$$z_{cl}^i = x_{cl}^i + \sqrt{-1}y_{cl}^i ,$$

it is helpful to take the Descartes decompositions of the unitary and normal generators,

$$u^i = v^i + \sqrt{-1}w^i = \frac{u^i + \bar{u}^i}{2} + \sqrt{-1}\frac{u^i - \bar{u}^i}{2\sqrt{-1}}$$

$$z^i = x^i + \sqrt{-1}y^i = \frac{z^i + \bar{z}^i}{2} + \sqrt{-1}\frac{z^i - \bar{z}^i}{2\sqrt{-1}} .$$

We can easily verify

$$\bar{v}^i = v^i, \; \bar{w}^i = w^i, \; [v^i, w^i] = 0, \; (v^i)^2 + (w^i)^2 = 1$$

$$\bar{x}^i = x^i, \; \bar{y}^i = y^i, \; [x^i, y^i] = 0 \,,$$

and recover $C^{alg}(T_\theta^n)$ from $C^{alg}(\mathbf{R}_\theta^{2n})$ as follows:

$$C^{alg}(T_\theta^n) \cong C^{alg}(\mathbf{R}_\theta^{2n})/(z^1 \bar{z}^1 - 1, \cdots, z^n \bar{z}^n - 1) \,.$$

2.2 Quantum Matrix Algebras

Next we define the unital $*$-algebra $M_\theta(2n; \mathbf{R})$. The elementary isomorphisms

$$M(2n; \mathbf{R}) \cong \mathbf{R}^{4n^2}$$

and

$$M(2n; \mathbf{R}) \cong End(\mathbf{R}^{2n}) \cong (\mathbf{R}^{2n})^* \otimes \mathbf{R}^{2n} \cong \mathbf{R}^{2n} \otimes (\mathbf{R}^{2n})^*$$

would justify the following inclusion

$$\iota : C^{alg}(M_\theta(2n; \mathbf{R})) \cong C^{alg}(\mathbf{R}_\Theta^{4n^2}) \hookrightarrow C^{alg}(\mathbf{R}_\theta^{2n}) \otimes C^{alg}(\mathbf{R}_{-\theta}^{2n}) \,.$$

Here

$$\Theta \in A(2n^2; \mathbf{R})$$

is determined by

$$\theta \in A(n; \mathbf{R}) \,.$$

So

Definition 3. *$M_\theta(2n; \mathbf{R})$ can be generated by $2n^2$ normal elements*

$$a_j^i, \; b_j^i \;\; (1 \le i, j \le n)$$

such that

$$\iota(a_j^i) = z^i \otimes z_j, \; \iota(b_j^i) = z^i \otimes \bar{z}_j \,,$$

$$z^i \in C^{alg}(\mathbf{R}_\theta^{2n}), \; z_j \in C^{alg}(\mathbf{R}_{-\theta}^{2n}), \; z_i := z^i$$

with commutation relations

$$a_j^i b_l^k = \lambda^{ik} \lambda_{jl} b_l^k a_j^i \;\; (\lambda^{ik} := \lambda_{ik}), \;\; etc \,.$$

Since $M(2n; \mathbf{R})$ fails to have group structure with respect to the ordinary multiplication of matrices, we cannot expect $C^{alg}(M_\theta(2n; \mathbf{R}))$ to have the corresponding Hopf algebra structure. But the essential obstruction is nothing but the antipode map. That is, we have no obstructions to define the corresponding bialgebra structure on it.

$$\Delta : C^{alg}(M_\theta(2n; \mathbf{R})) \longrightarrow (M_\theta(2n; \mathbf{R})) \otimes (M_\theta(2n; \mathbf{R}))$$

$$a^i_j \longmapsto a^i_k \otimes a^k_j + b^i_k \otimes \bar{b}^k_j$$

$$b^i_j \longmapsto a^i_k \otimes b^k_j + b^i_k \otimes \bar{a}^k_j$$

$$\varepsilon : C^{alg}(M_\theta(2n; \mathbf{R})) \longrightarrow \mathbf{C}$$

$$a^i_j \longmapsto \delta^i_j$$

$$b^i_j \longmapsto 0 .$$

$M(2n; \mathbf{R})$ has a natural action on $\mathbf{R}^{2n}$. It is not so hard to define the corresponding coaction of $C^{alg}(M_\theta(2n; \mathbf{R}))$ on $C^{alg}(\mathbf{R}^{2n}_\theta)$.

$$\alpha : M(2n; \mathbf{R}) \times \mathbf{R}^{2n} \longrightarrow \mathbf{R}^{2n}$$

$$\beta : \quad C^{alg}(\mathbf{R}^{2n}_\theta) \longrightarrow C^{alg}(M_\theta(2n; \mathbf{R})) \otimes C^{alg}(\mathbf{R}^{2n}_\theta)$$

$$z^i \longmapsto a^i_j \otimes z^j + b^i_j \otimes \bar{z}^j .$$

2.3 Quantum Orthogonal Groups and Unitary Groups

The dual objects of quantum linear Lie groups should be defined as the quatient algebras of $C^{alg}(M_\theta(2n; \mathbf{R}))$ by appropriate ideals.

$$C^{alg}(G_\theta) := C^{alg}(M_\theta(2n; \mathbf{R}))/\mathfrak{J}$$

Recall that $O(2n)$ is defined to be a subset of $M(2n; \mathbf{R})$ such that each element of it preserves the quadratic form $\sum_{i=1}^{n} z^i \bar{z}^i$. Thus it is quite natural to characterize $C^{alg}(O_\theta(2n))$ by the following proposition.

$$\pi : C^{alg}(M_\theta(2n; \mathbf{R})) \longrightarrow C^{alg}(O_\theta(2n)) := C^{alg}(M_\theta(2n; \mathbf{R}))/ \,^{\exists}I$$

$$such \ that$$

$$\beta' : \quad C^{alg}(\mathbf{R}^{2n}_\theta) \longrightarrow C^{alg}(O_\theta(2n)) \otimes C^{alg}(\mathbf{R}^{2n}_\theta)$$

$$\sum_{i=1}^{n} z^i \bar{z}^i \longmapsto 1 \otimes \sum_{i=1}^{n} z^i \bar{z}^i$$

See [1] for the precise description of ideal I.

One of the famous formulations of $U(n)$,

$$U(n) = \{g \in O(2n) \mid Jg = gJ\} ,$$

can be translated into the following dual formulation.

$$C^{alg}(U_\theta(n)) := C^{alg}(O_\theta(2n))/(\pi(b^i_j), \ \pi(\bar{b}^i_j))$$

3 Quantum Complex Projective Spaces

In this section we will construct quantum complex projective spaces as quantum homogeneous spaces of the quantum unitary groups appeared in previous section.

3.1 Restrictions and Coactions of Quantum Unitary Groups

Recall that the quotient space of the action

$$\alpha : U(n) \times (U(1) \times U(n-1)) \longrightarrow U(n)$$

is the complex projective space

$$P^{n-1}(\mathbf{C}) = U(n)/(U(1) \times U(n-1)) .$$

So it is natural that we consider the dual object for quantum complex projective space as invariant subalgebra of such a coaction as

$$\beta : C^{alg}(U_{\theta_n}(n)) \longrightarrow C^{alg}(U_{\theta_n}(n)) \otimes (C^{alg}(U_{\theta_1}(1)) \otimes C^{alg}(U_{\theta_{n-1}}(n-1))) .$$

$$C^{alg}(P_{\theta_n}^{n-1}(\mathbf{C})) := C^{alg}(U_{\theta_n}(n))^{(C^{alg}(U_{\theta_1}(1)) \otimes C^{alg}(U_{\theta_{n-1}}(n-1)))}$$

$$\theta_n \in A(n; \mathbf{R}), \quad \theta_1 = 0 \in A(1; \mathbf{R})$$

First we have to construct such a restriction as

$$\rho : C^{alg}(U_{\theta_k}(k)) \longrightarrow C^{alg}(U_{\theta_{k-1}}(k-1))$$

for the standard inclusion

$$\iota : U(k-1) \longrightarrow U(k) .$$

Lemma 1. *For the normal generators*

$$a_j^i, b_j^i \in C^{alg}(M_{\theta_k}(2k; \mathbf{R})),$$

let I and J to be ideals generated by

$$a_1^1 - 1, \quad b_1^1, \quad a_j^1, \quad a_1^i, \quad b_j^1, \quad b_1^i \quad for \quad 2 \le i, j \le k$$

and

$$a_k^k - 1, \quad b_k^k, \quad a_j^k, \quad a_k^i, \quad b_j^k, \quad b_k^i \quad for \quad 1 \le i, j \le k-1,$$

respectively. Then the images of quotient maps

$$\rho_k : C^{alg}(M_{\theta_k}(2k; \mathbf{R})) \longrightarrow C^{alg}(M_{\theta_k}(2k; \mathbf{R}))/I$$

and

$$\rho_k' : C^{alg}(M_{\theta_k}(2k; \mathbf{R})) \longrightarrow C^{alg}(M_{\theta_k}(2k; \mathbf{R}))/J$$

coincide

$$C^{alg}(M_{\theta_{k-1}}(2(k-1); \mathbf{R})) \quad and \quad C^{alg}(M_{\theta'_{k-1}}(2(k-1); \mathbf{R})),$$

respectively. Here, for a given antisymmetric martix θ_k of size k,

$$\exists \theta_{k-1}, \ \theta'_{k-1} \in A(k-1; \mathbf{R})$$

such that

$$\theta_k = \begin{pmatrix} \theta'_{k-1} & * \\ * & 0 \end{pmatrix}, \begin{pmatrix} 0 & * \\ * & \theta_{k-1} \end{pmatrix} \in A(k; \mathbf{R}) .$$

Moreover, the induced restrictions on C^{alg} of quantum unitary groups, denoted by the same notations above,

$$\rho_k : C^{alg}(U_{\theta_k}(k)) \longrightarrow C^{alg}(U_{\theta_{k-1}}(k-1))$$

and

$$\rho'_k : C^{alg}(U_{\theta_k}(k)) \longrightarrow C^{alg}(U_{\theta_{k-1}}(k-1))$$

are Hopf algebra homomorphisms.

Using these restrictions, we can get the restrictions corresponding to the standard inclusion

$$U(1) \times U(n-1) \hookrightarrow U(n) .$$

Proposition 1. *Let $\rho_{n,1}$ be*

$$\rho_{n,1} := ((\rho_2 \circ \cdots \circ \rho_n) \otimes \rho'_n) \circ \Delta .$$

Then

$$\rho_{n,1} : C^{alg}(U_{\theta_n}(n)) \longrightarrow (C^{alg}(U_{\theta_1}(1)) \otimes C^{alg}(U_{\theta_{n-1}}(n-1))$$

is a surjective $*$-Hopf algebra morphism, and

$$(id \otimes \rho_{n,1}) \circ \Delta : C^{alg}(U_{\theta_n}(n)) \longrightarrow C^{alg}(U_{\theta_n}(n))$$
$$\otimes (C^{alg}(U_{\theta_1}(1)) \otimes C^{alg}(U_{\theta_{n-1}}(n-1))$$

is a right coaction.

Now we reach the very definition of C^{alg} of quantum complex projective space.

Definition 4. *For the right comodule algebra $C^{alg}(U_{\theta_n}(n))$ is over*

$$(C^{alg}(U_{\theta_1}(1)) \otimes C^{alg}(U_{\theta_{n-1}}(n-1))),$$

we define $C^{alg}(P^{n-1}_{\theta_n}(\mathbf{C}))$ to be the invariant subalgebra of its right coaction.

$$C^{alg}(P^{n-1}_{\theta_n}(\mathbf{C})) := C^{alg}(U_{\theta_n}(n))^{(C^{alg}(U_{\theta_1}(1)) \otimes C^{alg}(U_{\theta_{n-1}}(n-1)))}$$

$$:= \{f \in C^{alg}(U_{\theta_n}(n)) \mid ((id \otimes \rho_{n,1}) \circ \Delta)(f) = f \otimes 1\}$$

Remark 1. It is not so difficult to give the definitions of

$$C^{alg}(Gr_\theta^{n,k}(\mathbf{C})),\ C^{alg}(F_\theta^{n,k_1,\ldots,k_d}(\mathbf{C}))\ and\ C^{alg}(St_\theta^{n,k}(\mathbf{C})),$$

which are dual of quantum complex Grassmannian manifold, complex flag manifold and complex Stiefel manifold, respectively. Of course, thier real versions can be also got similarly[2].

Remark 2. We can define odd dimensional quantum spheres as quantum homogeneous spaces of quantum unitary groups.

$$(id \otimes \rho_n) \circ \Delta : C^{alg}(U_{\theta_n}(n)) \longrightarrow C^{alg}(U_{\theta_n}(n)) \otimes C^{alg}(U_{\theta_{n-1}}(n-1))$$

$$C^{alg}(S_{\theta_n}^{2n-1}) := \{f \in C^{alg}(U_{\theta_n}(n)) \mid ((id \otimes \rho_n) \circ \Delta)(f) = f \otimes 1\}$$

On the other hand, we can find another definition of them in [1], which has the following expression:

$$C^{alg}\left(S_{\theta_n}^{2n-1}\right) := C^{alg}\left(\mathbf{R}_{\theta_n}^{2n}\right) / \left(\sum_{i=1}^{n} z^i \bar{z}^i - id\right)$$

Since $S^3 \cong SU(2)$ classically, it is expected that $C^{alg}(S_{\theta_2}^3) \cong C^{alg}(SU_{\theta_2}(2))$. But we cannot quantize $SU(n)$ in the context of θ-deformation [1]. In spite of this it is still interesting to consider the Hopf algebra structure on $C^{alg}(S_{\theta_2}^3)$ with respect to the above two expressions [3].

4 Main Results

The following-type theorems have been already proven for the case of some quantum groups in [1]. But the restrictions and coactions defined in this paper require a little longer proves than those shown in [1]. For more general cases, for example $Gr_{\theta_n}^{n,k}(\mathbf{C})$, the same theorems as for $P_{\theta_n}^{n-1}(\mathbf{C})$ are proven in [2], where the cyclic theory and K-theory of them will be also discussed.

The first is a splitting formura, which justifies the importance of the quantum tori in θ-deformation.

Theorem 1. *The noncommutativity between the generators of $C^{alg}(P_{\theta_n}^{n-1}(\mathbf{C}))$ is absorbed in quantum tori. That is:*

$$C^{alg}(P_{\theta_n}^{n-1}(\mathbf{C})) = (C^{alg}(P^{n-1}(\mathbf{C})) \otimes C^{alg}(T_\theta^n)$$

$$\otimes\ C^{alg}(T_{-\theta}^n))^{(\sigma \otimes \sigma) \times (\tau \otimes \tau)^{-1}}.$$

Here,

$$\sigma \times \sigma : \quad T^n \times T^n \quad \longrightarrow \quad Aut(C^{alg}(M_\theta(2n; \mathbf{R})))$$

$$(s, t) \quad \longmapsto \quad \sigma_s \otimes \sigma_t$$

$$\sigma_s \otimes \sigma_t : C^{alg}(M_\theta(2n; \mathbf{R})) \longrightarrow \quad C^{alg}(M_\theta(2n; \mathbf{R}))$$

$$a_j^i \quad \longmapsto exp(2\pi\sqrt{-1}(s_i + t_j))\, a_j^i$$

$$b_j^i \quad \longmapsto exp(2\pi\sqrt{-1}(s_i - t_j))\, b_j^i$$

$$\tau \quad : \quad T^n \quad \longrightarrow \quad Aut(C^{alg}(T_\theta^n))$$

$$t \quad \longmapsto \quad \tau_t$$

$$\tau_t \quad : \quad u^i \quad \longrightarrow \quad exp(2\pi\sqrt{-1}t_i)\, u^i$$

The next is on the nondegeneracy of dimension of quantum projective spaces. This theorem also states a typical phenomenon in θ-deformation.

Theorem 2.

$$dim_H(C^{alg}(P_{\theta_n}^{n-1}(\mathbf{C}))) = dim(P^{n-1}(\mathbf{C}))$$

Here, dim_H denotes the Hochschild dimension of algebras, which is the last degree of nontrivial Hochschild homology of algebras. Using the projective resolutions of $C^\infty(H_\theta), C^\infty(H)$ and $C^\infty(T_\theta^n)$ tensored with de Rham algebra of them, and then calculate the spectral sequences.

References

1. A. Connes and M. Dubois-Violette: *Noncommutative finite-dimensional mani-folds. I. Spherical manifolds and related examples*, Comm. Math. Phys. **230**, 3 (2002) pp 539–579.
2. S. Kamimura, *Quantum homogeneous spaces deformed by antisymmetric matri-ces*, preprint.
3. S. Kamimura, *Two-parametric deformation of $SU(2)$ or S^3*, preprint.
4. T. Masuda, Y. Nakagami and J. Watanebe: *Non-commutative differential geo-metry on the quantum $SU(2)$ I -an algebraic viewpoint-*, K-theory **4** (1990), pp 157–180.
5. T. Masuda, Y. Nakagami and J. Watanebe: *Non-commutative differential geo-metry on the quantum two sphere of Podoles I -an algebraic viewpoint-*, K-theory **5** (1991), pp 151–175.

Part II

Poisson Geometry
and Deformation Quantization

The Part II deals with several topics from Poisson geometry and its quantum counterpart as formulated using deformation quantization. Poisson manifolds generalize the classical phase space quite drastically but find also applications far beyond purely "mechanical" theories like e.g. in the Poisson-sigma models and hence in (quantum) field theories. Deformation quantization on the other hand is one of the most successful quantization schemes when the classical phase space is no longer flat. However, also here one has applications far beyond the purely "quantum mechanical" world since the star products are e.g. used for noncommutative field theories by defining what the underlying noncommutative space-time should be.

The first contribution discusses symplectic connections of Ricci type, i.e. those whose curvature tensor is determined by the Ricci tensor alone. It is shown how such connections can locally be obtained by a phase space reduction out of a flat symplectic connection.

In the second contribution, the role of gauge transformations by closed two-forms in Poisson geometry and the relations with Dirac structures and Morita equivalence of Poisson manifolds are discussed.

In the third contribution, the classification of star products quantizing arbitrary quadratic Poisson structures on the plane is given.

In the fourth contribution, the authors obtain a universal deformation formula for three-dimensional solvable Lie groups allowing them to define new examples of strict deformation quantizations.

The fifth contribution gives a general geometrical framework for noncommutative field theories by deforming arbitrary vector bundles and relates these deformations to Morita equivalence of star products and their corresponding Picard groupoids.

The sixth contribution deals with secondary characteristic classes for Lie algebroids arising from a suitably defined adjoint representation of the Lie algebroid.

Local Models for Manifolds
with Symplectic Connections of Ricci Type[*]

M. Cahen

Université Libre de Bruxelles, Campus Plaine, CP 218, 1050 Brussels, Belgium
mcahen@ulb.ac.be

Summary. We show that any symplectic manifold (M, ω) of dimension $2n$ $(n \geq 2)$ admitting a symplectic connection of Ricci type can locally be constructed by a reduction procedure from the Euclidean space $\mathbb{R}^{2n+2}$ endowed with a constant symplectic structure and the standard flat connection. We also prove that on the bundle of symplectic frames $B(M)$ over M, there exists a 1-form with values in the algebra $sp(n + 1, \mathbb{R})$ which locally satisfies a Maurer-Cartan type equation.

1 Introduction

On any smooth, finite dimensional, paracompact manifold M, there exists a smooth riemannian metric g. The space of riemannian metrics on M, $\mathcal{E}(M)$, is infinite dimensional. One may impose restrictions to the metric; for example by means of a variational principle. If the functional is chosen to be

$$\int_M \rho_g d\mu_g$$

where ρ_g is the scalar curvature of g and $d\mu_g$ is the standard measure associated to g, the critical points are the so called Einstein metrics. Riemannian geometers have studied the existence of Einstein metrics on a given manifold M; in the case there is existence they have looked at the moduli space of Einstein metrics on M, i.e. the space of Einstein metrics modulo the action of the diffeomorphism group of M.

On any smooth, finite dimensional, paracompact manifold M, there does not exist a smooth symplectic form ω. The manifold must be even dimensional, orientable; but these 2 conditions are far from sufficient as exemplified by the spheres S^{2n} $(n \geq 2)$ which do not admit a symplectic structure. We shall thus consider a symplectic manifold (M, ω). A symplectic connection ∇ is a linear connection which is torsion free and for which ω is parallel. The space of symplectic connections on (M, ω), $\mathcal{E}(M, \omega)$ is infinite dimensional. One may impose restrictions to the connection; for example by means of a variational principle. Let the functional be chosen to be

---

[*] This paper describes work done in collaboration with Simone Gutt and Lorenz Schwachhoefer. Our research was partially supported by an Action de Recherche Concertée de la Communauté française de Belgique.

M. Cahen: *Local Models for Manifolds with Symplectic Connections of Ricci Type*, Lect. Notes Phys. **662**, 77–87 (2005)
www.springerlink.com

$$\int_M r^2 \omega^n \, ,$$

where $\dim M = 2n$ and r denotes the Ricci tensor of the connection ∇ (i.e. $r(X,Y) = \mathrm{tr}[Z \to R(X,Z)Y]$, where X, Y, Z are vector fields on M and $R(X,Z)$ is the curvature endomorphism associated to X and Z for the connection ∇). Finally r^2 is the scalar defined as follows. Let ρ be the endomorphism

$$\omega(X, \rho Y) \underset{\mathrm{def}}{=} r(X,Y)$$

Then

$$r^2 \underset{\mathrm{def}}{=} \mathrm{tr}\, \rho^2$$

Remark that $\mathrm{tr}\, \rho = 0$ as ρ_x belongs to the symplectic algebra of $(T_x M, \omega_x)$.

The Euler Lagrange equations of this functional are:

$$\underset{X,Y,Z}{\oplus} (\nabla_X r)(Y, Z) = 0 \, .$$

where $\oplus$ denotes the sum over cyclic permutations of the indicated quantities.

A connection ∇ satisfying these field equations is said to be **preferred**.

By analogy with the riemannian situation we can formulate:

Problem 1 Can one describe the moduli space of preferred connections on (M, ω), i.e. the space of preferred connections on (M, ω) modulo the action of the symplectic diffeomorphism group.

The following has been proven in [2].

Theorem 1 Let (M, ω) be a compact symplectic surface; let ∇ be a complete preferred symplectic connection. Then

(i)   if $M = S^2$, ∇ is the Levi Civita connection associated to a metric of constant positive curvature
(ii) if $M = T^2$, the connection ∇ is flat
(iii) if M is a surface of genus $g \geq 2$, ∇ is the Levi Civita connection associated to a metric of constant negative curvature.

In dimension $2n \geq 4$ very little is known. Fortunately a subclass of preferred connections may be described with some detail. Let me first define the subclass.

Let (M, ω) be a symplectic manifold and ∇ a symplectic connection. At a point $x \in M$, the curvature tensor R_x of ∇ is a tensor of type $\binom{0}{4}$ having the following symmetries

$$R_x(X, Y, Z, T) \underset{\mathrm{def}}{=} \omega(R(X,Y)Z, T)$$

(i) $\quad R_x(X,Y,Z,T) = -R_x(Y,X,Z,T)$
(ii) $\quad R_x(X,Y,Z,T) = R_x(X,Y,T,Z)$
(iii) $\quad \underset{X,Y,Z}{\oplus} R_x(X,Y,Z,T) = 0.$

From (i) and (ii), $R_x \in \Lambda^2 T_x^* M \otimes \odot^2 T_x^* M$, where $\odot^k V$ is the symmetrized k-tensor product of the vector space V. Recall Koszul's exact sequences:

$$0 \leftrightarrows \odot^4 V \leftrightarrows V \otimes \odot^3 V \underset{a}{\overset{s}{\leftrightarrows}} \Lambda^2 V \otimes \odot^2 V \underset{a}{\overset{s}{\leftrightarrows}} \Lambda^3 V \otimes V \leftrightarrows \Lambda^4 V \leftrightarrows 0$$

where

$$a(u_1 \wedge \ldots \wedge u_p \otimes v_1 \ldots v_q) \underset{\text{def}}{=} \sum_{i=1}^{q} u_1 \wedge \ldots \wedge u_p \wedge v_i \otimes v_1 \ldots \hat{v}_i \ldots v_q$$

$$s(u_1 \wedge \ldots \wedge u_p \otimes v_1 \ldots v_q) \underset{\text{def}}{=} \sum_{j=1}^{p} u_1 \wedge \ldots \wedge \hat{u}_j \wedge \ldots \wedge u_p \otimes u_j v_1 \ldots v_q (-1)^{p-j} .$$

Then

$$a^2 = s^2 = 0$$

$$(as + sa)|_{\Lambda^p V \otimes \odot^q V} = (p+q)id|_{\Lambda^p V \otimes \odot^q V}$$

Since

$$(aR_x)(X,Y,Z,T) = \underset{X,Y,Z}{\oplus} R_x(X,Y,Z,T) = 0$$

we see from (iii) that the space $\mathcal{R}_x$ of curvature tensors at x is

$$\mathcal{R}_x = \ker a \subset \Lambda^2 T_x^* M \otimes \odot^2 T_x^* M$$

The group $Sp(T_x M, \omega_x)$ acts on $\mathcal{R}_x$. Under this action the space $\mathcal{R}_x$ decomposes in 2 stable subspaces :

$$\mathcal{R}_x = \mathcal{E}_x \oplus \mathcal{W}_x$$

The action of $Sp(T_x M, \omega_x)$ on each of the subspaces is irreducible [6]. These subspaces may be described as follows.

Let $t \in \odot^2 T_x^* M$; the map $j : \odot^2 T_x^* M \to \mathcal{R}_x : t \to as(\omega_x \otimes t)$ is injective and $Sp(T_x M, \omega_x)$ equivariant. The image $j \odot^2 T_x^* M$ is the stable subspace $\mathcal{E}_x$.

The symplectic form ω_x induces a non degenerate scalar product on $\Lambda^2 T_x^* M \otimes \odot^2 T_x^* M$; its restriction to $\mathcal{E}_x$ is also non degenerate. Hence :

$$\mathcal{R}_x = \mathcal{E}_x \oplus \underset{\text{not}}{\mathcal{E}_x^{\perp} \cap \mathcal{R}_x} = \mathcal{E}_x \oplus \mathcal{W}_x$$

If r_x denotes as above the Ricci tensor associated to R_x, one checks that the Ricci tensor associated to $j(r_x)$ is $-2(n+1)r_x$. Hence the decomposition of the curvature tensor R_x into its $\mathcal{E}_x$ component (denoted E_x) and its $\mathcal{W}_x$ component (denoted W_x) reads:

$$R_x = E_x + W_x$$

$$E_x(X,Y,Z,T) = -\frac{1}{2(n+1)}\big[2\omega_x(X,Y)r_x(Z,T) + \omega_x(X,Z)r_x(Y,T)$$

$$+\omega_x(X,T)r_x(Y,Z) - \omega_x(Y,Z)r_x(X,T) - \omega_x(Y,T)r_x(X,Z)\big]$$

A connection ∇ is said to be **of Ricci type** if, at each point x, $W_x = 0$.

Observe that in dimension 2 $(n = 1)$, the space $\mathcal{W}$ vanishes identically. Thus, in a certain sense, the condition for a connection to be of Ricci type, generalizes in higher dimension the surface situation.

Lemma 1 *Let (M,ω) be a symplectic manifold of dimension $2n$ $(n \geq 2)$ and let ∇ be a symplectic connection of Ricci type. Then ∇ is a preferred connection.*

This leads to

Problem 2 Can one describe the moduli space of symplectic connections of Ricci type on the symplectic manifold (M,ω).

This paper is a contribution to the solution of problem 2. More precisely, we show that any symplectic manifold (M,ω) of dimension $2n$ $(n \geq 2)$ admitting a symplectic connection of Ricci type can locally be constructed by a reduction procedure from the Euclidean space $\mathbb{R}^{2n+2}$ endowed with a constant symplectic structure and the standard flat connection. We also prove that on the bundle of symplectic frames $B(M)$ over M, there exists a 1-form with values in the algebra $sp(n+1,\mathbb{R})$ which locally satisfies a Maurer-Cartan type equation.

2 Some Properties of the Curvature of a Ricci-Type Connection

Let (M,ω) be a symplectic manifold of dim $2n$ $(n \geq 2)$ and let ∇ be a Ricci type symplectic connection. Then

Lemma 2 *[3] The curvature endomorphism reads*

$$R(X,Y) = -\frac{1}{2(n+1)}[-2\omega(X,Y)\rho - \rho Y \otimes \underline{X} + \rho X \otimes \underline{Y} - X \otimes \underline{\rho Y} + Y \otimes \underline{\rho X}]$$

$$(1)$$

where $\underline{X}$ denotes the 1-form $i(X)\omega$ (for X a vector field on M) and where ρ is the endomorphism associated to the Ricci tensor:

$$r(U, V) \underset{\text{def}}{=} \omega(U, \rho V) .$$ (2)

Furthermore:

(i) *there exists a vector field u such that*

$$\nabla_X \rho = -\frac{1}{2n+1}[X \otimes \underline{u} + u \otimes \underline{X}] ;$$ (3)

(ii) *there exists a function f such that*

$$\nabla_X u = -\frac{2n+1}{2(n+1)}\rho^2 X + fX ;$$ (4)

(iii) *there exists a real number K such that*

$$tr\rho^2 + \frac{4(n+1)}{2n+1}f = K ;$$

(iv) *the hamiltonian vector field associated to f, X_f, reads*

$$X_f = -\frac{1}{n+1}\rho u$$

and its covariant derivative reads

$$\nabla_Y X_f = -\frac{1}{(n+1)(2n+1)}u\omega(u, Y) + \frac{2n+1}{2(n+1)^2}\rho^3 Y - \frac{1}{n+1}f\rho Y .$$

3 Manifolds with Ricci Type Connections

We describe a construction of symplectic manifolds admitting a connection of Ricci type.

Let $0 \neq A \in sp(n+1, \mathbb{R})$ and denote by Σ_A, the closed hypersurface $\Sigma_A \subset \mathbb{R}^{2n+2}$ with equation:

$$\Omega(x, Ax) = 1$$

where Ω is the standard symplectic form on $\mathbb{R}^{2n+2}$; in order for Σ_A to be non empty we replace, if necessary, A, by $-A$.

Let $\tilde{\nabla}$ be the standard flat symplectic affine connection on $\mathbb{R}^{2n+1}$. If X, Y are vector fields tangent to Σ_A put:

$$(\nabla_X Y)_x = (\dot{\nabla}_X Y)_x - \Omega(AX, Y)x;$$

this defines a torsion free linear connection along Σ_A.

The vector field Ax is an affine vector field for this connection; it is clearly complete. Denote by H the 1-parametric group of diffeomorphisms of Σ_A generated by this vector field.

Since the vector field Ax is nowhere 0 on Σ_A, for any $x_0 \in \Sigma_A$, there exists a neighborhood $U_{x_0}(\subset \Sigma_A)$, a ball $D \subset \mathbb{R}^{2n}$ of radius r_0, centered at the origin, and interval $I(\subset H)$ symmetric with respect to the neutral element of H and a diffeomorphism

$$\chi : D \times I \to U_{x_0}$$

such that $\chi(0,1) = x_0$ and $\chi(y,h) = h \cdot \chi(y,1)$ (where $\cdot$ denotes the action of H on Σ_A). We shall denote

$$\pi : U_{x_0} \to D \quad \pi = p_1 \otimes \chi^{-1} .$$

If we view Σ_A as a constraint manifold in $\mathbb{R}^{2n+2}$, D is a local version of the Marsden-Weinstein reduction of Σ_A.

If $x \in \Sigma_A$, $T_x \Sigma_A = \rangle Ax \langle^{\perp}$; let $\mathcal{H}_x (\subset T_x \Sigma_A) = \rangle x, Ax \langle^{\perp}$; then

$$T_x \mathbb{R}^{2n+2} = (\mathcal{H}_x \oplus \mathbb{R}Ax) \oplus \mathbb{R}x$$

and π_{*x} defines an isomorphism between $\mathcal{H}_x$ and the tangent space $T_y D$ for $y = \pi(x)$. A vector belonging to $\mathcal{H}_x$ will be called horizontal.

A symplectic form on D, $\omega^{(1)}$ is defined by

$$\omega_y^{(1)}(X, Y) = \Omega_x(\bar{X}, \bar{Y}) \quad y = \pi(x) \tag{5}$$

where $\bar{X}$ (resp. $\bar{Y}$) denotes the horizontal lift of X (resp. Y). A symplectic connection $\nabla^{(r)}$ on D is defined by

$$\overline{\nabla_X^{(r)} Y}(x) = \nabla_{\bar{X}} \bar{Y}(x) + \Omega(\bar{X}, \bar{Y})Ax \tag{6}$$

Proposition 1 *[1] The manifold $(D, \omega^{(1)})$ is a symplectic manifold and $\nabla^{(r)}$ is a symplectic connection of Ricci type. Furthermore*

$$\overline{\rho^{(r)} X}(x) = -2(n+1)\overline{A\bar{X}}$$

$$\bar{u}(x) = -2(n+1)(2n+1)\overline{A^2 x}$$

$$(\pi^* f)(x) = 2(n+1)(2n+1)\Omega(A^2 x, Ax)$$

$$K = 4(n+1)^2 \mathrm{tr} A^2$$

If $0 \neq A$ is an element of $sp(n+1, \mathbb{R})$ and $A^2 = \lambda I$ for a certain $\lambda \in \mathbb{R}$ then the natural map $\Sigma_A \to M^{(r)}$ endows Σ_A with a structure of circle or line bundle over $M^{(r)}$. Furthermore $(M^{(r)}, \omega^{(r)}, \nabla^{(r)})$ is a symplectic symmetric space. In fact all symmetric spaces, whose canonical connection is of Ricci type are of this type. The only compact simply connected one is $\mathbb{P}_n(\mathbb{C})$. [4]

The following analysis will show that the examples described above are crucial.

Let (M, ω) be a symplectic manifold of dim $2n$ $(n \geq 2)$ and let ∇ be a symplectic connection of Ricci type. Let (N, α) be a smooth $(2n + 1)$-dimensional contact manifold (i.e. α is a smooth 1-form such that $\alpha \wedge (d\alpha)^n \neq 0$ everywhere). Let X be the corresponding Reeb vector field (i.e. $i(X)d\alpha = 0$ and $\alpha(X) = 1$). Assume there exists a smooth submersion $\pi : N \to M$ such that $d\alpha = 2\pi^*\omega$. Then at each point $x \in N$, $\text{Ker}\,(\pi_{*x}) = \mathbb{R}X$. Furthermore $\mathcal{L}_X \alpha = 0$. Observe that such a contact manifold (N, α) exists if (M, ω) is an exact symplectic manifold. Indeed choose $N = M \times \mathbb{R}$ and if $\omega = d\lambda$ let $\alpha = p_1^* \lambda + dt$.

This implies in particular that such a construction may be performed above any contractible open set $U \subset M$. In what follows we will make local constructions and not remind at each step the locality requirement.

If U is a vector field on N we can define its "horizontal lift" $\overline{U}$ on N by:

$$(i) \quad \pi_* \overline{U} = 0 \qquad (ii) \quad \alpha(\overline{U}) = 0 \,.$$

Let us denote by ν the 2-form $\pi^* \omega$ on N. Define a connection $\dot{\nabla}$ on N by:

$$\dot{\nabla}_{\overline{U}} \overline{V} = \overline{\nabla_U V} - \nu(\overline{U}, \overline{V}) X$$

$$\dot{\nabla}_X \overline{U} = \dot{\nabla}_{\overline{U}} X = -\frac{1}{2(n+1)} \overline{\rho U}$$

$$\dot{\nabla}_X X = -\frac{1}{2(n+1)(2n+1)} \overline{u}$$

where ρ is the Ricci endomorphism of (M, ∇) and where u is the vector field on M appearing in formula (3) of Lemma 2. Then $\dot{\nabla}$ is a torsion free connection on N and the Reeb vector field X is an affine vector field for this connection. The curvature of this connection has the following form:

$$\dot{R}(\overline{U}, \overline{V})\overline{W} = \frac{1}{2(n+1)} [\nu(\overline{\rho V}, \overline{W})\overline{U} - \nu(\overline{\rho U}, \overline{W})\overline{V}]$$

$$\dot{R}(\overline{U}, \overline{V})X = \frac{1}{2(n+1)(2n+1)} [\nu(\overline{u}, \overline{V})\overline{U} - \nu(\overline{u}, \overline{U})\overline{V}]$$

$$\dot{R}(\overline{U}, X)\overline{V} = \frac{1}{2(n+1)(2n+1)} \nu(\overline{u}, \overline{V})\overline{U} + \frac{1}{2(n+1)} \nu(\overline{U}, \overline{\rho V})X$$

$$\dot{R}(\overline{U}, X)X = \frac{1}{2(n+1)(2n+1)} [-\pi^* f\, \overline{U} + \nu(\overline{U}, \overline{u})X]$$

where f is the function appearing in formula (4) of Lemma 2. Since ν has constant rank $2n$ on M, there exists a symplectic manifold (P, μ) of dimension $2n + 2$ and a smooth embedding $i : N \to P$ such that $i^*\mu = \nu$. Furthermore this is essentially unique [7]. This construction can be realized as follows. The cotangent bundle T^*N contains a 1-dimensional subbundle, generated at each point $x \in N$ by α_x. Denote by P this subbundle which is clearly $N \times \mathbb{R}$; denote by s the variable along $\mathbb{R}$ and let $\theta = e^{2s} p_1^*\alpha$ $(p_1 : P \to N)$. Choose

$$\mu = d\theta = 2e^{2s} \, ds \wedge \alpha + e^{2s} \, d\alpha$$

and let $i : N \to P$ $x \mapsto (x, 0)$. Obviously $i^*\mu = \nu$. We now define a connection ∇^1 on P as follows. If Z is a vector field along N, we denote by the same letter the vector field on P such that

$$(i) \;\; Z_{i(x)} = i_{*x}Z \qquad (ii) \;\; [Z, \partial_s] = 0 \,.$$

The formulas for ∇^1 are:

$$\nabla^1_Z Z' = \dot{\nabla}_Z Z' + \gamma(Z, Z')\partial_s$$

where

$$\gamma(Z, Z') = \gamma(Z', Z)$$

$$\gamma(X, X) = \frac{1}{2(n+1)(2n+1)}\pi^*f$$

$$\gamma(X, \overline{U}) = -\frac{1}{2(n+1)(2n+1)}\nu(\overline{u}, \overline{U})$$

$$\gamma(\overline{U}, \overline{V}) = -\frac{1}{2(n+1)}\nu(\overline{U}, \overline{\rho V})$$

and

$$\nabla^1_Z \partial_s = \nabla^1_{\partial_s} Z = Z$$

$$\nabla^1_{\partial_s} \partial_s = \partial_s \,.$$

Theorem 2 *The connection ∇^1 on (P, μ) is symplectic and has zero curvature.*

Proposition 2 *Let $\psi(s)$ be a smooth function on P. Then ψ has vanishing third covariant differential if and only if*

$$\partial_s^2\psi - 2\partial_s\psi = 0 \,. \tag{7}$$

In particular the function e^{2s} has this property.

Corollary 1 *In a chart of P in which ∇^1 is the standard flat connection on $\mathbb{R}^{2n+2}$, the function ψ solution of (7) is a polynomial function of degree at most 2.*

Corollary 2 *Let (P, μ, ∇^1) be as above and let Σ be the constrained submanifold defined by $\psi(s) = 1$. Then*

(i) *The vector field X along Σ is up to sign the Hamiltonian vector field associated to ψ, i.e. $i(X)\mu = -d\psi$.*

(ii) *Let Z (resp. Z') be a vector field tangent to Σ; let Y be the vector field transversal to Σ such that $i(Y)\mu = \alpha$. Then the formula*

$$\nabla^2_Z Z' = \nabla^1_Z Z' + (\nabla^1(\nabla^1\psi))(Z, Z')Y$$

defines a torsion free affine connection ∇^2 along Σ and the vector field X is an affine vector field for this connection.

(iii) *Let H be the 1-parametric group generated by X. Then Σ/H can be identified with M and is the classical Marsden Weinstein reduction of (P, μ) for the constraint Σ. If H_x is the subspace of $T_x\Sigma$ which is the kernel of α_x one has*

$$T_x\Sigma = \mathbb{R}X \oplus H_x .$$

The 2-form $\tilde{\omega}$ on M defined by

$$\tilde{\omega}_y(Z_1, Z_2) = 2d\alpha_x(\overline{Z_1}, \overline{Z_2}) ,$$

where $y = \pi(x)$ ($\pi : \Sigma \to \Sigma/H$) and $\overline{Z}$ ($\in T_x\Sigma$) is such that $\pi_*(\overline{Z}) = Z$ and $\alpha(\overline{Z}) = 0$, coincides with ω.

(iv) *The connection ∇^3 on M defined by*

$$\nabla^3_{Z_1}\overline{Z_2} = \nabla^2_{\overline{Z_1}}\overline{Z_2} - \alpha(\nabla^2_{\overline{Z_1}}\overline{Z_2})X$$

coincides with the connection ∇, of Ricci-type, we started with.

Corollary 3 (i) *If the function ψ is a polynomial of degree 1, the curvature of (M, ∇) vanishes.*

(ii) *If the function ψ is a homogeneous polynomial of order 2, there exists a unique element $A \in sp(n+1, \mathbb{R}; \mu)$ such that*

$$(\nabla(\nabla\psi))(Z, Z') = \mu(Z, AZ') .$$

Theorem 3 *The local geometry of the symplectic manifold (M, ω) endowed with a Ricci-type symplectic connection ∇ is entirely determined by the element A of $sp(n+1, \mathbb{R})$ constructed in corollary (3).*

Remark 1 It may be proven that the "local" moduli space is isomorphic to the orbit space of $sp(n+1, \mathbb{R})$ under the adjoint action combined with dilations [8].

4 A Bundle Construction

We shall end this contribution by exhibiting a Maurer-Cartan-type 1-form on the bundle of symplectic frames over a manifold (M, ω) endowed with a symplectic connection ∇ of Ricci-type. Let $B(M) \xrightarrow{\pi} M$ be the principal bundle of symplectic frames over M. Denote by $\tilde{u} : B(M) \to \mathbb{R}^{2n}$ the $Sp(n, \mathbb{R})$ equivariant function given by $\tilde{u}(\xi) = \xi^{-1}u(x)$ where $\pi(\xi) = x$. Similarly denote by $\tilde{\rho} : B(M) \to sp(n, \mathbb{R})$ the $Sp(n, \mathbb{R})$ equivariant function given by $\tilde{\rho}(\xi) = \xi^{-1}\rho(x)\xi$; we view $sp(n, \mathbb{R}) \subset \text{End } \mathbb{R}^{2n}$; the symmetry of the Ricci tensor implies that $\tilde{\rho}(\xi)$ belongs to the symplectic algebra.

Define the $Sp(n, \mathbb{R})$ equivariant map $\tilde{A} : B(M) \to sp(n+1, \mathbb{R})$

$$\tilde{A}(\xi) = \begin{pmatrix} 0 & \dfrac{(\pi^* f)(\xi)}{2(n+1)(2n+1)} & \dfrac{-\tilde{u}(\xi)}{2(n+1)(2n+1)} \\ 1 & 0 & 0 \\ 0 & \dfrac{-\tilde{u}(\xi)}{2(n+1)(2n+1)} & \dfrac{-\tilde{\rho}(\xi)}{2(n+1)} \end{pmatrix}$$

where $\tilde{u}(\xi) = i(u(\xi))\dot{\Omega}$. We have chosen a basis of the symplectic vector space $\mathbb{R}^{2n+2}$ relative to which the symplectic form has matrix

$$\Omega = \begin{pmatrix} 0 & 1 & 0 \\ -1 & 0 & 0 \\ 0 & 0 & \dot{\Omega} \end{pmatrix} \qquad \dot{\Omega} = \begin{pmatrix} 0 & I_n \\ -I_n & 0 \end{pmatrix}.$$

The symplectic group $Sp(n, \mathbb{R}) := Sp(\mathbb{R}^{2n}, \dot{\Omega})$ injects into $Sp(n+1, \mathbb{R}) := Sp(\mathbb{R}^{2n+2}, \Omega)$ as the set of matrices

$$\begin{pmatrix} I_2 & 0 \\ 0 & A \end{pmatrix} \qquad A \in Sp(n, \mathbb{R}).$$

Lemma 3 *There exist a 1-form $\tilde{B}$ on $B(M)$, with values in $sp(n+1, \mathbb{R})$ such that:*

(i) $d\tilde{A} = [\tilde{B}, \tilde{A}];$

(ii) $\tilde{B}(Z^*) = \begin{pmatrix} 0 & 0 & 0 \\ 0 & 0 & 0 \\ 0 & 0 & -Z \end{pmatrix}$, $Z \in sp(n, \mathbb{R})$ and Z^* denoting the fundamen-

tal vector field associated to Z.

$$\tilde{B}(\overline{X}) = \begin{pmatrix} 0 & \dfrac{\dot{\Omega}(\tilde{u}, \tilde{X})}{2(n+1)(2n+1)} & \dfrac{\widetilde{\rho(X)}}{2(n+1)} \\ 0 & 0 & \tilde{X} \\ -\tilde{X} & \dfrac{\widetilde{\rho(X)}}{2(n+1)} & 0 \end{pmatrix}$$

where $\overline{X}$ *is the horizontal lift of* X;
(iii) $R_h^* \tilde{B} = h^{-1} \tilde{B} h \qquad \forall h \in Sp(n, \mathbb{R})$;
(iv) $d\tilde{B} - [\tilde{B}, \tilde{B}] = 2\tilde{A}\pi^*\omega$.

Remark 2 If one restricts $B(M)$ to a contractible open set U, the 2-form $\omega_{|_U}$ is exact. The 1-form $\tilde{B}$ is not uniquely defined by the condition (i) of the lemma; indeed, one can replace $\tilde{B}$ by $\tilde{B}' = \tilde{B} + \pi^*\lambda\tilde{A}$. If one chooses λ such that $d\lambda = 2\omega$, the 1-form $\tilde{B}'$ satisfies a Maurer-Cartan equation.

References

1. P. Baguis, M. Cahen, A construction of symplectic connections through reduction. L.M.P. 57 (2001), pp. 149–160.
2. F. Bourgeois and M. Cahen, A variational principle for symplectic connections. *J. Geom. Phys.* **30** (1999) 233–265.
3. M. Cahen, S. Gutt, J. Horowitz and J. Rawnsley, Homogeneous symplectic manifolds with Ricci-type curvature, *J. Geom. Phys.* **38** (2001) 140–151.
4. M. Cahen, S. Gutt and J. Rawnsley, Symmetric symplectic spaces with Ricci-type curvature, in Conférence Moshe Flato 1999, vol 2, G. Dito et D. Sternheimer (eds), Math. Phys. Studies 22 (2000) 81–91.
5. M. Cahen, L. Schwachöfer: in preparation
6. M. De Visscher, Mémoire de licence, Bruxelles, 1999. See also I. Vaisman, Symplectic Curvature Tensors *Monats. Math.* **100** (1985) 299–327.
7. V. Guillemin, S. Sternberg, *Symplectic techniques in physics.* Cambridge University text 1984.
8. S. Kobayashi and K. Nomizu, *Foundations of differential geometry. Vol II.* John Wiley & Sons, New York–London, 1963.

On Gauge Transformations
of Poisson Structures

H. Bursztyn

Department of Mathematics, University of Toronto, Toronto, Ontario M5S 3G3, Canada
henrique@math.toronto.edu

Summary. We discuss various questions in Poisson geometry centered around the notion of gauge transformations associated with 2-forms. The topics in this note include the relationship between gauge transformations and Morita equivalence of Poisson manifolds, gauge transformations of Lie bialgebroids and Poisson groupoids and integration of Dirac structures.

1 Introduction

Connections between Poisson geometry and topological sigma-models have led to the notion of Poisson structures "twisted" by closed 3-forms, see e.g. [23, 33]. Gauge transformations of Poisson structures associated with 2-forms were used in [38] as a tool to study this more general type of Poisson geometry. Roughly speaking, a gauge transformation modifies a given Poisson structure by adding to its leafwise symplectic structure the pullback of a globally defined 2-form. In this paper, we discuss several topics in Poisson geometry revolving around this operation.

The relationship between gauge transformations and Xu's Morita equivalence of Poisson manifolds [44] is suggested by their similar role as Poisson analogues of Morita equivalence of algebras [32, 35], a key notion of equivalence in noncommutative geometry and string theory, see e.g. [15, 37]. On one hand, Xu's purely geometric Morita theory for Poisson manifolds strongly resembles the one for $(C^*\text{-})$algebras, see e.g. [25]; in fact, at least in some examples, there are quantization procedures concretely relating them [26], though a general correspondence is yet to be found [27]. On the other hand, in the framework of formal deformation quantization [3, 24], two formal Poisson structures are quantized to Morita equivalent algebras only when they are gauge equivalent with respect to an integral 2-form [9, 10, 22]. The connection between Xu's Morita equivalence and gauge transformations was unraveled in [8], and we will recall it in Sect. 3.

Xu's geometric Morita theory for Poisson manifolds is closely related to the theory of symplectic groupoids [16, 40, 45]. Since symplectic groupoids are objects "integrating" Poisson structures (in a sense analogous to the integration of Lie algebras to Lie groups), clarifying how gauge transformations relate to Xu's Morita equivalence naturally leads one to consider the

H. Bursztyn: *On Gauge Transformations of Poisson Structures*, Lect. Notes Phys. **662**, 89–112 (2005)
www.springerlink.com

"integration" of gauge transformations to the level of symplectic groupoids. We will address this issue in two ways: first, in Sect. 3, we revisit the results of [8]; then, in Sect. 4, we present a more general approach extending the notion of gauge transformation to Lie bialgebroids and studying its global counterpart on Poisson groupoids.

The most natural framework for the study of gauge transformations, to be recalled in Sect. 2, is that of *Dirac structures* [17], a common generalization of Poisson structures, closed 2-forms and regular foliations. It turns out that the study of gauge transformations on symplectic groupoids sheds light on an interesting question in Poisson geometry: What are the global objects "integrating" Dirac manifolds? The solution to this problem is given by the *presymplectic groupoids* of [7], which we recall in Sect. 5 along with the closely related notion of *presymplectic realization*. We also illustrate, following [7, Sect. 7] (see also [46]), interesting connections between these objects and certain generalized notions of hamiltonian actions and momentum maps in symplectic geometry [1, 2].

As outlined above, this paper mostly reviews the results in [7, 8]; the only exception is Sect. 4, which contains a more general approach to [8, Thm. 4.1] leading to interesting objects (multiplicative Dirac structures) generalizing Poisson and presymplectic groupoids [5].

Acknowledgments: This note is an expanded version of a talk given at the International Workshop on Quantum Field Theory and Noncommutative Geometry, held at Tohoku University, Sendai, in November of 2002. I thank Yoshi Maeda and Satoshi Watamura for their kind invitation and warm hospitality. I would also like to thank the Institute for Pure and Applied Mathematics, UCLA, for its hospitality while this paper was being written, and Marius Crainic, Yvette Kosmann-Schwarzbach and Stefan Waldmann for helpful comments on the manuscript.

2 Gauge Transformations and Dirac Structures

2.1 Gauge Transformations of Poisson Structures

Let P be a smooth manifold. To fix our notation and terminology, recall that a bivector field $\pi \in \Gamma^\infty(\wedge^2 TP)$ is a **Poisson structure** on P if the bracket $\{f, g\} := \pi(df, dg)$ defines a Lie bracket on $C^\infty(P)$; this condition is equivalent to $[\pi, \pi]_s = 0$, where $[\cdot, \cdot]_s$ is the Schouten bracket. We refer the reader to [12] for definitions and details.

Let (P, π) be a Poisson manifold and consider a *closed* 2-form $B \in \Omega^2(P)$. We will also refer to closed 2-forms as **presymplectic**. Let us consider the addition of the pullback of B to the leafwise symplectic form of π. We call B π-**admissible** if the resulting leafwise 2-form is again symplectic (note that it is always closed, but not necessarily nondegenerate). In this case, this new symplectic foliation on P is associated with a global Poisson structure, which we denote by $\tau_B(\pi)$. The map $\pi \mapsto \tau_B(\pi)$ is a **gauge transformation** of π associated with B.

If B is not π-admissible, then the procedure described in the previous paragraph results in a *presymplectic foliation* of P, which does not correspond to any Poisson structure. We will see, however, that there is a more general geometric structure underlying it.

A convenient way of describing gauge transformations is by means of the bundle maps $\widetilde{\pi} : T^*P \to TP$ and $\widetilde{B} : TP \to T^*P$, defined by $\widetilde{\pi}(\alpha) := \pi(\cdot, \alpha)$ and $\widetilde{B}(X) := B(X, \cdot)$. In these terms, B is π-admissible if and only if

$$\text{Id} + \widetilde{B}\widetilde{\pi} : T^*P \longrightarrow T^*P \tag{1}$$

is invertible. In this case, $\tau_B(\pi)$ is completely determined by the condition that its associated bundle map satisfies

$$\widetilde{\tau_B(\pi)} = \widetilde{\pi}(\text{Id} + \widetilde{B}\widetilde{\pi})^{-1} . \tag{2}$$

In particular, if π is nondegenerate, then so is $\tau_B(\pi)$, and (2) simply says that

$$\widetilde{\tau_B(\pi)}^{-1} = \widetilde{\pi}^{-1} + \widetilde{B} .$$

So any two symplectic structures on a manifold P are gauge equivalent.

2.2 Poisson Structures as Subbundles of $TP \oplus T^*P$

A key step in unraveling the geometric structure underlying the presymplectic foliations of the previous section is noticing that Poisson structures on P can be completely described as certain subbundles of $E := TP \oplus T^*P$: for each Poisson structure π on P, we set

$$L_\pi := \text{graph}(\widetilde{\pi}) = \{(\widetilde{\pi}(\alpha), \alpha) \mid \alpha \in T^*P\} \subset E . \tag{3}$$

The result of a gauge transformation of π is represented, in these terms, as the subbundle

$$\tau_B(L_\pi) := \{(\widetilde{\pi}(\alpha), \alpha + \widetilde{B}(\widetilde{\pi}(\alpha))) \mid \alpha \in T^*P\} \subset E . \tag{4}$$

As a subbundle, (4) is well defined for any B, though it will be the graph associated with another Poisson structure if and only if $\tau_B(L_\pi) \cap TP = \{0\}$ at all points of P, which is equivalent to B being π-admissible. In this case,

$$\tau_B(L_\pi) = L_{\tau_B(\pi)} = \text{graph}(\widetilde{\tau_B(\pi)}) .$$

A natural question now is finding an intrinsic characterization of the subbundles of E of the forms (3) and (4). In order to detect them, we need to consider the following extra structure on the "ambient" bundle E:

(1) The symmetric pairing $\langle\cdot,\cdot\rangle : \Gamma^\infty(E) \times \Gamma^\infty(E) \to C^\infty(P)$,

$$\langle(X,\alpha),(Y,\beta)\rangle := \beta(X) + \alpha(Y) ; \tag{5}$$

(2) The bracket $[\cdot,\cdot] : \Gamma^\infty(E) \times \Gamma^\infty(E) \to \Gamma^\infty(E)$,

$$[(X,\alpha),(Y,\beta)] := ([X,Y], \mathcal{L}_X\beta - i_Y d\alpha) . \tag{6}$$

The bracket in 2) is the standard **Courant bracket** on E [17], written in its non-skew-symmetric version of [28, 38]. The bundle E, together with $\langle\cdot,\cdot\rangle$ and $[\cdot,\cdot]$, is an example of a *Courant algebroid* [28]. The next result follows from [17].

Proposition 1. *A subbundle $L \subset TP \oplus T^*P$ is of the form $L_\pi = \mathrm{graph}(\widetilde{\pi})$ for a bivector field π if and only if*

(i) $\mathrm{rank}(L) = \dim(P)$;
(ii) $TP \cap L = \{0\}$ at all points of P;
(iii) L is isotropic with respect to $\langle\cdot,\cdot\rangle$, i.e., $\langle(X,\alpha),(Y,\beta)\rangle = 0$ whenever $(X,\alpha),(Y,\beta) \in L$;

In this case, π is a Poisson structure if and only if

(iv) $\Gamma^\infty(L)$ *is closed under the Courant bracket* $[\cdot,\cdot]$.

Conditions i) and ii) say that L is the graph of a bundle map $T^*P \to TP$, which is always associated with a $(2,0)$-tensor $\pi \in \Gamma^\infty(TP \otimes TP)$; condition iii) is then equivalent to π being skew symmetric, i.e., a bivector field. The fact that iv) amounts to the integrability condition $[\pi,\pi]_s = 0$ is shown in [17].

Remark 1. Since the symmetric pairing $\langle\cdot,\cdot\rangle$ has signature zero, conditions i) and ii) of Proposition 1 are equivalent to L being *maximally isotropic with respect to* $\langle\cdot,\cdot\rangle$.

A direct computation shows that subbundles of the form (4) satisfy all the conditions in Proposition 1 except for ii), which holds if and only if B is π-admissible. This leads us to the geometric object we are after.

2.3 Dirac Structures and Lie Algebroids

A **Dirac structure** on a smooth manifold P is a subbundle of $E = TP \oplus T^*P$ satisfying conditions i), iii) and iv) of Prop. 1; in other words, a Dirac structure is a subbundle L of E which is maximally isotropic with respect to $\langle\cdot,\cdot\rangle$ and for which $\Gamma^\infty(L)$ is closed under the Courant bracket $[\cdot,\cdot]$. The set of Dirac structures on P is denoted by $\mathrm{Dir}(P)$.

We can naturally generalize gauge transformations from Poisson to Dirac structures [38]: for $L \in \mathrm{Dir}(P)$ and $B \in \Omega^2(P)$ closed, we set

$$\tau_B(L) := \{(X, \alpha + \tilde{B}(X)) \mid (X, \alpha) \in L\}.$$

In this extended sense, **gauge transformations** define an action of the abelian group of closed 2-form on $\mathrm{Dir}(P)$.

Of course, Poisson structures are Dirac structures intersecting TP trivially. Let us illustrate Dirac structures with two other examples.

Example 1. (Closed 2-forms)

Maximally isotropic subbundles L of $(E, \langle \cdot, \cdot \rangle)$ satisfying

$$L \cap T^*P = \{0\} \tag{7}$$

at all points of P are precisely the graphs associated with 2-forms $\omega \in \Omega^2(P)$; in this case, condition $iv)$ of Prop. 1 translates into $d\omega = 0$. So closed 2-forms correspond to Dirac structures intersecting T^*P trivially.

Example 2. (Regular foliations)

Let F be a subbundle of TP, and let F° denote the subbundle of T^*P annihilating F. Then $L = F \oplus F^\circ$ is clearly isotropic with respect to $\langle \cdot, \cdot \rangle$, and $\mathrm{rank}(L) = \dim(P)$. In this example, a direct computation shows that the "Courant-bracket condition", $iv)$ of Prop. 1, becomes the usual integrability condition of Frobenius: if $X, Y \in \Gamma^\infty(F)$, then $[X, Y] \in \Gamma^\infty(F)$, where $[\cdot, \cdot]$ is the Lie bracket of vector fields. So regular foliations are examples of Dirac structures.

Analogously to Poisson manifolds, Dirac manifolds are closely related to the world of Lie algebroids, see e.g. [12].

A **Lie algebroid** over P is a vector bundle $A \to P$ together with a Lie algebra bracket $[\cdot, \cdot]$ on $\Gamma^\infty(A)$ and a bundle map $a : A \to TP$, called the **anchor**, satisfying the Leibniz identity

$$[\xi, f\eta] = f[\xi, \eta] + \mathcal{L}_{a(\xi)}\eta, \quad \xi, \eta \in \Gamma^\infty(A).$$

As a result, a induces a Lie algebra homomorphism $\Gamma^\infty(A) \to \mathcal{X}(P)$. Whenever there is no risk of confusion, we will write $\mathcal{L}_{a(\xi)}$ simply as $\mathcal{L}_\xi$.

If P is a point, we recover the definition of a Lie algebra; another example of a Lie algebroid is $A = TP$, with $a = \mathrm{Id}$. In general, the image $a(A) \subseteq TP$ defines a generalized integrable distribution, determining a foliation of P; the leaves of this foliation are called **orbits** of A.

A Poisson structure π on P always induces a Lie algebroid structure on T^*P, see e.g. [12, Sect. 17.3]: the Lie bracket on $\Omega^1(P)$ is

$$[\alpha, \beta] := \mathcal{L}_{\tilde{\pi}(\beta)}\alpha - \mathcal{L}_{\tilde{\pi}(\alpha)}\beta - d(\pi(\alpha, \beta)), \tag{8}$$

and the anchor is $-\tilde{\pi}$; the orbits of T^*P on P are precisely the symplectic leaves of π.

This picture generalizes to Dirac structures as follows. Let

$$\rho : TP \oplus T^*P \to TP \quad \text{and} \quad \rho^* : TP \oplus T^*P \to T^*P$$

be the natural projections. Then the restriction of the Courant bracket to a Dirac subbundle L defines a Lie algebra bracket on $\Gamma^\infty(L)$, which together with the map $\rho|_L : L \to TP$ makes L into a Lie algebroid over P. When $L = \text{graph}(\tilde{\pi})$ for a Poisson structure π, the map $\rho^*|_L : L \to T^*P$ establishes an isomorphism of Lie algebroids, where T^*P has the Lie algebroid structure described in the previous paragraph.

Just as the orbits of T^*P carry a symplectic structure when P is a Poisson manifold, the orbits of a general Dirac structure L define a *presymplectic foliation* of P: the leafwise 2-form θ associated with L is defined, at each $x \in P$, by

$$\theta_x(X, Y) = \alpha(Y), \tag{9}$$

where $X, Y \in \rho(L)_x \subset T_xP$ and α is any element in T_x^*P satisfying $(X, \alpha) \in L_x$. One can check that (9) is well defined at each point of $\rho(L)$ and determines a smooth closed leafwise 2-form. Clearly this presymplectic foliation is symplectic if and only if the underlying Dirac structure comes from a Poisson structure.

Remark 2. A direct computation shows that, for any closed 2-form B, the map $\tau_B : E \to E$, $(X, \alpha) \mapsto (X, \alpha + \tilde{B}(X))$, preserves the Courant bracket. It immediately follows that gauge-equivalent Dirac structures correspond to isomorphic Lie algebroids: if $L \subset E$ is a Dirac subbundle, then $\tau_B : L \to \tau_B(L)$ is an isomorphism.

2.4 Dirac Maps

In order to define Dirac maps, we will reformulate the notion of Poisson maps in terms of subbundles of $TP \oplus T^*P$.

Recall that if (P_1, π_1), (P_2, π_2) are Poisson manifolds, then a map $f : P_1 \to P_2$ is a **Poisson map** if $f^* : C^\infty(P_2) \to C^\infty(P_1)$ preserves Poisson brackets. In terms of the bundle maps $\tilde{\pi_1}$ and $\tilde{\pi_2}$, this condition is equivalent to

$$(\tilde{\pi_2})_{f(x)} = T_x f \circ (\tilde{\pi_1})_x \circ (T_x f)^*, \quad \text{for } x \in P_1, \tag{10}$$

which can be alternatively written as

$$L_{\pi_2} = \{(Tf \circ \tilde{\pi_1} \circ (Tf)^*(\alpha), \alpha) \mid \alpha \in T^*P_2\}. \tag{11}$$

Since $X = \tilde{\pi_1}((Tf)^*\alpha)$ if and only if $(X, (Tf)^*\alpha) \in L_{\pi_1}$, we can rewrite (11) just in terms of the subbundles L_{π_1} and L_{π_2}:

$$L_{\pi_2} = \{(Tf(X), \alpha) \mid X \in TP_1, \ \alpha \in T^*P_2, \ (X, (Tf)^*(\alpha)) \in L_{\pi_1}\}.$$

This condition is now easily generalized to Dirac structures: if (P_1, L_1) and (P_2, L_2) are Dirac manifolds, then a map $f : P_1 \to P_2$ is a **Dirac map** if and only if

$$L_2 = \{(Tf(X), \alpha) \mid X \in TP_1, \ \alpha \in T^*P_2, \ (X, (Tf)^*\alpha) \in L_1\} \ .$$

The reader is referred to [8] for more on Dirac maps.

Remark 3. In the terminology of [8], the maps just defined are **forward Dirac maps**, since they generalize pushforward of bivector fields. There is an analogous notion of **backward Dirac map** generalizing pullback of 2-forms [8], which will not play a role in this paper.

2.5 Twisted Courant Brackets

The way of incorporating a closed "background" 3-form $\phi \in \Omega^3(P)$ [38] into the formalism of Dirac structures is by modifying the standard Courant bracket on E to

$$[(X, \alpha), (Y, \beta)]_\phi := ([X, Y], \mathcal{L}_X\beta - i_Y d\alpha + \phi(X, Y, \cdot)) \ . \tag{12}$$

The bundle E together with $\langle \cdot, \cdot \rangle$ and $[\cdot, \cdot]_\phi$ still satisfies the axioms of a Courant algebroid [38], and we refer to $[\cdot, \cdot]_\phi$ as the ϕ-**twisted Courant bracket** on E.

Just as before, we define ϕ-**twisted Dirac structures** as maximally isotropic subbundles of E with respect to $\langle \cdot, \cdot \rangle$ whose sections are closed under the $[\cdot, \cdot]_\phi$. Then ϕ-**twisted Poisson structures** are bivectors $\pi \in \mathcal{X}^2(P)$ whose graphs are Dirac structures; a direct computation shows that this is equivalent to π satisfying

$$[\pi, \pi]_s = 2(\wedge^3\widetilde{\pi})(\phi) \ .$$

Similarly, a 2-form ω gives rise to a ϕ-twisted Dirac structure if and only if

$$d\omega + \phi = 0 \ ,$$

in which case it is called a ϕ-**twisted presymplectic form**.

A ϕ-twisted Dirac structure induces a singular foliation on P with ϕ-twisted presymplectic leaves.

Example 3. (Cartan-Dirac structures on Lie groups [38, Ex. 4.2])
Let G be a Lie group with Lie algebra $\mathfrak{g}$. Suppose $(\cdot, \cdot)_\mathfrak{g}$ is a bi-invariant metric on G, which we use to identify TG and T^*G. On $TG \oplus TG$, we consider the maximally isotropic subbundle

$$L = \left\{ \left(v_r - v_l, \frac{1}{2}(v_l + v_r) \right), \ v \in \mathfrak{g} \right\} \ ,$$

where v_l (resp. v_r) is obtained from $v \in \mathfrak{g}$ by left (resp. right) translation. A direct computation shows that L is a ϕ-twisted Dirac structure, where ϕ is the associated bi-invariant Cartan 3-form, defined on Lie algebra elements by

$$\phi(u, v, w) = \frac{1}{2}(u, [v, w])_{\mathfrak{g}} \ .$$

We call L the **Cartan-Dirac structure** on G associated with $(\cdot, \cdot)_{\mathfrak{g}}$ [7]. In this example, the presymplectic leaves of L coincide with the conjugacy classes of G.

3 Gauge Transformations, Symplectic Groupoids and Morita Equivalence

Following [8], as a step to clarify the relationship between gauge transformations, symplectic groupoids and Xu's Morita equivalence, we discuss gauge transformations on dual pairs.

3.1 Gauge Transformations of Dual Pairs

Let (S, ω_S) be a symplectic manifold, and let (P_1, π_1) and (P_2, π_2) be Poisson manifolds. A **dual pair** [39] is a pair of Poisson maps

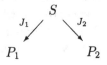

for which the J_1- and J_2-fibers are the symplectic orthogonal of one another. A dual pair is called **full** if J_1 and J_2 are surjective submersions and **complete** if the maps J_1 and J_2 are complete. (Recall that a Poisson map $J : Q \to P$ is **complete** if whenever $f \in C^\infty(P)$ has compact support, the hamiltonian vector field $X_{J^* f} \in \mathfrak{X}(Q)$ is complete.)

The following result, proven in [8], describes the effect of gauge transformations on dual pairs.

Theorem 1. *Let* $(P_1, \pi_1) \xleftarrow{J_1} (S, \omega_S) \xrightarrow{J_2} (P_2, \pi_2)$ *be a [full] dual pair, and let* B_i *be a closed 2-form on* P_i, $i = 1, 2$. *Let* $\widehat{\omega}_S = \omega_S + J_1^* B_1 + J_2^* B_2$. *Then* $\widehat{\omega}_S$ *is a symplectic form if and only if* B_i *is* π_i-*admissible*, $i = 1, 2$, *in which case*

$$(S, \widehat{\omega}_S) \tag{13}$$

$$
\begin{array}{ccc}
 & (S, \widehat{\omega}_S) & \\
 & J_1 \swarrow \quad \searrow J_2 & \\
(P_1, \tau_{B_1}(\pi_1)) & & (P_2, \tau_{B_2}(\pi_2))
\end{array}
$$

is again a [full] dual pair.

In general, $\tau_{B_1}(L_{\pi_1})$ and $\tau_{B_2}(L_{\pi_2})$ are just Dirac structures, and $\widehat{\omega}_S$ is a presymplectic form. In this more general setting, the maps J_1 and J_2 are Dirac maps, and the diagram (13) is an example of a **pre-dual pair** in the sense of [8].

Before we move on, let us recall some necessary facts about Lie and symplectic groupoids.

3.2 Lie and Symplectic Groupoids

Lie groupoids are the global counterparts of Lie algebroids. In order to fix our notation, let us recall that a Lie groupoid over a manifold P consists of another manifold $\mathcal{G}$ together with surjective submersions $t, s : \mathcal{G} \to P$, called **target** and **source**, a partially defined **multiplication** $m : \mathcal{G}^{(2)} \to \mathcal{G}$, where $\mathcal{G}^{(2)} = \{(g, h) \in \mathcal{G} \times \mathcal{G}, \; s(g) = t(h)\}$ is the set of **composable pairs**, a **unit section** $\varepsilon : P \to \mathcal{G}$, and an **inversion** $i : \mathcal{G} \to \mathcal{G}$, all related by the appropriate axioms[1] (see e.g. [12, Sect. 13]). To simplify the notation, we will often identify an element x in P with its image $\varepsilon(x) \in \mathcal{G}$.

Note that, if P is a point, a Lie groupoid becomes just a Lie group.

A Lie groupoid $\mathcal{G}$ over P has an associated Lie algebroid $A(\mathcal{G}) = \mathrm{Ker}(Ts)|_P \to P$, with anchor

$$a = Tt|_P : A(\mathcal{G}) \to TP \,,$$

and bracket induced by the Lie bracket of vector fields on $\mathcal{G}$.

An **integration** of a Lie algebroid $A \to P$ is a Lie groupoid $\mathcal{G}$ together with an identification $A(\mathcal{G}) \cong A$. Unlike Lie algebras, Lie algebroids are not always integrable, see [19] and references therein. However, if A is integrable, then there exists a canonical Lie groupoid $\mathcal{G}(A)$ with simply-connected s-fibers[2], unique up to isomorphism, called the *Weinstein groupoid* in [19].

Example 4. (Pair and fundamental groupoids)

Let P be a manifold. Then $\mathcal{G} = P \times P$ acquires a groupoid structure if we set $s(x, y) = y$, $t(x, y) = x$ and $m((x, y)(y, z)) = (x, z)$. This is called the **pair groupoid**. In this case, $A(\mathcal{G}) = TP$. In general, the s-simply-connected integration of TP is $\Pi(P)$, the **fundamental groupoid** of P, which naturally covers $P \times P$.

Example 5. (Tangent and cotangent groupoids)

Let $\mathcal{G}$ be a Lie groupoid over P. Then $T\mathcal{G}$ is a Lie groupoid over TP with source (resp. target) $Ts : T\mathcal{G} \to TP$ (resp. $Tt : T\mathcal{G} \to TP$); the multiplication
- is defined by

---

[1] A groupoid is as a small category in which all morphisms are isomorphisms; in this sense, P corresponds to the set of objects and $\mathcal{G}$ to the arrows, and the compatibility axioms between the structure maps follow.

[2] Simply connected will always mean connected with trivial fundamental group.

$$U \bullet V = Tm(U, V) ,$$

where m is the multiplication on $\mathcal{G}$, see [29]. Note that $(T\mathcal{G})^{(2)} = T(\mathcal{G}^{(2)})$.

The cotangent bundle $T^*\mathcal{G}$ is a Lie groupoid over $A(\mathcal{G})^*$, see [16, 29]: The source and target maps are defined by

$$\tilde{s}(\mu_g)(\xi) = \mu_g(Tl_g(\xi - Tt(\xi))) \text{ and } \tilde{t}(\mu_g)(\eta) = \mu_g(Tr_g(\eta)) ,$$

where $\mu_g \in T_g^*\mathcal{G}$, $\xi \in A(\mathcal{G})_{s(g)}$ and $\eta \in A(\mathcal{G})_{t(g)}$; l_g (resp. r_g) is the left (resp. right) multiplication by $g \in \mathcal{G}$. The composition $\circ$ on $T^*\mathcal{G}$ is defined by

$$(\mu_g \circ \nu_h)(U_g \bullet V_h) = \mu_g(U_g) + \nu_h(V_h), \text{ for } (V_g, U_h) \in T_{(g,h)}\mathcal{G}^{(2)} . \qquad (14)$$

A 2-form ω on a Lie groupoid $\mathcal{G}$ is called **multiplicative** if the graph of m is an *isotropic* submanifold of $(\mathcal{G}, \omega) \times (\mathcal{G}, \omega) \times (\mathcal{G}, -\omega)$; this is equivalent to the condition

$$m^*\omega = \mathrm{pr}_1^*\omega + \mathrm{pr}_2^*\omega , \qquad (15)$$

where $\mathrm{pr}_i : \mathcal{G}^{(2)} \to \mathcal{G}$, $i = 1, 2$, are the natural projections.

If π is a Poisson structure on P inducing an integrable Lie algebroid structure on T^*P, then the associated groupoid $\mathcal{G}(P) := \mathcal{G}(T^*P)$ carries a canonical multiplicative symplectic structure ω [13, 20]. A groupoid together with a multiplicative symplectic form is called a **symplectic groupoid** [16, 40].

Remark 4. The groupoids $\mathcal{G}(P)$ associated with Poisson manifolds coincide with reduced phase-spaces of Poisson sigma-models, in such a way that the canonical symplectic forms arise through (infinite-dimensional) symplectic reduction [13].

On the other hand, if $(\mathcal{G}, \omega)$ is a symplectic groupoid over P, then the compatibility (15) alone implies several interesting properties. In particular,

(i) s- and t-fibers are symplectically orthogonal to one another;
(ii) $\varepsilon : P \to \mathcal{G}$ is a lagrangian embedding, and i is an anti-symplectomorphism;
(iii) P inherits a Poisson structure π, uniquely determined by the condition that $t : \mathcal{G} \to P$ (resp. $s : \mathcal{G} \to P$) is a Poisson map (resp. anti-Poisson map).
(iv)   The symplectic form ω induces a map

$$\mathrm{Ker}(Ts)|_P \xrightarrow{\sim} T^*P, \quad \xi \mapsto i_\xi\omega|_{TP} \qquad (16)$$

which establishes an isomorphism of Lie algebroids

$$A(\mathcal{G}) \cong T^*P . \qquad (17)$$

In particular, $\dim(\mathcal{G}) = 2\dim(P)$.

An **integration** of a Poisson manifold (P, π) is a symplectic groupoid $\mathcal{G}$ over P for which the induced Poisson structure on P coincides with π. The isomorphism (17) also makes $\mathcal{G}$ into an integration of T^*P in the sense of Lie algebroids.

Example 6. (Symplectic manifolds)

A symplectic manifold (M, ω_M) is always integrable. A symplectic groupoid over it is the pair groupoid $M \times M$ equipped with the symplectic structure $\omega = \omega_M \times (-\omega_M)$. In this case, $\mathcal{G}(M) = \Pi(M)$ equipped with the symplectic form obtained by pulling back ω by the cover map $\Pi(M) \to M \times M$.

Example 7. (Dual of Lie algebras)

Let $\mathfrak{g}$ be a Lie algebra. Then $\mathfrak{g}^*$, with the Lie-Poisson structure, is always integrable. Indeed, if G is a Lie group integrating $\mathfrak{g}$, then a symplectic groupoid over $\mathfrak{g}^*$ is the semi-direct product [12] $G \ltimes \mathfrak{g}^*$, with respect to the coadjoint action. The symplectic structure on this groupoid is induced by the canonical symplectic form on T^*G via the identification $T^*G \cong G \times \mathfrak{g}^*$. Note that this is a particular case of the cotangent groupoid of Example 5 when P is a point.

Various other examples of symplectic groupoids can be found in [16, 20]; obstructions to integrability are discussed in detail in [20].

3.3 Gauge Transformations and Morita Equivalence of Poisson Manifolds

We call two Poisson manifolds (P_1, π_1) and (P_2, π_2) **gauge equivalent up to Poisson diffeomorphism** if there exists a Poisson diffeomorphism $f : (P_1, \pi_1) \to (P_2, \tau_B(\pi_2))$ for some closed 2-form B on P_2. The following properties are immediate:

(1) Two symplectic manifolds are gauge equivalent up to Poisson diffeomorphism if and only if they are symplectomorphic;
(2) Poisson manifolds which are gauge equivalent up to Poisson diffeomorphism have isomorphic foliations (though generally different leafwise symplectic structures);
(3) Gauge equivalence up to Poisson diffeomorphism preserves the Lie algebroids associated with the Poisson structures; as a result, it preserves all Poisson cohomology groups [12, 38].

Following [8], we now recall how gauge equivalence relates to Xu's Morita equivalence.

Two Poisson manifolds (P_1, π_1), (P_2, π_2) are called **Morita equivalent** [44] if there exists a complete, full dual pair

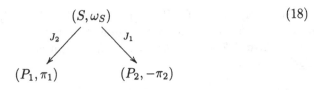

$$(18)$$

with J_i-simply-connected fibers, $i = 1, 2$; we call this diagram a **Morita bimodule**. Let us describe some properties of Morita equivalence, in analogy with gauge equivalence:

(1) For symplectic manifolds, the fundamental group is a complete Morita invariant [44]; in particular, any simply-connected symplectic manifold is Morita equivalent to a point.
(2) Morita equivalent Poisson manifolds have isomorphic spaces of symplectic leaves, see e.g. [8, 20, 45].
(3) Morita equivalence preserves first Poisson cohomology groups [18, 21].

Remark 5. (i) A symplectic groupoid $\mathcal{G}$ over P is a self-Morita equivalence of P. In fact, one can prove that a Poisson manifold is Morita equivalent to itself if and only if it can be integrated to a symplectic groupoid [42].
(ii) The regularity conditions in the definition of a Morita bimodule allow them to be "composed" through a certain "tensor product" operation [44, 45]. This makes Morita equivalence a transitive relation and endows the self-Morita equivalences of a Poisson manifold with an interesting group structure [11].
So, despite its name, Morita equivalence only becomes an equivalence relation when restricted to the class of integrable Poisson manifolds.

The comparison of properties 1), 2) and 3) for gauge and Morita equivalences suggests that they should be related and that the former should be stronger than the latter. We now show that this is in fact the case.

Let (P, π) be an integrable Poisson manifold with s-simply-connected symplectic groupoid $(\mathcal{G}, \omega)$. Let B be a π-admissible closed 2-form on P. Theorem 1 implies that

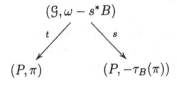

is a full dual pair, which is in fact complete [8, Thm. 5.1]. Since s- and t-fibers are simply connected by assumption, it follows that

Theorem 2. *The Poisson manifolds (P, π) and $(P, \tau_B(\pi))$ are Morita equivalent. As a consequence, gauge equivalence up to Poisson diffeomorphism implies Morita equivalence.*

Let us make some remarks on the converse.

Remark 6. (i) A comparison between properties 1) for gauge and Morita equivalences immediately implies that the converse of Theorem 2 does not hold: just consider two nonisomorphic symplectic manifolds with the same fundamental group. As shown in [8, Ex. 5.2], by studying certain symplectic fibrations, one can also find examples of Morita equivalent Poisson structures on the *same* manifold which are not gauge equivalent up to Poisson diffeomorphism.

(ii) There is, however, a generic class of Poisson structures on 2-dimensional surfaces for which gauge and Morita equivalences coincide. Let Σ be a compact connected oriented surface. We call a Poisson structure π on Σ **topologically stable** if the zero set of π consists of n smooth disjoint, closed curves on Σ, for some $n \geq 0$, where π vanishes at most linearly [34]. Building on [8, 34], it is proven in [11] that two surfaces equipped with topologically stable Poisson structures are Morita equivalent if and only if they are gauge equivalent up to Poisson diffeomorphism. Furthermore, one can use this fact to construct a complete set of gauge/Morita invariants for these structures, which can all be assembled into a labelled graph [11].

3.4 Gauge Transformations of Symplectic Groupoids

Let $(\mathcal{G}, \omega)$ be a symplectic groupoid associated with the Poisson manifold (P, π). By Remark 2, the Lie algebroids associated to Poisson structures gauge equivalent to π are all isomorphic. Therefore $\mathcal{G}$ integrates all of them as a Lie groupoid. Since gauge-equivalent Poisson structures are generally not isomorphic, gauge transformations must affect ω. We will now recall how.

Since $(P, \pi) \xleftarrow{t} (\mathcal{G}, \omega) \xrightarrow{s} (P, -\pi)$ is a full dual pair, Theorem 1 implies that, if B is a closed π-admissible 2-form on P, then

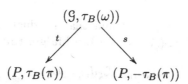

is also a full dual pair, where

$$\tau_B(\omega) := \omega + t^*B - s^*B . \tag{19}$$

This gives an indication of the next result.

Theorem 3. $(\mathcal{G}, \tau_B(\omega))$ *is a symplectic groupoid integrating* $(P, \tau_B(\pi))$.

A direct computation, see [8, Thm. 4.1], shows that $\tau_B(\omega)$ is multiplicative, so that $(\mathcal{G}, \tau_B(\omega))$ is a symplectic groupoid; it integrates $\tau_B(\pi)$ since $t : (\mathcal{G}, \tau_B(\omega)) \to (P, \tau_B(\pi))$ is a Poisson map.

We will prove a more general result in the next section.

4 Gauge Transformations of Lie Bialgebroids and Poisson Groupoids

4.1 Lie Bialgebroids and Poisson Groupoids

A common generalization of symplectic groupoids and Poisson-Lie groups was introduced by Weinstein in [41]: a bivector field $\pi_{\mathcal{G}}$ on a Lie groupoid $\mathcal{G}$ is called **multiplicative** if the graph of the multiplication m is a *coisotropic*[3] submanifold of $(\mathcal{G}, \pi_{\mathcal{G}}) \times (\mathcal{G}, \pi_{\mathcal{G}}) \times (\mathcal{G}, -\pi_{\mathcal{G}})$; a **Poisson groupoid** is a Lie groupoid together with a multiplicative Poisson structure.

As in the case of symplectic groupoids, several properties follow directly from the definition:

(i) $\{s^*C^\infty(P), t^*C^\infty(P)\} = 0$ in $C^\infty(\mathcal{G})$;

(ii) $\varepsilon : P \to \mathcal{G}$ is a coisotropic embedding, and i is an anti-Poisson map;

(iii) There is a unique Poisson structure π on P for which which t is a Poisson map (and s is anti-Poisson).

Example 8. (Pair groupoid)

If (P, π) is any Poisson manifold, then the pair groupoid $P \times P$ endowed with the Poisson structure $\pi \times (-\pi)$ is a Poisson groupoid.

It follows from this example that any Poisson structure can be realized as the base of a Poisson groupoid (even if it is not integrable) and there can be many Poisson groupoids inducing it.

The first hint about the nature of the infinitesimal counterpart of a Poisson groupoid $\mathcal{G}$ is the observation [41] that, since P sits in $\mathcal{G}$ as a coisotropic submanifold, its conormal bundle, which can be identified with $A(\mathcal{G})^*$, has itself an induced Lie algebroid structure. So a Poisson groupoid is associated with a pair of Lie algebroids $(A(\mathcal{G}), A(\mathcal{G})^*)$; the compatibility between them was explained in [29].

Suppose A is a Lie algebroid and that its dual A^* also carries a Lie algebroid structure. Then (A, A^*) is a **Lie bialgebroid** if

$$d_*[\xi, \eta] = \mathcal{L}_\xi d_*\eta - \mathcal{L}_\eta d_*\xi \tag{20}$$

for all $\xi, \eta \in \Gamma^\infty(A)$. Here $d_* : \Gamma^\infty(\wedge^k A) \to \Gamma^\infty(\wedge^{k+1} A)$ is the differential associated with the Lie algebroid structure of A^*, and $[\cdot, \cdot]$ is the Schouten-type bracket of A [12, 29],

$$[\cdot, \cdot] : \Gamma^\infty(\wedge^k A) \times \Gamma^\infty(\wedge^m A) \to \Gamma^\infty(\wedge^{k+m-1} A) \, ,$$

extending the Lie bracket on $\Gamma^\infty(A)$. The Lie derivatives $\mathcal{L}_\xi$ and $\mathcal{L}_\eta$ in (20) mean $[\xi, \cdot]$ and $[\eta, \cdot]$, respectively.

---

[3] If (P, π) is a Poisson manifold, then a submanifold $N \subset P$ is coisotropic if, at each point, $\alpha, \beta \in TN^\circ$ implies that $\pi(\alpha, \beta) = 0$.

The pair $(A(\mathcal{G}), A(\mathcal{G})^*)$ on a Poisson groupoid $\mathcal{G}$ is a Lie bialgebroid [29]; conversely, if (A, A^*) is a Lie bialgebroid and A is integrable, then there is a unique Poisson structure on $\mathcal{G}(A)$ making it into a Poisson groupoid with Lie bialgebroid (A, A^*) [30].

Note that a Lie bialgebroid over a base point is a Lie bialgebra. In this case, the associated Poisson groupoids are Poisson-Lie groups.

Example 9. (Poisson structures and symplectic groupoids)

If the Poisson structure on a Poisson groupoid $\mathcal{G}$ is nondegenerate, then the associated symplectic form makes it into a symplectic groupoid. The Lie bialgebroid in this case is the pair $(A(\mathcal{G}), A(\mathcal{G})^*) = (T^*P, TP)$, where T^*P has the Lie algebroid structure coming from π (the induced Poisson structure on P), and TP has the canonical Lie algebroid structure. Conversely, if π is a Poisson structure on P, then (T^*P, TP) is a Lie bialgebroid, and a Poisson groupoid associated with it is a symplectic groupoid integrating P.

If (A, A^*) is a Lie bialgebroid with anchors $a : A \to TP$ and $a_* : A^* \to TP$ then

$$a \circ a_*^* : T^*P \to TP \tag{21}$$

defines a Poisson structure on P [29]; here $a_*^* : T^*P \to A$ is the dual of a_*. We will refer to (21) as the Poisson structure **induced** by (A, A^*). If $\mathcal{G}$ is an associated Poisson groupoid, then this Poisson structure is the unique one making t into a Poisson map (see property *iii*) after the definition of Poisson groupoids).

4.2 Gauge Transformations

Let (P, π) be a Poisson manifold, and let $(A, A^*) = (T^*P, TP)$ be the Lie bialgebroid associated with π, as in Example 9; since $A = T^*P$ and $A^* = TP$, we denote the associated Lie bracket on $\Omega^1(P)$ by $[\cdot, \cdot]$ and the usual Lie bracket of vector fields by $[\cdot, \cdot]_*$.

Suppose B is a closed π-admissible 2-form on P, and let

$$\varphi_B = (\mathrm{Id} + \widetilde{B}\widetilde{\pi}) : T^*P \to T^*P, \quad \text{and} \quad \psi_B = \varphi_B^* = (\mathrm{Id} + \widetilde{\pi}\widetilde{B}) : TP \to TP.$$

Then the Lie algebroid structure on T^*P induced by $\tau_B(\pi)$, denoted by $(T^*P)_B$, has anchor $a_B = \widetilde{\pi}\varphi_B^{-1}$ and bracket

$$[\alpha, \beta]^B := \varphi_B[\varphi_B^{-1}(\alpha), \varphi_B^{-1}(\beta)].$$

Although $(T^*P)_B$ is isomorphic to T^*P, the Lie bialgebroids (T^*P, TP) and $((T^*P)_B, TP)$ are *not* isomorphic in general. Note that the Lie bialgebroid $((T^*P)_B, TP)$ is isomorphic to $(T^*P, (TP)_B)$, where $(TP)_B$ has anchor $(a_*)_B = \psi_B^{-1}$ and bracket

$$[X, Y]_*^B := \psi_B[\psi_B^{-1}(X), \psi_B^{-1}(Y)].$$

We now generalize this operation for arbitrary Lie bialgebroids (details to appear in [5]).

Let (A, A^*) be a Lie bialgebroid, with anchors $a : A \to TP, a_* : A^* \to TP$, and brackets $[\cdot, \cdot], [\cdot, \cdot]_*$, respectively. Let π be the induced Poisson structure on P.

Let $B \in \Omega^2(P)$ be a π-admissible 2-form. One can check that π-admissibility is equivalent to the invertibility of the map

$$\psi_B := (\mathrm{Id} + a^* \widetilde{B} a_*) : A^* \to A^* . \tag{22}$$

Here $a^* : T^*P \to A^*$ is the dual of a. Then a **gauge transformation** of (A, A^*) associated with B is the pair of Lie algebroids

$$\tau_B(A, A^*) := (A, (A^*)_B),$$

where the anchor and bracket of $(A^*)_B$ are given by

$$(a_*)_B = a_*(\psi_B)^{-1}, \quad [u, v]_*^B := \psi_B[\psi_B^{-1}(u), \psi_B^{-1}(v)]_* . \tag{23}$$

Using that B is closed, one can check that the pair $\tau_B(A, A^*) = (A, (A^*)_B)$ is a Lie bialgebroid, but this will also follow from Theorem 4 below. The bundle map associated with the Poisson structure induced by $\tau_B(A, A^*)$ on P is

$$a(\mathrm{Id} + a_*^* \widetilde{B} a)^{-1} a_*^* : T^*P \to TP , \tag{24}$$

and a simple computation shows that this map coincides with $\widetilde{\pi}(\mathrm{Id} + \widetilde{B}\widetilde{\pi})^{-1}$. Hence the Poisson structure induced by $\tau_B(A, A^*)$ is just $\tau_B(\pi)$.

Since a gauge transformation of a Lie bialgebroid (A, A^*) only affects A^*, its effect on the associated Poisson groupoids is only a change of the Poisson structure. This picture is clarified by the next result.

Theorem 4. *Let $(\mathcal{G}, \pi_\mathcal{G})$ be a Poisson groupoid over P, with Lie bialgebroid (A, A^*), and induced Poisson structure π on P. Let $B \in \Omega^2(P)$ be a closed 2-form, and let $B_\mathcal{G} := t^*B - s^*B \in \Omega^2(\mathcal{G})$. Then B is π-admissible if and only if $B_\mathcal{G}$ is $\pi_\mathcal{G}$-admissible, in which case $(\mathcal{G}, \tau_{B_\mathcal{G}}(\pi_\mathcal{G}))$ is a Poisson groupoid with Lie bialgebroid $\tau_B(A, A^*)$.*

The first assertion follows from an extension of [8, Lem. 2.12] saying that Tt induces an isomorphism between the kernels of the leafwise presymplectic forms associated with the Dirac structures $\tau_{B_\mathcal{G}}(L_{\pi_\mathcal{G}})$ and $\tau_B(L_\pi)$.

Before we prove the second assertion, let us recall that a bivector $\pi_\mathcal{G}$ on a Lie groupoid $\mathcal{G}$ is multiplicative if and only if $\widetilde{\pi}_\mathcal{G} : T^*\mathcal{G} \to T\mathcal{G}$ is a morphism of Lie groupoids [30], where $T^*\mathcal{G}$ and $T\mathcal{G}$ are the groupoids of Example 5; in this case, the induced map of identity sections $A(\mathcal{G})^* \to TP$ is the anchor of the Lie algebroid structure on $A(\mathcal{G})^*$.

Lemma 1. *Let $B \in \Omega^2(P)$ be closed, and let $B_\mathcal{G} = t^*B - s^*B$. Then $\widetilde{B}_\mathcal{G} : T\mathcal{G} \to T^*\mathcal{G}$ is a groupoid morphism (which is equivalent to $B_\mathcal{G}$ being multiplicative).*

Proof. A direct computation using the definitions in Example 5 shows that $\widetilde{B_\mathcal{G}}$ commutes with source and target maps.

Now suppose $U_g, V_h \in T\mathcal{G}^{(2)}$. Evaluating $\widetilde{B_\mathcal{G}}(U_g) \circ \widetilde{B_\mathcal{G}}(V_h)$ at $U'_g \bullet V'_h$ yields

$$\widetilde{B_\mathcal{G}}(U_g)(U'_g) + \widetilde{B_\mathcal{G}}(V_h)(V'_h) = t^* B_\mathcal{G}(U_g, U'_g) - s^* B_\mathcal{G}(U_g, U'_g)$$
$$+ t^* B_\mathcal{G}(V_h, V'_h) - s^* B_\mathcal{G}(V_h, V'_h) \,.$$

But $U_g, V_h \in T\mathcal{G}^{(2)}$ means that $Ts(U_g) = Tt(V_h)$, so we obtain

$$\widetilde{B_\mathcal{G}}(U_g) \circ \widetilde{B_\mathcal{G}}(V_h)(U'_g \bullet V'_h) = t^* B_\mathcal{G}(U_g, U'_g) - s^* B_\mathcal{G}(V_h, V'_h) \,. \tag{25}$$

On the other hand,

$$\widetilde{B_\mathcal{G}}(U_g \bullet V_h)(U'_g \bullet V'_h) = t^* B_\mathcal{G}(Tm(U_g, V_h), Tm(U'_g, V'_h)) - \tag{26}$$
$$s^* B_\mathcal{G}(Tm(U_g, V_h), Tm(U'_g, V'_h)) \,.$$

But $TtTm(U, V) = Tt(U)$ and $TsTm(U, V) = Ts(V)$ on any groupoid. Plugging these identities into (26), we get (25) and, as a result, $\widetilde{B_\mathcal{G}}(U_g \bullet V_h) = \widetilde{B_\mathcal{G}}(U_g) \circ \widetilde{B_\mathcal{G}}(V_h)$. $\square$

We can now complete the proof of Theorem 4. Since $\widetilde{\pi_\mathcal{G}} : T^*\mathcal{G} \to T\mathcal{G}$ and $\widetilde{B_\mathcal{G}} : T\mathcal{G} \to T^*\mathcal{G}$ are groupoid morphisms, so is the composition $\widetilde{\pi_\mathcal{G}}(\mathrm{Id} + \widetilde{B_\mathcal{G}}\widetilde{\pi_\mathcal{G}})^{-1}$. Hence $\tau_{B_\mathcal{G}}(\pi_\mathcal{G})$ is a multiplicative Poisson structure.

A simple computation shows that

$$\widetilde{\tau_{B_\mathcal{G}}(\pi_\mathcal{G})}|_{A^*} = a_*(\mathrm{Id} + a^*\widetilde{B}a_*)^{-1} \,,$$

which coincides with $(a_*)_B$; finally, a longer but direct computation shows that the bracket induced by $\tau_{B_\mathcal{G}}(\pi_\mathcal{G})$ on $\Gamma^\infty(A^*)$ is $[\cdot, \cdot]^B_*$.

Note that Theorem 3 now follows as a corollary.

Remark 7. (Dirac groupoids)

What is the object resulting from a gauge transformation on a Poisson groupoid if B is *not* admissible? In general, $\tau_{B_\mathcal{G}}(L_{\pi_\mathcal{G}})$ is just a Dirac structure on $\mathcal{G}$, which can be shown to be *multiplicative* in the sense that it is a subgroupoid of the product groupoid $T\mathcal{G} \times T^*\mathcal{G}$;[4] furthermore, the Dirac structure $\tau_B(L_\pi)$ on P is *induced* from $\tau_{B_\mathcal{G}}(L_{\pi_\mathcal{G}})$ in the sense that

$$t : (\mathcal{G}, \tau_{B_\mathcal{G}}(L_{\pi_\mathcal{G}})) \to (P, \tau_B(L_\pi))$$

is a Dirac map. As we will discuss in the next section, this last property does not follow from multiplicativity in general; it requires a certain "nondegeneracy condition" on the Dirac structure. These observations motivate the

---

[4] Note that bivector fields and 2-forms on $\mathcal{G}$ are multiplicative in the usual sense if and only if their graphs are subgroupoids of $T\mathcal{G} \times T^*\mathcal{G}$.

investigation of Lie groupoids carrying multiplicative Dirac structures. The particular case of multiplicative presymplectic forms was studied thoroughly in [7] (see the next section); the general case will be treated elsewhere [5].

The infinitesimal versions of multiplicative Dirac structures on $\mathcal{G}$ encompass Lie bialgebroids and Dirac structures on P. The associated integration problem should be handled by means of an extension of the methods in [30], where the correspondence between Lie algebroid and groupoid morphisms should now be phrased in terms of subalgebroids and subgroupoids of the associated product groupoids. Finally, one can also consider "background" closed 3-forms; for example, if B is not closed, then $\tau_{B_\mathcal{G}}(\pi_\mathcal{G})$ becomes twisted with respect to $s^*\phi - t^*\phi$, $\phi = dB$, and its infinitesimal counterpart $\tau_B(A, A^*)$ is closely related to the objects in [36], indicating a generalization of [14] to twisted Poisson groupoids.

5 Integration of Dirac Structures

5.1 Back to Gauge Equivalence of Symplectic Groupoids

Let us discuss the objects obtained by gauge transformation of symplectic groupoids in more detail. Let $(\mathcal{G}, \omega)$ be a symplectic groupoid, $B \in \Omega^2(P)$ a closed 2-form, and let us consider the Dirac structure $\tau_B(L_\pi)$ and the presymplectic form $\tau_B(\omega) = \omega + t^*B - s^*B$. As pointed out in Remark 7, $\tau_B(L_\pi)$ and $\tau_B(\omega)$ satisfy the following properties:

(i) $\tau_B(\omega)$ is multiplicative;
(ii) $t : (\mathcal{G}, \tau_B(\omega)) \to (P, \tau_B(L_\pi))$ is a Dirac map (while s is anti-Dirac);
(iii) $A(\mathcal{G}) \cong \tau_B(L_\pi)$.

In order to extract the properties defining the global objects integrating Dirac manifolds from this example, we explore property $iii)$ a bit further, as it reveals a certain compatibility between the kernel of $\tau_B(\omega)$ and the groupoid structure.

The identification $A(\mathcal{G}) \cong \tau_B(L_\pi)$ can be divided into two steps: first, $A(\mathcal{G}) \cong T^*P \cong L_\pi$; second, $L_\pi \cong \tau_B(L_\pi)$. For the first part, the isomorphism is given by

$$A(\mathcal{G}) \longrightarrow L_\pi, \quad \xi \mapsto (\tilde{\pi}(i_\xi\omega|_{TP}), i_\xi\omega|_{TP}).$$

Claim. $\tilde{\pi}(i_\xi\omega|_{TP}) = Tt(\xi)$.

To prove the claim, note that since $t : \mathcal{G} \to P$ is a Poisson map, $Tt(\xi) = \tilde{\pi}(\alpha)$ if and only if $Tt^*\alpha = i_\xi\omega$. Using the multiplicativity (15) of ω, one shows that $(Tt)^*(i_\xi\omega|_{TP}) = i_\xi\omega$, which implies the result (see [7, Sect. 3]).

Composing this map with a gauge transformation we get

$$A(\mathcal{G}) \xrightarrow{\sim} \tau_B(L_\pi), \quad \xi \mapsto (Tt(\xi), i_\xi\tau_B(\omega)|_{TP}), \tag{27}$$

which is the desired isomorphism of Lie algebroids.

5.2 Presymplectic Groupoids

Let us now consider a Lie groupoid $\mathcal{G}$ equipped with a multiplicative closed 2-form ω. The analogue of the map (27) is

$$A(\mathcal{G}) \to TP \oplus T^*P, \quad \xi \mapsto (Tt(\xi), i_\xi(\omega)|_{TP}) . \tag{28}$$

Guided by the discussion in Subsection 5.1, we would like this map to be injective and its image to be a Dirac structure. A direct computation shows that the map (28) is injective if and only if

$$\mathrm{Ker}(\omega_x) \cap \mathrm{Ker}(T_x s) \cap \mathrm{Ker}(T_x t) = \{0\} \tag{29}$$

for all $x \in P$; in this case, the fibers of its image have the same dimension as P (which is necessary if it is to be a Dirac structure) if and only if $\dim(\mathcal{G}) = 2\dim(P)$. One can check that the compatibility between the Lie bracket on $\Gamma^\infty(A(\mathcal{G}))$ and the Courant bracket on $TP \oplus T^*P$ follows from the multiplicativity of ω [7, Prop. 3.5]. So the image of (28) is a Dirac structure on P if and only if (29) holds and $\dim(\mathcal{G}) = 2\dim(P)$. It turns out that this is all we need.

A **presymplectic groupoid** is a Lie groupoid $\mathcal{G}$ with $\dim(\mathcal{G}) = 2\dim(P)$ equipped with a closed multiplicative 2-form ω satisfying (29). In the presence of a closed 3-form $\phi \in \Omega^3(P)$, we require ω to be **relatively ϕ-closed**, i.e.,

$$d\omega = s^*\phi - t^*\phi,$$

and $(\mathcal{G}, \omega, \phi)$ is called a **ϕ-twisted presymplectic groupoid**. It is proved in [7] that [twisted] presymplectic groupoids are the global counterparts of [twisted] Dirac structures.

Theorem 5. *Let $(\mathcal{G}, \omega)$ be a [ϕ-twisted] presymplectic groupoid. Then the image of the map (28) is a [ϕ-twisted] Dirac structure L on P, isomorphic to $A(\mathcal{G})$ as a Lie algebroid via (28) itself; furthermore, $t : (\mathcal{G}, \omega) \to (P, L)$ is a Dirac map (and s is anti-Dirac) and L is the unique Dirac structure on P with this property. (We call $(\mathcal{G}, \omega)$ an **integration** of L.)*

Conversely, let L be a [ϕ-twisted] Dirac structure on P whose associated Lie algebroid is integrable, and let $\mathcal{G}(L)$ be its s-simply connected integration. Then there exists a unique 2-form ω making $(\mathcal{G}(L), \omega)$ into a presymplectic groupoid integrating L.

The following examples illustrate the correspondence established in the theorem.

Example 10. (Poisson structures)

A presymplectic groupoid $\mathcal{G}$ over P induces a Poisson structure on P if and only if ω is symplectic. So we obtain the well-known correspondence between Poisson manifolds and symplectic groupoids. In the presence of a closed 3-form, Thm. 5 recovers the correspondence between twisted symplectic groupoids and twisted Poisson structures proved in [14].

Example 11. (Closed 2-forms)

As in the case of symplectic manifolds, presymplectic manifolds can be integrated by their fundamental groupoids with presymplectic form defined just as in Example 6.

Example 12. (Cartan-Dirac structures and the AMM-groupoid)

Let G be a Lie group with bi-invariant metric $(\cdot, \cdot)_{\mathfrak{g}}$. The **AMM groupoid** [4] over G is the action groupoid $\mathcal{G} = G \ltimes G$ with respect to the conjugation action equipped with the 2-form [2]

$$\omega_{(g,x)} = \frac{1}{2}\left((\mathrm{Ad}_x p_g^*\lambda, p_g^*\lambda)_{\mathfrak{g}} + (p_g^*\lambda, p_x^*(\lambda + \bar{\lambda}))_{\mathfrak{g}}\right)$$

where p_g and p_x denote the projections onto the first and second components of $G \times G$, and λ and $\bar{\lambda}$ are the left and right Maurer-Cartan forms. One can check that the AMM-groupoid integrates L, the Cartan-Dirac structure on G associated with $(\cdot, \cdot)_{\mathfrak{g}}$. If G is simply connected, then $(G \ltimes G, \omega)$ is isomorphic to the canonical s-simply connected integration of L; in general, one must pull-back ω to $\widetilde{G} \ltimes G$, where $\widetilde{G}$ is the universal cover of G [7, Thm. 7.6].

See [7, Sect. 8] for a detailed account of integration of Dirac structures associated with regular foliations.

5.3 Presymplectic Realizations and Hamiltonian Actions

A **symplectic realization** of a Poisson manifold (P, π) is a Poisson map from a symplectic manifold (S, ω_S) to P (see e.g. [12]). Let us briefly recall how these objects connect symplectic groupoids to the theory of hamiltonian actions.

If $J : S \to P$ is a symplectic realization, then it induces a map $a_S : T^*P \to TS$ assigning to each $\alpha \in T^*_{J(y)}P$ a vector $Z \in T_y S$ uniquely defined by

$$i_Z \omega_S = J^*\alpha .$$

This map satisfies the compatibility $TJ \circ a_S = \widetilde{\pi}$ and induces a Lie algebra homomorphism $\Omega^1(P) \to \mathfrak{X}^1(S)$; in fact, a_S defines an *action* of the Lie algebroid T^*P on S [20]. When J is complete, this action can be integrated to an action of the groupoid $\mathcal{G}(P)$ on S with moment J [20, 31]. We denote this action by $m_S : \mathcal{G}(P) \times_P S \to S$; furthermore, this action is *symplectic* in the sense that

$$m_S^*\omega_S = \mathrm{pr}_{\mathcal{G}}^*\omega + \mathrm{pr}_S^*\omega_S , \tag{30}$$

where $\mathrm{pr}_{\mathcal{G}} : \mathcal{G}(P) \times_P S \to \mathcal{G}$ and $\mathrm{pr}_S : \mathcal{G}(P) \times_P S \to S$ are the natural projections. In this context, symplectic realizations are the infinitesimal versions of symplectic groupoid actions.

Example 13. (Hamiltonian actions)

Let G be a simply connected Lie group with Lie algebra $\mathfrak{g}$. A symplectic realization $J : S \to \mathfrak{g}^*$ induces an action of the Lie algebroid $T^*\mathfrak{g}^* = \mathfrak{g} \ltimes \mathfrak{g}^*$ (this semi-direct product is with respect to the coadjoint action), which is equivalent to an action of $\mathfrak{g}$ on S for which J is $\mathfrak{g}$-equivariant [31]. It is easy to check that the $\mathfrak{g}$-action on S is hamiltonian with momentum map J.

If the realization J is complete, then the Lie algebroid action integrates to an action of $\mathcal{G}(\mathfrak{g}^*) = G \ltimes \mathfrak{g}^*$, which is equivalent to a hamiltonian G-action on S for which J is an equivariant momentum map.

Poisson actions can be similarly described as symplectic realizations of duals of Poisson-Lie groups, see [43] and references therein.

The natural way to extend the notion of symplectic realization to the realm of Dirac manifolds is to consider Dirac maps from presymplectic manifolds. So let (P, L) be a Dirac manifold, let (S, ω_S) be a presymplectic manifold and let $J : S \to P$ be a Dirac map. Let us try to define an action of L on S analogous to the one for Poisson structures: if $(X, \alpha) \in L_{J(y)}$, then J being a Dirac map means that there exists a vector $Z \in T_y S$ satisfying

$$T_y J(Z) = X , \text{ and } J^*\alpha = i_Z \omega_S ; \tag{31}$$

note, however, that these equations determine Z uniquely if and only if

$$\mathrm{Ker}(TJ) \cap \mathrm{Ker}(\omega_S) = \{0\} . \tag{32}$$

In this case, the map

$$a_S : L \to TS , \quad (X, \alpha) \mapsto Z ,$$

defines an action of the Lie algebroid L on S, generalizing the property of symplectic realizations.

Bringing closed 3-forms into the picture, we arrive at the following definition: a **presymplectic realization** of a ϕ-twisted Dirac manifold (P, L) is a Dirac map $J : (S, \omega_S) \to (P, L)$, where ω_S is a $J^*\phi$-twisted presymplectic form (i.e., $d\omega_S + J^*\phi = 0$), such that $\mathrm{Ker}(TJ) \cap \mathrm{Ker}(\omega_S) = \{0\}$.

An immediate consequence of this definition is

Corollary 1. *Any presymplectic realization* $J : (S, \omega_S) \to (P, L)$ *of a ϕ-twisted Dirac structure is canonically equipped with an infinitesimal action of the Lie algebroid of L.*

Furthermore, if the realization $J : S \to P$ is complete[5] and L is integrable, then there is an induced smooth action of the presymplectic groupoid $(\mathcal{G}(L), \omega)$ on S satisfying (30).

---

[5] In the sense that the vector field associated with $l \in \Gamma^\infty(L)$ is complete whenever l has compact support.

Having the notion of presymplectic realizations, we can repeat the discussion in Example 13 replacing $\mathfrak{g}^*$ by other Dirac manifolds.

Example 14. (Quasi-hamiltonian actions [2])

Let G be a simply connected Lie group equipped with the Cartan-Dirac structure L associated with a bi-invariant metric $(\cdot, \cdot)_{\mathfrak{g}}$. Suppose $J : S \to G$ is a presymplectic realization of G; here S is equipped with a $J^*\phi$-twisted presymplectic form ω_S, where ϕ is the Cartan 3-form. In this case, we can identify $L \cong \mathfrak{g} \ltimes G$, where the semi-direct product is with respect to the action by conjugation. As in Example 13, an action of L on S is equivalent to an action $a_S : \mathfrak{g} \to TS$ preserving ω_S for which J is $\mathfrak{g}$-equivariant. In terms of a_S, the conditions that J is a Dirac map and (32) are expressed equivalently by [7]

1. $i_{a_S(v)}(\omega_S) = \frac{1}{2}J^*(\lambda + \bar{\lambda}, v)_{\mathfrak{g}}$, where λ is the left invariant Maurer-Cartan 1-form;
2. $\mathrm{Ker}(\omega_S)_y = \{a_S(v)_y : v \in \mathrm{Ker}(\mathrm{Ad}_{J(y)} + 1)\}$, at each $y \in S$.

These properties define the infinitesimal form of quasi-hamiltonian spaces [2].

As in Examples 13, if J is complete, then we have an action of the AMM-groupoid $\mathcal{G} = G \ltimes G$ on S; this action is equivalent to a G-action on S for which J is equivariant. This makes S into a quasi-hamiltonian G-space with group-valued moment map J [2].

Replacing symplectic realizations by Poisson maps from arbitrary Poisson manifolds in Example 13 yields hamiltonian actions on Poisson manifolds; analogously, a Dirac manifold together with a Dirac map (satisfying a suitable nondegeneracy condition like (32)) into a Lie group equipped with a Cartan-Dirac structure is equivalent to a hamiltonian quasi-Poisson manifold [1]. This topic will be treated in more detail in a separate paper [6].

References

1. Alekseev, A., Kosmann-Schwarzbach, Y., Meinrenken, E.: *Quasi-Poisson manifolds.* Canad. J. Math. **54** (2002), 3–29.
2. Alekseev, A., Malkin, A., Meinrenken, E.: *Lie group valued moment maps.* J. Differential Geom. **48** (1998), 445–495.
3. Bayen, F., Flato, M., Frønsdal, C., Lichnerowicz, A., Sternheimer, D.: *Deformation Theory and Quantization.* Ann. Phys. **111** (1978), 61–151.
4. Behrend. K., Xu, P., Zhang, B.: *Equivariant gerbes over compact simple Lie groups.* C. R. Acad. Sci. Paris **333** (2003), 251–256.
5. Bursztyn, H.: *Multiplicative Dirac structures.* In preparation.
6. Bursztyn, H., Crainic, M.: *Dirac structures, moment maps and quasi-Poisson manifolds.* Preprint.
7. Bursztyn, H., Crainic, M., Weinstein, A., Zhu, C.: *Integration of twisted Dirac brackets*, Duke Math. J., to appear. Math.DG/0303180.

8. Bursztyn, H., Radko, O.: *Gauge equivalence of Dirac structures and symplectic groupoids.* Ann. Inst. Fourier (Grenoble) **53** (2003), 309–337.
9. Bursztyn, H., Waldmann, S.: *The characteristic classes of Morita equivalent star products on symplectic manifolds.* Comm. Math. Phys. **228** (2002), 103–121.
10. Bursztyn, H., Waldmann, S.: In preparation.
11. Bursztyn, H., Weinstein: *Picard groups in Poisson geometry,* Moscow Math. J., to appear. Math.SG/0304048.
12. Cannas da Silva, A., Weinstein, A.: *Geometric models for noncommutative algebras.* American Mathematical Society, Providence, RI, 1999.
13. Cattaneo, A., Felder, G.: *Poisson sigma models and symplectic groupoids.* In: *Quantization of singular symplectic quotients,* Progr. Math. 198, 61–93, Birkhäuser, Basel, 2001.
14. Cattaneo, A., Xu, P.: *Integration of twisted Poisson structures.* J. Geom. Phys., to appear. Math.SG/0302268.
15. Connes, A., Douglas, M.R., Schwarz, A.: *Noncommutative geometry and matrix theory: compactification on tori.* J. High Energy Phys. **1998**.2, Paper 3, 35 pp. (electronic).
16. Coste, A., Dazord, P., Weinstein, A.: *Groupoïdes symplectiques.* In: *Publications du Département de Mathématiques. Nouvelle Série. A, Vol. 2,* i–ii, 1–62. Univ. Claude-Bernard, Lyon, 1987.
17. Courant, T.: *Dirac manifolds.* Trans. Amer. Math. Soc. **319** (1990), 631–661.
18. Crainic, M.: *Differentiable and algebroid cohomology, van Est isomorphisms, and characteristic classes.* Math.DG/0006064, to appear in Comment. Math. Helv.
19. Crainic, M., Fernandes, R.: *Integrability of Lie brackets.* Ann. of Math. (2) **157** (2003), 575–620.
20. Crainic, M., Fernandes, R.: *Integrability of Poisson brackets.* Math.DG/0210152.
21. Ginzburg, V.L., Lu, J.-H.: *Poisson cohomology of Morita-equivalent Poisson manifolds.* Internat. Math. Res. Notices **10** (1992), 199-205.
22. Jurco, B., Schupp, P., Wess, J.: *Noncommutative line bundle and Morita equivalence.* Lett. Math. Phys. **61** (2002), 171–186.
23. Klimcik, C., Strobl, T.: *WZW-Poisson manifolds.* J. Geom. Phys. **4** (2002), 341-344.
24. Kontsevich, M.: *Deformation Quantization of Poisson Manifolds, I.* Preprint q-alg/9709040.
25. Landsman, N.P.: *Mathematical Topics between Classical and Quantum Mechanics. Springer Monographs in Mathematics.* Springer-Verlag, Berlin, Heidelberg, New York, 1998.
26. Landsman, N.P.: *Operator algebras and Poisson manifolds associated to groupoids.* Comm. Math. Phys. **222** (2001), 97–116.
27. Landsman, N.P.: *Quantization as a functor,* math-ph/0107023.
28. Liu, Z.-J., Weinstein, A., Xu , P.: *Manin triples for Lie bialgebroids.* J. Diff. Geom. **45** (1997), 547–574.
29. Mackenzie, K., Xu , P.: *Lie bialgebroids and Poisson groupoids.* Duke Math. J. **73** (1994), 415–452.
30. Mackenzie, K., Xu , P.: *Integration of Lie bialgebroids.* Topology **39** (2000), 445–467.

31. Mikami, K., Weinstein, A.: *Moments and reduction for symplectic groupoid actions.* Publ. RIMS, Kyoto Univ. **24** (1988), 121–140.

32. Morita, K.: *Duality for modules and its applications to the theory of rings with minimum condition.* Sci. Rep. Tokyo Kyoiku Daigaku Sect. A **6** (1958), 83–142.

33. Park, J.-S.: *Topological open p-branes.* In: *Symplectic geometry and Mirror symmetry (Seoul, 2000)*, 311-384, World Sci. Publishing, Rover Edge, NJ, 2001.

34. Radko, O.: *A classification of topologically stable Poisson structures on a compact oriented surface.* J. Symp. Geometry. **1** (2002), 523-542.

35. Rieffel, M.A.: *Morita equivalence for C^*-algebras and W^*-algebras,* J. Pure Appl. Algebra **5** (1974), 51–96.

36. Roytenberg, D.: *Quasi-Lie bialgebroids and twisted Poisson manifolds.* Lett. Math. Phys. **61** (2002), 123-137.

37. Schwarz, A.: *Morita equivalence and duality.* Nuclear Phys. B **534**.3 (1998), 720–738.

38. Ševera, P., Weinstein, A.: *Poisson geometry with a 3-form background.* Prog. Theor. Phys. Suppl. **144** (2001), 145–154.

39. Weinstein, A.: *The local structure of Poisson manifolds.* J. Diff. Geom. **18** (1983), 523–557.

40. Weinstein, A.: *Symplectic groupoids and Poisson manifolds.* Bull. Amer. Math. Soc. (N.S.) **16** (1987), 101–104.

41. Weinstein, A.: *Coisotropic calculus and Poisson groupoids.* J. Math. Soc. Japan **40** (1988), 705–727.

42. Weinstein, A.: *Affine Poisson structures.* Internat. J. Math. **1** (1990), 343–360.

43. Weinstein, A.: *The geometry of momentum.* Math.SG/0208108

44. Xu, P.: *Morita equivalence of Poisson manifolds.* Comm. Math. Phys. **142** (1991), 493–509.

45. Xu, P.: *Morita equivalent symplectic groupoids.* In: *Symplectic geometry, groupoids, and integrable systems (Berkeley, CA, 1989)*, 291-311. Springer, New York, 1991.

46. Xu, P.: *Morita equivalence and momentum maps.* Math.SG/0307319.

Classification of All Quadratic Star Products on a Plane* **

N. Miyazaki

Department of Mathematics, Faculty of Economics, Keio University, 4-1-1, Hiyoshi, Yokohama, 223-8521, JAPAN
miyazaki@math.hc.keio.ac.jp

Summary. In this paper we classify all quadratic star products on a plane by using Hochschild cohomology and Poisson cohomology.

1 Introduction

Deformation quantization, or moreprecisely, star products as a deformation of the usual product of functions on a phase space for an understanding of quantum mechanics were introduced in [1] (see also [23]). The existence and classification problems of star products have been solved by succesive steps from special classes of symplectic manifolds to general Poisson manifolds. At present, it is well-known that the set of all equivalent classes of hermitian star products on a symplectic manifold M is parametrized by the space of formal power series of parameter $\hbar$ with coefficients in the second de Rham cohpomology space of M (cf. [8, 19], Appendix in the present paper):

Theorem 1.

$$\{[*] : \quad * \text{ is a hermitian star product on } M\}$$

$$\cong \{ \text{ Poincaré-Cartan classes } \} \cong \frac{[\omega]_{dR}}{\nu} \bowtie H_{dR}^2(M)[[\nu]] .$$

where $p \bowtie V$ means attaching a vector space V to a point p, and $\nu = \hbar i$.

On the other hand, it is difficult to clarify the parameter spaces for general Poisson manifolds. In the present paper, using *Hochschild cohomology*, we shall classify all quadratic star products on a plane.

Theorem 2. *The moduli space*

$$\{[*] : \quad * \text{ is a star product on } \mathbb{R}^2 \text{ w.r.t. quadratic Poisson structure } \}$$

is parametrized by

---

* **Key words**: Hochschild cohomology, cyclic cohomology, deformation quantization, star product, etc.

** 2000 *Mathematics Subject Classification*: 53D55, 58B34, 58J40, 16E40

N. Miyazaki: *Classification of All Quadratic Star Products on a Plane*, Lect. Notes Phys. **662**, 113–126 (2005)
www.springerlink.com

$$\left(\frac{[0]}{\hbar} \bowtie \chi_{\mathbb{C}}^2(\mathbb{R}^2)[[\hbar]]\right) \bigcup \left(\frac{[xy\partial_x \wedge \partial_y]}{\hbar} \bowtie \mathbb{C}^2[[\hbar]]\right)$$

$$\bigcup \left(\frac{[x^2 + y^2\partial_x \wedge \partial_y]}{\hbar} \bowtie \mathbb{C}^2[[\hbar]]\right) \bigcup \left(\frac{[y^2\partial_x \wedge \partial_y]}{\hbar} \bowtie C_{\mathbb{C}}^\infty(\mathbb{R}_x)[[\hbar]]\right),$$

where [ ] *denotes a Poisson cohomology class, and* χ^2 *means the space of all bivector fields.*

Before introducing deformation quantization, in a context which seemed unrelated to deformation theory, pseudodifferential operators were introduced and also became a hot subject in mathematics thanks to the Atiyah-Singer index theorem, which expresses an analytic Fredholm index in topological terms. The symbol calculus of pseudodifferential operators playing a crucial role in its proof gives a nontrivial example of a star product. Moreover there have been many gneralizations of the original index theorems. For instance, algebraic versions developed by Connes in the context of noncommutative geometry. Under the above situation, Hochschild homology and cohomology played important roles in study of deformation quantization [6], [15] and [18], especially it is effectively used for noncommutative index theorem. For more informations on these subjects, readers should consult excellent references [23] and [5].

For example, Hochschild homology of star algebra on a symplectic manifold is isomorphic to the space of formal power series of the Planck constant with coefficients in the usual de Rham cohomology [2]:

Theorem 3. *Suppose that* M *is a* $2l$-*dimensional symplectic manifold. Then*

$$HH_n(C^\infty(M)[[\hbar]], *) \cong H_{dR}^{2l-n}(M)[[\hbar]],$$
$$HC_n(C^\infty(M)[[\hbar]], *) \cong \oplus_{i=0}^\infty H_{dR}^{2l-n+2i}(M)[[\hbar]].$$

where HC denotes cyclic homology.

For general Poisson manifolds the computation is very complicated because of the lack of powerful methods to compute it. If the rank of a Poisson structure π is everywhere constant on a manifold M, (M, π) is said to be regular. In this case we can show a similar result as in symplectic manifolds.

If (M, π) is not regular, certain difficulties will arise in the computation of Hochschild homology and cohomology of star algebra. A typical example of such manifolds is the Lie-Poisson manifold which is the dual space of a finite dimensional Lie algebra. There are some results on the computation of Hochschild homology and cohomology of star algebra on the Lie-Poisson manifold. In the present paper, we are also concerned with Hochschild homology of star algebra on a quadratic Poisson manifold.

2 Formal Deformation Quantization

2.1 Star Product

As mentioned in Introduction, the framework of deformation quantization is a quantization scheme introduced in [1]. In this approach, algebras of quantum observables are defined by formal deformation of classical observables as follows:

Definition 1. *A formal deformation quantization of Poisson manifold* (M, π) *is a family of product* $* = *_\hbar$ *(depending on the Planck constant* $\hbar$*) on the space of formal power series of parameter* $\hbar$ *with coefficients in* $C^\infty(M)$*,* $C^\infty(M)[[\hbar]]$ *defined by*

$$f *_\hbar g = fg + \hbar \pi_1(f, g) + \cdots + \hbar^n \pi_n(f, g) + \cdots, \quad \forall f, g \in C^\infty(M)[[\hbar]]$$

satisfying

1. *$*$ is associative,*
2. *$\pi_1(f, g) = \frac{1}{2\sqrt{-1}} \{f, g\}$,*
3. *each π_n $(n \geq 1)$ is a $\mathbb{C}[[\hbar]]$-bilinear and bidifferential operator,*

where $\{,\}$ is the Poisson bracket defined by the Poisson structure π.

Deformed algebra (resp. deformed algebra structures) are called a *star algebra* (resp. *star-products*).

Note that on a symplectic vector space $\mathbb{R}^{2n}$, there exists the "canonical" deformation quantization, so-called the *Moyal product formula*: Let f, g be smooth functions of a Darboux coordinate (x, y) on $\mathbb{R}^{2n}$ and $\nu = i\hbar$. Using the binomial theorem, we set

$$f(\overleftarrow{\partial_x} \wedge \overrightarrow{\partial_y})^m g = \sum_{|\alpha|+|\beta|=m} \frac{m!}{\alpha! \beta!} (-1)^{|\beta|} f(\overleftarrow{\partial_x} \overrightarrow{\partial_y})^\alpha (\overleftarrow{\partial_y} \overrightarrow{\partial_x})^\beta g,$$

and

$$\exp\left[\frac{\nu}{2} \overleftarrow{\partial_x} \wedge \overrightarrow{\partial_y}\right] = \sum_n \frac{1}{m!} \left(\frac{\nu}{2}\right)^m (\overleftarrow{\partial_x} \wedge \overrightarrow{\partial_y})^m.$$

Under these notation, the Moyal product formula is written in the following way:

$$f * g = f \exp\left[\frac{\nu}{2} \overleftarrow{\partial_x} \wedge \overrightarrow{\partial_y}\right] g.$$

2.2 The Formality Theorem

In this subsection, we recall Kontsevich formality theorem [10].

Differential Graded Lie Algebra of T_{poly}-Fields

Let M be a smooth manifold. Set $T_{poly}(M) = \oplus_{k \geq -1} \Gamma(M, \wedge^{k+1} TM)$, and let $[\cdot, \cdot]_S$ be the *Schouten* bracket:

$$[X_0 \wedge \cdots \wedge X_m, Y_0 \wedge \cdots \wedge Y_n]_S = \sum_{i,j} (-1)^{i+j+m} [X_i, Y_j] \cdots \wedge \hat{X}_i \wedge \cdots \wedge \hat{Y}_j \wedge \cdots ,$$

where $X_i, Y_i \in \Gamma(M, TM)$. Then, the triple

$$(T_{poly}(M)[[\hbar]], d = 0, [\cdot, \cdot] = [\cdot, \cdot]_S)$$

forms a differential graded Lie algebra. It is well-known that for any bivector $\pi \in \Gamma(M, \wedge^2 TM)$, π is a Poisson structure if and only if

$$[\pi, \pi]_S = 0 . \tag{1}$$

Differential Grade Lie Algebra of D_{poly}-Fields

Let $(A, \bullet)$ be an associative algebra and set $C(A) = \oplus_{k \geq -1} C^k$, $C^k = Hom(A^{\otimes k+1}; A)$. For $\varphi_i \in C^{k_i}$ $(i = 1, 2)$, we set

$$\varphi_1 \circ \hat{\varphi}_2 (a_0 \otimes a_1 \otimes \cdots \otimes a_{k_1 + k_2})$$

$$= \sum_{i=0}^{k} (-1)^{ik_2} \varphi_1 (a_0 \otimes \cdots \otimes a_{i-1} \tag{2}$$

$$\otimes \varphi_2 (a_i \otimes \cdots \otimes a_{i+k_2}) \otimes a_{i+k_2+1} \otimes \cdots \otimes a_{k_1+k_2}) .$$

Then the *Gerstenhaber* bracket is defined in the following way:

$$[\varphi_1, \varphi_2]_G = \varphi_1 \circ \hat{\varphi}_2 - (-1)^{k_1 k_2} \varphi_2 \circ \hat{\varphi}_1 \tag{3}$$

and Hochschild coboundary operator $\delta = \delta_\bullet$ with respect to $\bullet$ is defined by $\delta_\bullet(\varphi) = (-1)^k [\bullet, \varphi]$ $(\varphi \in C^k)$. Then it is known that the triple

$$(C(A), d = \delta_\bullet, [\cdot, \cdot] = [\cdot, \cdot]_G)$$

is a differential graded Lie algebra.

Let M be a smooth manifold. Set $\mathcal{F} = C^\infty(M)$, and $D_{poly}(M)^n(M)$ equals a space of all multidifferential operators from $\mathcal{F}^{\otimes n+1}$ into $\mathcal{F}$. Then $D_{poly}(M)[[\hbar]] = \oplus_{n \geq -1} D_{poly}^n(M)[[\hbar]]$ is a subcomplex of $C(\mathcal{F}[[\hbar]])$. Furthermore, the triple $(D_{poly}(M)[[\hbar]], \delta, [\cdot, \cdot]_G)$ is a differential graded Lie algebra.

Proposition 1. *Let B be a bilinear operator and $f \star g = f \cdot g + B(f, g)$. Then it is known that $\star$ is associative if and only if*

$$\delta B + \frac{1}{2} [B, B]_G = 0. \tag{4}$$

Next we recall the moduli space $\mathcal{MC}(C(V[1]))$. Suppose that $V = \oplus_{k \in \mathbb{Z}} V^k$ is a graded vector space, and $[1]$ is the shift-functor, that is, $V[1]^k = V^{k+1}$. $V[1] = \oplus_k V[1]^k$ is called a shifted graded vector space of V. We set $C(V) = \oplus_{n \geq 1} Sym^n(V)$ where $Sym^n(V) = T^n(V)/\{\cdots \otimes (x_1 x_2 - (-1)^{k_1 k_2} x_2 x_1) \otimes \cdots ; x_i \in V^{k_i}\}$. For $b \in V[1]$, set $e^b = 1 + b + \frac{b \otimes b}{2!} + \cdots \in C(V[1])$.

Definition 2. $\ell(e^b) = 0$ is called a Batalin-Vilkovisky-Maurer-Cartan equation, where $\ell = d + (-1)^{deg} \circ [\circ, \bullet]$.

Using this equation, we define the moduli space as follows:

Definition 3.

$$\widehat{\mathcal{MC}}(C(V[1])) = \{b; \ell(e^b) = 0\}, \tag{5}$$

$$\mathcal{MC}(C(V[1])) = \widehat{\mathcal{MC}}(C(V[1]))/ \sim , \tag{6}$$

where V stands for $T_{poly}(M)[[\hbar]]$ and $D_{poly}(M)[[\hbar]]$, and $\sim$ means the gauge equivalencet (cf. [10]).

Note that (1) and (4) can be seen as the Batalin-Vilkovisky-Maurer-Cartan equations.

With these preliminaries, we can state precise version of Kontsevich formality theorem:

Theorem 4. There exists a map $\mathcal{U}$ such that

$$\mathcal{U} : \mathcal{MC}(C(T_{poly}(M)[[\hbar]][1])) \cong \mathcal{MC}(C(D_{poly}(M)[[\hbar]][1])) .$$

Thus, for any Poisson manifold (M, ω) there exists a formal deformation quantization.

3 Proof of Theorem 2

3.1 Classification of the Moyal-Type Quadratic Star Products

As seen in the previous section, Theorem 4 gives a star product on any Poisson manifold. However, in general, it is difficult to classify all star products with respect to quadratic Poisson structures explisitly. In this subsection, we classify all star products with respect to quadratic Poisson structures on a **plane**. First we note that a direct computation gives the following:

Proposition 2. As for commutative vector fields $D_1, \cdots, D_l$ on a manifold M, a product $*$ defined by

$$f * g = f \exp\left[\frac{\nu}{2} \sum_{i \in I, j \in J} \overleftarrow{D_i} \wedge \overrightarrow{D_j} \right] g$$

gives a deformation quantization of $\pi := \sum_{i \in I, j \in J} D_i \wedge D_j$ which defines a Poisson structure on the manifold M.

In what follows, we will refer to a product arising in this way as the *Moyal-type star product*. Using this lemma, we shall classify the Moyal-type quadratic star products. Remark that in order to classify all quadratic Poisson structures on a plane, we have only to classify $sl_2(\mathbb{R})$ by Jordan form (cf. [11]). By this procedure, we have the classification of all quadratic Poisson structures π on a plane up to a linear isomorphism:

1. $\pi = 0$,
2. $\pi = (x^2 + y^2)\partial_x \wedge \partial_y = r\partial_r \wedge \partial_\theta$ $(x = r\cos\theta, y = r\sin\theta)$,
3. $\pi = xy\partial_x \wedge \partial_y$,
4. $\pi = y^2\partial_x \wedge \partial_y$,

Combining Proposition 2 with what mentioned above we have the following.

Proposition 3. *The following is a complete list of all Moyal-type star products $*$ on a plane with respect to quadratic Poisson structures π:*

$$f * g = \begin{cases} f \cdot g & (\text{ if } \pi = 0), \\ f \exp[\frac{\nu}{2}\overleftarrow{r\partial_r} \wedge \overrightarrow{\partial_\theta}]g & (\text{if } \pi = (x^2 + y^2)\partial_x \wedge \partial_y), \\ f \exp[\frac{\nu}{2}\overleftarrow{x\partial_x} \wedge \overrightarrow{y\partial_y}]g & (\text{if } \pi = xy\partial_x \wedge \partial_y), \\ f \exp[\frac{\nu}{2}\overleftarrow{\partial_x} \wedge \overrightarrow{y^2\partial_y}]g & (\text{if } \pi = y^2\partial_x \wedge \partial_y). \end{cases} \tag{7}$$

In what follows, the above formula (7) (resp. deformed algebra) will be refered to as the *"Moyal-type quadratic star product"* (resp. *"Moyal-type quadratic star algebra"*) associated with a quadratic Poisson structure.

3.2 Classification of All Star Products of Quadratic Poisson Structures on a Plane

First we consider the moduli spaces defined in §2. The second cohomology has the geometrical meaning as follows: A variational computation of the Batalin-Vilkovisky-Maurer-Cartan equation ((1), (4) and Definition 2) gives the following:

$$T_{[\pi]}\mathcal{MC}(T_{poly}(M)[[\hbar]]) \cong H^2_{LP,\pi}(M)[[\hbar]],$$

where $T_{[\pi]}\mathcal{MC}(T_{poly}(M)[[\hbar]])$ denotes the tangent space of the moduli space of $\mathcal{MC}(T_{poly}(M)[[\hbar]])$ at a point $[\pi]$ and $H^2_{LP,\pi}(M)$ denotes the Lichnerowicz-Poisson cohomology. Similarly, we see

$$T_{[*]}\mathcal{MC}(D_{poly}(M)[[\hbar]]) \cong HH^2(C^\infty(M)[[\hbar]], *),$$

where $[*]$ is a point defined by $[*] = \mathcal{U}([\pi])$ and $HH^{\cdot}$ means Hochschild cohomology. Summing up what mentioned above gives

Lemma 1.

$$H^2_{LP,\pi}(M)[[\hbar]] \cong HH^2(C^\infty(M)[[\hbar]], *).$$

Since (1) and the definition of equivalence (cf. [8]):

$$T : C^\infty(M)[[\hbar]] \to C^\infty(M)[[\hbar]] : \mathbb{C}[[\hbar]]\text{-differential operator,}$$

the classification problem is reduced to compute the second Hochschild cohomology with respect to a quadratic star product. Furthermore, by Proposition 2, we may assume that $*$ is one of products defined by (7). Thus, combining Proposition 1 with results in [14], we obtain the following.

Proposition 4. *The moduli space of each Moyal-type quadratic star product*

$$\{[*] : \quad \text{star product on } \mathbb{R}^2 \text{ w.r.t. a quadratic Poisson structure } \pi\}$$

is parametrized by

$$
\begin{array}{ll}
\frac{[0]}{\hbar} \bowtie \chi_{\mathbb{C}}^2(\mathbb{R}^2)[[\hbar]] & (\pi = 0)\,, \\[4pt]
\frac{[xy\partial_x \wedge \partial_y]}{\hbar} \bowtie \mathbb{C}^2[[\hbar]] & (\pi = xy\partial_x \wedge \partial_y)\,, \\[4pt]
\frac{[x^2 + y^2 \partial_x \wedge \partial_y]}{\hbar} \bowtie \mathbb{C}^2[[\hbar]] & (\pi = (x^2 + y^2)\partial_x \wedge \partial_y)\,, \\[4pt]
\frac{[y^2 \partial_x \wedge \partial_y]}{\hbar} \bowtie C_{\mathbb{C}}^\infty(\mathbb{R}_x)[[\hbar]] & (\pi = y^2\partial_x \wedge \partial_y)\,,
\end{array}
$$

where $[\ ]$ *denotes a Poisson cohomology class,* $p \bowtie V$ *means attaching a vector space V to a point p, and χ^2 means the space of all bivector fields.*

Summing up what mentioned above gives Theorem 2.

Note that using the cohomology version of Connes' periodicity exact sequence, we can calculate cyclic cohomology of star algebras on a plane with singular Poisson structures.

3.3 Hochschild Homology

From this section, we are mainly concerened with a quadratic Poisson structure $\pi = y^2\partial_x \wedge \partial_y$. As mentioned in the previous subsection, $HH_2(C^\infty(M)$ $[[\hbar]], *)$ (resp. $H_2^{KP,\pi}(M)[[\hbar]]$) can be seen as a cotangent space of the moduli space $\mathcal{MC}(C(D_{poly}(M)[[\hbar]][1]))$ (resp. $\mathcal{MC}(C(T_{poly}(M)[[\hbar]][1]))$) introduced in the previous section. Thus it is natural to expect that

$$H_n^{KP,\pi}(M)[[\hbar]] \cong HH_n(C^\infty(M)[[\hbar]], *)\,.$$

First we consider the fundamental formula on which our arguments will be based.

Proposition 5.

$$H_n^{KP,\pi}(M)[[\hbar]] \cong HH_n(C^\infty(M)[[\hbar]], *)\,,$$

where $H_n^{KP,\pi}$ denotes the Koszul-Poisson homology.

Sketch of the Proof for Proposition 5. Now let us start with setting up our notation of Hochschild homology of $\mathcal{A}$. The Hochschild homology of algebra $\mathcal{A}$ is defined as a homology of the following complex:

$$C_* : \cdots \xrightarrow{b} \mathcal{A} \otimes \mathcal{A}^{\otimes n} \xrightarrow{b} \mathcal{A} \otimes \mathcal{A}^{\otimes n-1} \xrightarrow{b} \cdots ,$$

where Hochschild boundary operator b is defined by setting

$$\begin{aligned}
&b(a_0 \otimes a_1 \otimes \cdots \otimes a_n) \\
&= \sum_{i=0}^{n-1} (-1)^i a_0 \otimes \cdots \otimes (a_i * a_{i+1}) \otimes \cdots \otimes a_n \\
&\quad + (-1)^n (a_n * a_0) \otimes a_1 \otimes \cdots \otimes a_{n-1} .
\end{aligned}$$

We introduce a modified *Koszul-Poisson* complex $\hat{C}_*$ defined by replacing the usual Poisson bracket by the commutation bracket with respect to star-product. Using the grading C_* (resp. $\hat{C}_*$) and $\hbar$-adic filtration denoted by F_* (resp. $\hat{F}_*$), we consider the spectral sequences E (resp. $\hat{E}_*$) for the homology of C_* (resp. $\hat{C}_*$). Furthermore we also need an *anti-symmetrization operator* ϵ of the modified Koszul-Poisson complex into the Hochschild complex as a compatible mapping of the filtered complex as follows:

$$\begin{aligned}
&\epsilon(a_0 \otimes da_1 \wedge \cdots \wedge da_n) \\
&= \sum_{\sigma \in S_n} \mathrm{sgn}\sigma \, a_0 \otimes a_{\sigma^{-1}(1)} \otimes \cdots \otimes a_{\sigma^{-1}(n)} .
\end{aligned}$$

By this operator ϵ we can compare the modified Koszul-Poisson complex with the Hochschild complex. According to the Hochschild-Kostant-Rosenberg theorem (cf. [9]), we see that the E_1 -term of the Hochshild complex coincides with one of the modified Koszul-Poisson complex. By virtue of the spectral sequence comparsion theorem, we see that Hochschild homology of formal deformation quantization is described by the formal power series of $\hbar$ with coefficient in the original Koszul-Poisson homology, see also [13].

Lemma 2. $HH_n(\mathcal{A}) = Ker\tilde{\delta}_n / Im\tilde{\delta}_{n+1}[[\hbar]]$, *where* $\tilde{\delta}_n = [i_\pi, d] : C^\infty(M) \otimes \wedge^n dC^\infty(M) \to C^\infty(M) \otimes \wedge^{n-1} dC^\infty(M)$ *is the original Koszul-Poisson boundary operator.*

Hence the problem is reduced to calculating the Koszul-Poisson homology.

As to a star product obtained from a quadratic Poisson structure $\pi = y^2 \partial_x \wedge \partial_y$ on a plane. A direct calculation gives the next proposition.

Proposition 6. *Suppose that* $*$ *is the quadratic star product associated with* $\pi = y^2 dx \wedge dy$. *Then*

$$HH_0(C^\infty(\mathbb{R}^2)[[\hbar]], *) \cong \frac{C_C^\infty(\mathbb{R}^2)}{\{(y^2)\frac{d\zeta}{dx \wedge dy}|\zeta \in \Lambda^1\}}[[\hbar]] ,$$

$$HH_1(C^\infty(\mathbb{R}^2)[[\hbar]], *) \cong \frac{dC_C^\infty(\mathbb{R}^2)}{\{d(y^2)f|f \in C_C^\infty(\mathbb{R}^2)\}}[[\hbar]] ,$$

$$HH_2(C^\infty(\mathbb{R}^2)[[\hbar]], *) \cong \{0\} .$$

Proof. Let $\omega = y^2$. Since

$$\tilde{\delta}(adx \wedge dy) = \{-d(\omega a)\},$$

and

$$\tilde{\delta}(bdx + cdy) = \omega(\partial_x c - \partial_y b),$$

$\tilde{\delta}(adx \wedge dy) = \frac{\hbar}{i}\{-d(\omega a)\}$, we see

$$
\begin{aligned}
&HH_1(\mathcal{A})\\
&= \frac{\{bdx+cdy:\omega(\partial_x c-\partial_y b)=0,b,c\in C_C^\infty(\mathbb{R}^2)\}}{\{d(\omega a):a\in C_C^\infty(\mathbb{R}^2)\}}[[\hbar]]\\
&= \frac{\{bdx+cdy:d(bdx+cdy)=0,b,c\in C_C^\infty(\mathbb{R}^2)\}}{\{d(\omega a):a\in C_C^\infty(\mathbb{R}^2)\}}[[\hbar]] \qquad (8)\\
&= \frac{\{dc:c\in C_C^\infty(\mathbb{R}^2)\}}{\{d(\omega a):a\in C_C^\infty(\mathbb{R}^2)\}}[[\hbar]]\\
&= \frac{dC_C^\infty(\mathbb{R}^2)}{d\langle\omega\rangle}[[\hbar]] ,
\end{aligned}
$$

where the 2nd equality of (8) follows from $\omega \neq 0$. By the same way, we get the first and third equalities of Proposition 6. $\square$

Remark that combining this results with Connes' periodicity exact sequence, we have informations on cyclic homology.

4 Appendix

In this section, we review the theory of Weyl manifold established in [18], [19], [26].

Definition 4. *A* Weyl algebra $\mathcal{W}$ *is an algebra formally generated by generators*

$$Z_0 = \nu,\ Z_1 = X_1,\ldots, Z_n = X_n, Z_{n+1} = Y_1,\ldots, Z_{2n} = Y_n , \qquad (9)$$

satisfying the following relations:

$$[X_i, X_j] = [Y_i, Y_j] = 0,\quad [\nu, X_i] = [\nu, Y_j] = 0,\quad [X_i, Y_j] = -\nu\delta_{ij} , \qquad (10)$$

where $[a, b] = a * b - b * a$.

From the point of view of symplectic geometry, a *Weyl manifold* $\mathcal{W}_M$ is an algebra bundle over a symplectic manifold M, whose transition functions satisfy a special properties.

Starting with the Weyl algebra $\mathcal{W}$, we consider a locally trivial algebra bundle $\mathcal{W}_M$ with the fiber $\mathcal{W}$ in the following way: given a suitable open covering $M = \bigcup_\alpha U_\alpha$, we take the atlas $\{\mathcal{W}_{U_\alpha} = U_\alpha \times \mathcal{W}\}$. On each local trivialization $\mathcal{W}_{U_\alpha}$, we define a concept of local Weyl function, as a certain class of sections of $\mathcal{W}_{U_\alpha}$. $\mathcal{W}_M$ is called a contact Weyl manifold if each coordinate transformation of two local trivializations preserves the class of Weyl functions. In order to give the precise definition of the Weyl manifold, we first need the of definition of Weyl function.

Definition 5. *A section s having the folloing form:*

$$s = f^{\#} := \sum \frac{1}{\alpha!\beta!} \partial_x^{\alpha} \partial_{\xi}^{\beta} f(x, \xi) X^{\alpha} \Xi^{\beta} \quad (f \in C^{\infty}(U)[[\nu]])$$

is called a Weyl function. *We denote the set of all Weyl functions on* $\mathcal{W}_U$ *by* $\mathcal{F}(\mathcal{W}_U)$.

By definition of the product of Weyl algebra, Weyl continuation and the Moyal product formula, we can easily see the following formula.

Proposition 7. $(f *_M g)^{\#} = f^{\#} * g^{\#}, \quad (f, g \in C^{\infty}(U)[[\nu]])$.

Next we define a class of transition functions.

Definition 6. *A map* $\Phi : \mathcal{W}_U \to \mathcal{W}_U$ *is called a* Weyl diffeomrphism *if*

1. $\Phi_z(\mathcal{W}_z) = \mathcal{W}_{\phi(z)}$ $(\forall z \in U)$, *and* $\Phi_z|_{\mathcal{W}_z}$ *is a* ν-preserving Weyl algebra isomorphism, where ϕ is a base map indeced by Φ,
2. $\Phi_z(\bar{a}) = \overline{\Phi_z(a)}$,
3. $\Phi^*(\mathcal{F}(\mathcal{W}_{\phi(U)})) = \mathcal{F}(\mathcal{W}_U)$.

We here give the precise definition of Weyl manifolds [18].

Definition 7. *A family* $\mathcal{W}_M = \{\mathcal{W}_{U_{\alpha}}, \hat{\Phi}_{\alpha\beta} : \mathcal{W}_{U_{\alpha\beta}} \to \mathcal{W}_{U_{\beta\alpha}}\}$ *is called a* Weyl manifold *over a symplectic manifold* M *if each* $\hat{\Psi}_{\alpha\beta}$ *is a Weyl diffeomorphism.*

The following is first shown by Omori-Maeda-Yoshioka.

Theorem 5. *For any symplectic manifold* M, *there exists a Weyl manifold over* M.

In order to explain the classification of deformation quantization, we introduce a contact Weyl algebra in the following way.

Definition 8. *A* contact Weyl algebra $\mathcal{C}$ *is an algebra formally generated by generators* (9) *and* τ *satisfying the relations* (10) *and* $[\tau, \nu] = 2\nu^2$, $[\tau, Z_i] = \nu Z_i$ *where* $[a, b] = a * b - b * a$.

For this algebra, just as a tirivial Weyl algebra bundle over an open subset U, we denote $\mathcal{C} \times U$ by $\mathcal{C}_U$.

Definition 9. *A map* $\Psi : \mathcal{C}_U \to \mathcal{C}_U$ *is called a* contact Weyl diffeomrphism *if*

1. Ψ *is an algebra bundle isomorphism,*
2. $\Psi_z(\mathcal{W}_z) = \mathcal{W}_{\psi(z)}$ $(\forall z \in U)$, *and* $\Psi_z|_{\mathcal{W}_z}$ *is a Weyl diffeomorphism, where* ψ *is a base map indeced by* Ψ,
3. $\Psi_z(\bar{a}) = \overline{\Psi_z(a)}$,
4. $\Psi^*(\tilde{\tau}) = \tilde{\tau} + f^{\#}$ $(\exists f \in C^{\infty}(U)[[\nu]])$ *where* $\tilde{\tau} = \tau + \sum \omega_{ij} z^i Z^j$.

It is known in [18] that

Lemma 3. *For any Weyl diffeomorphism Φ, there exists a canonical lifting-contact Weyl diffeomorphism Ψ such that $\Psi|_{W_U} = \Phi$. Ψ is denoted by Φ^c.*

However, it is impossible to make a contact Weyl algebla bundle glueing local trivial contact Weyl algebra bundles by contact Weyl diffeomorphisms unlike a Weyl manifold, that is, $\mathcal{C}_M = \{C_{U_\alpha}, \Psi_{\alpha\beta} : \mathcal{C}_{U_{\alpha\beta}} \to \mathcal{C}_{U_{\beta\alpha}}\}$ only forms a **stack**.

Assume that a contact Weyl diffeomorphism $\Psi = \exp r^\#$ on $\mathcal{U}$ satisfies $\Psi|_{W_U} = 1$, then r must not contain variables x_i, y_j, thus we obtain the following.

Proposition 8. *Assume that a map Ψ is a contact Weyl diffeomorphism.*

1. *If the diffeomorphism φ on the base map induced by Ψ is the identity, there exists uniquely a Weyl function $g^\#(\nu^2)$ such that $\Psi = \exp[ad(\frac{1}{\nu}g^\#(\nu^2))]$.*
2. *$\Psi|_{W_U} = 1$ if and only if $\exists c(\nu^2) \in \mathbb{R}[[\nu^2]]$ such that $\Psi = \exp[\frac{1}{\nu}ad(c(\nu^2))]$.*

From this proposition, we can define the Poincaré-Cartan class in the following way [19]. Assume that $W_M = \{W_{U_\alpha}, \Phi_{\alpha\beta}\}$ is a Weyl manifold. Then

$$\Phi_{\alpha\beta}\Phi_{\beta\gamma}\Phi_{\gamma\alpha} = 1 \ .$$

According to Proposition 8, we have

$$\Phi_{\alpha\beta}\Phi_{\beta\gamma}\Phi_{\gamma\alpha} = \exp\left[\frac{1}{\nu}(c_{\alpha\beta\gamma}(\nu^2))\right], \quad (\exists c_{\alpha\beta\gamma}(\nu^2) \in \mathbb{R}[[\nu^2]]) \ .$$

We can show that $\{c_{\alpha\beta\gamma}\}$ is a Čech 2-cocycle, and then it defines Čech 2-class.

Definition 10. *We refer to this cocycle (resp. class) as the* Poincaré-Cartan *cocycle (resp. class) (PC-cocycle (resp. class) for short) and denote it by $c(W_M)$.*

For the PC-class we have the following.

Theorem 6. *For any $c = c^{(0)} + \sum_{i=1}^{\infty} c^{(2i)}\nu^{2i}$ $(c^{(2i)} \in \check{H}^2(M;\mathbb{R}))$ such that $[c^{(0)}]$ is the class of symplectic 2-form, there exists a family of contact Weyl diffeomorphisms $\{\Psi_{\alpha\beta} : \mathcal{C}_{U_{\alpha\beta}} \to \mathcal{C}_{U_{\beta\alpha}}\}$, such that $\Psi_{\alpha\beta\gamma}|_{W_{U_{\alpha\beta\gamma}}} = 1$, where $\Psi_{\alpha\beta\gamma} := \Psi_{\alpha\beta}\Psi_{\beta\gamma}\Psi_{\gamma\alpha}$, and $\{c_{\alpha\beta\gamma}(\nu^2)\}$ defines Čech 2 cohomology class which coincides with c. Moreover there is one to one correspondence between the set of Poincaré-Cartan classes and the set of isomorphism classes of Weyl manifolds.*

Proof. The proof is already known in [19], but we give an outline of it here for readers. First we show that for any cocycle $\{c_{\alpha\beta\gamma}(\nu^2)\}$, there exists a Weyl manifold such that $c(\mathcal{W}_M) = [c_{\alpha\beta\gamma}(\nu^2)]$. Suppose that $c = \sum_{k\geq 0} \nu^{2k} c^{(2k)} \in \check{H}^2(M)[[\nu^2]]$ is given. According to the existence theorem of Weyl manifold in [18], we may start with a Weyl manifold $\mathcal{W}_M^{(0)}$ with a Poincaré-Cartan cocycle $\{c_{\alpha\beta\gamma}^{(0)}\}$, and changing patching Weyl diffeomorphisms we construct a Weyl manifold with a Poincaré-Cartan class c. Let $\Phi_{\alpha\beta}^* : \mathcal{F}(\mathcal{W}_{U_{\alpha\beta}}) \to \mathcal{F}(\mathcal{W}_{U_{\beta\alpha}})$ be the patching Weyl diffeomorphism of $\mathcal{W}_M^{(0)}$ and let $\Psi_{\alpha\beta}^*$ be its extension as a contact Weyl diffeomorphism (cf. Proposition 3). Let $\{c_{\alpha\beta\gamma}^{(2k)}\}$ be a Čech cocycle involved in $c^{(2k)}$. Note that since the sheaf cohomology of C^∞-functions $H^2(M; \mathcal{E}) = \{0\}$, there is $h_{\alpha\beta}^{(2)} \in C^\infty(U_{\alpha\beta})$ on each $U_{\alpha\beta}$ such that

$$- c_{\alpha\beta\gamma}^{(2)} = h_{\alpha\beta}^{(2)} + \varphi^* h_{\beta\gamma}^{(2)} + \varphi_{\alpha\gamma}^* h_{\gamma\alpha}^{(2)} . \tag{11}$$

If we use $\dot{\Psi}_{\alpha\beta}^* = \Psi_{\alpha\beta}^* e^{ad\nu\tilde{h}_{\beta\alpha}}$ as patching diffeomorphism for every $V_\alpha \cap V_\beta \neq \phi$, then using the formula $\Psi_{\alpha\beta}^* e^{ad(h)} = e^{ad(\Psi_{\alpha\beta}^* h)} \Psi_{\alpha\beta}^*$ for $h \in \mathcal{F}(\mathcal{W}_{U_{\alpha\beta}})$, we see

$$\dot{\Phi}_{\alpha\beta}^* \dot{\Phi}_{\beta\gamma}^* \dot{\Psi}_{\gamma\alpha}^* = \Psi_{\alpha\beta}^* \Psi_{\beta\gamma}^* \Psi_{\gamma\alpha}^* e^{ad(\nu\Psi_{\alpha\beta}^* \tilde{h}_{\beta\alpha})} e^{ad(\nu\Psi_{\alpha\gamma}^* \tilde{h}_{\gamma\beta})} e^{ad(\nu\Psi_{\alpha\alpha}^* \tilde{h}_{\alpha\gamma})} ,$$

where we set $\tilde{h}_{\alpha\beta} = (h_{\alpha\beta}^{(2)})^\# + \nu^2 r_{\alpha\beta}^\#$ for a function $r_{\alpha\beta} \in C^\infty(U_{\alpha\beta})[[\nu^2]]$. By (11), we have

$$e^{ad(\nu\Psi_{\alpha\beta}^* \tilde{h}_{\beta\alpha})} e^{ad(\nu\Psi_{\alpha\gamma}^* \tilde{h}_{\gamma\beta})} e^{ad(\nu\Psi_{\alpha\alpha}^* \tilde{h}_{\alpha\gamma})} = e^{\nu^2 c_{\alpha\beta\gamma}^{(2)} ad\nu^{-1}} \quad mod\ \nu^4 . \tag{12}$$

By working on the term ν^4, $\nu^6, \ldots$, we can tune up $r_{\alpha\beta}$ by recursively, so that

$$e^{ad(\nu\Psi_{\alpha\beta}^* \tilde{h}_{\beta\alpha})} e^{ad(\nu\Psi_{\alpha\gamma}^* \tilde{h}_{\gamma\beta})} e^{ad(\nu\Psi_{\alpha\alpha}^* \tilde{h}_{\alpha\gamma})} = e^{\nu^2 c_{\alpha\beta\gamma}^{(2)} ad\nu^{-1}} . \tag{13}$$

It follows that $\{\dot{\Psi}_{\alpha\beta}^*\}$ defines a Weyl manifold $\dot{\mathcal{W}}_M$ with the Poincaré-Cartan class $c^{(0)} + \nu^2 c^{(2)}$. Replacing $\Psi_{\alpha\beta}^*$ by $\dot{\Psi}_{\alpha\beta}^*$ and repeating a similar argument as above, we can replace the condition mod ν^4 in (12) by mod ν^6. Repeating this procedure, we have a Weyl manifold $\mathcal{W}_M$ such that $c(\mathcal{W}_M) = c \in \check{H}(M)[[\nu^2]]$.

Let $\{c_{\alpha\beta\gamma}\}$, $\{c_{\alpha\beta\gamma}'\}$ be Poincaré-Cartan cocycles of $\{\mathcal{C}_U\}$, $\{\mathcal{C}_U'\}$ respectively, which give same Poincaré-Cartan classes. Then, there exists $b_{\alpha\beta} \in \mathbf{R}[[\nu^2]]$ on every $V_\alpha \cap V_\beta \neq \phi$ such that $b_{\alpha\beta} = -b_{\beta\alpha}$ and $c_{\alpha\beta\gamma}' - c_{\alpha\beta\gamma} = b_{\alpha\beta} + b_{\beta\gamma} + b_{\gamma\alpha}$. Note that $b_{\beta\gamma}$ may be replaced by $b_{\beta\gamma} + c_{\beta\gamma}$ such that $c_{\alpha\beta} + c_{\beta\gamma} + c_{\gamma\alpha} = 0$. Since $e^{b_{\alpha\beta} ad(\nu^{-1})}$ is an automorphism, we can replace $\Psi_{\alpha\beta}$ by $\dot{\Psi}_{\alpha\beta} = \Psi_{\alpha\beta} e^{b_{\alpha\beta} ad(\nu^{-1})}$. Since $e^{b_{\alpha\beta} ad(\nu^{-1})}$ is the identity on $\mathcal{F}(\mathcal{W}_{U_{\alpha\beta}})$, this replacement does not change the isomorphism class of $\mathcal{F}(\mathcal{W}_M)$, but it changes the Poincaré-Cartan cocycle from $\{c_{\alpha\beta\gamma}\}$ to $\{c_{\alpha\beta\gamma}'\}$. This means that the map from the set of PC-cocycles into $\mathbf{W_M}$, the set of isomorphism classes of Weyl manifolds induces a map F from $\mathbf{PC_M}$, the set of PC classes into $\mathbf{W_M}$.

Next we construct the inverse map $\Psi : \mathcal{W}'_M \to \mathcal{W}_M$ which induces the identity on the base manifold. Equivalently Ψ^* defines an algebra isomorphism of $\mathcal{F}(\mathcal{W}_M)$ onto $\mathcal{F}(\mathcal{W}'_M)$. The isomorphism is given by a family $\{\Psi^*_\alpha\}$ of isomorphisms:

$$\Psi^*_\alpha : \mathcal{F}(\mathcal{W}_{U_\alpha}) \to \mathcal{F}(\mathcal{W}'_{U_\alpha}) ,$$

each of which induces the identity map on the base space U_α such that

$$\Psi^*_\alpha \Psi^*_{\alpha\beta} \Psi^{*-1}_\beta = \acute{\Psi}^*_{\alpha\beta} . \tag{14}$$

If we extend Ψ^*_α to a contact Weyl diffeomorphism, then the above replacement makes no change of Poincaré-Cartan cocycle. We use the same notation Ψ^*_α for this contact Weyl diffeomorphism. By (14), and Proposition 8, we have

$$\Psi^*_\alpha \Psi^*_{\alpha\beta} \Psi^{*-1}_\beta e^{b_{\alpha\beta} ad(\nu^{-1})} = \acute{\Psi}^*_{\alpha\beta} . \tag{15}$$

on contact Weyl algebra bundle. However this type of replacement changes the Poincaré-Cartan cocycle within the same cohomology class. This means that there is a map from $\mathbf{W_M}$ into $\mathbf{PC_M}$, and which is obviously the inverse map of F. □

Remark that there exists a contact weyl algebra bundle with a connection ∇^Q such that its first Chern class coincides with Poincaé-Cartan class (cf. [26]). Then $\nabla^Q|_{\mathcal{W}_M}$ gives a flat connection on $\mathcal{W}_M$ and there is a one to one correspondence σ between the space of parallel sections with respect to $\nabla|_{\mathcal{W}_M}$ and $C^\infty(M)[[\hbar]]$. Combining this map with fiberwise star product, we can define a star product: $f * g = \sigma(\sigma^{-1}(f) *_{\text{fiberwise}} \sigma^{-1}(g))$. Hence we obtain Theorem 1.

Acknowledgement

The author thanks Professors Giuseppe Dito, Yoshiaki Maeda, Hitoshi Moriyoshi, Nobutada Nakanishi, Toshikazu Natsume, Ryszard Nest, Hideki Omori, Daniel Sternheimer, Boris Tsygan, Satoshi Watamura and Akira Yoshioka for their valuable comments.

This work was partially supported by Grant-in-Aid for Scientific Research (#13740049, #15740045), Ministry of Education, Science and Culture, Japan, and also Keio Gijuku Academic Development Funds.

References

1. Bayen, F., Flato, M., Fronsdal, C., Lichnerowicz, A. and Sternheimer, D. *Deformation theory and quantization I*, Ann. of Phys. 111 (1978), 61–110.
2. Brylinski, J.L. and Getzler, E. *The homology of algebras of pseudo-differential symbols and the noncommutative residue.* K-theory 1 (1987), 385–403.

3. Brusztyn, H. and Waldmann, S. *Deformation quantization of Hermitian vector bundles,* math.QA/0009170

4. De Wilde, M. and Lecomte, P.B. *Existence of star-products and formal deformations of the Poisson Lie algebra of arbitrary symplectic manifolds,* Lett. Math. Phys. 7 (1983), 487–496.

5. Dito, G. and Sternheimer, D. *Deformation Quantization: Genesis, Developments and Metamorphoses* math.QA/0201168

6. Elliot, G.A., Natsume, T. and Nest, R. *The Atiyah-Singer index theorem as passage to the classical limit in Quantum mechanics,* Commun. Math. Phys. 182 (1996), 505–533

7. Fedosov, B.V. *A simple geometrical construction of deformation quantization,* J. Differential Geom. 40 (1994), 213–238.

8. Gutt, S. and Rawnsley, J. *Equivalence of star products on a symplectic manifold; an introduction of Deligne's Cech cohomology classes,* J. Geom. Phys. 29 (1999), 347–392.

9. Hochschild, G., Kostant, B. and Rosenberg, A. *Differential forms on regular affine algebras,* Trans. Amer. Math. Soc. 102 (1962), 383–408.

10. Kontsevich, M. *Deformation quantization of Poisson manifolds,* q-alg/9709040.

11. Liu, Z. and Xu, P. *On quadratic Poisson structures,* Lett. Math. Phys. 26(1992), no.1 33–42.

12. Maeda, Y. and Kajiura, H. *Introduction to deformation quantization,* Lectures in Math. Sci. The Univ. of Tokyo, 20 (2002), Yurinsya

13. Miyazaki, N. *Hochschild homology and cohomology of star algebra on a plane with a singular Poisson structure,* submitted.
   preparation.

14. Nakanishi, N. *Poisson cohomology of plane quadratic Poisson structures,* Publ. RIMS, Kyoto Univ. 33 (1997), 73–89.

15. Natsume, T. and Moriyoshi, H. *Operator algebras and Geometry,* Math. Soc. Japan, Memoir vol. 2 (2001)

16. Nest, R. and Tsygan, B. *Algebraic index theorem,* Commun. Math. Phys. 172 (1995), 223–262

17. Omori, H. *Infinite-dimensional Lie groups,* AMS, MMONO 158, (1995)

18. Omori, H., Maeda, Y. and Yoshioka, A. *Deformation quantization and Weyl manifolds,* Advances in Mathematics 85 (1991), 224–255.

19. Omori, H., Maeda, Y., Miyazaki, N. and Yoshioka, A. *Poincaé-Cartan class and deformation quantization of Kähler manifolds,* Commun. Math. Phys. 194 (1998), 207–230.

20. Rieffel, M.A. *Morita equivalence for operator algebras* 285–298, Amer. Math. Soc. Providence, R. i.,1982

21. Rieffel, M.A., *Projective modules over higher-dimensional noncommutative tori,* Canad. J. Math. 40.2 (1988), 257–338

22. Rosenberg, J. *K-theory under deformation quantization,* q-alg/9607021

23. Sternheimer, D. *Deformation quantization twenty years after,* q-alg/9809056.

24. Voronov, T. *Quantization,Poisson bracket and beyond,* Contemp. Math. 315

25. Weinstein, A. *Deformation quantization.* Seminaire Bourbaki, Vol. 1993/94. Asterisque No. 227, (1995), Exp. No. 789, 5, 389–409.

26. Yoshioka, A. *Contact Weyl manifold over a symplectic manifold,* in "Lie groups, Geometric structures and Differential equations", Adv. St. in Pure Math. 37(2002), 459–493

Universal Deformation Formulae
for Three-Dimensional Solvable Lie Groups

P. Bieliavsky[1], P. Bonneau[2], and Y. Maeda[3]

[1] Université Libre de Bruxelles, Belgium
   pbiel@ulb.ac.be
[2] Université de Bourgogne, France
   bonneau@u-bourgogne.fr
[3] Keio University, Japan
   maeda@math.keio.ac.jp

Summary. We apply methods from strict quantization of solvable symmetric spaces to obtain universal deformation formulae for actions of every three-dimensional solvable Lie group. We also study compatible co-products by generalizing the notion of smash product in the context of Hopf algebras. We investigate in particular the dressing action of the 'book' group on $SU(2)$. This work is aimed to be applied in a string theoretical context to produce noncommutative deformations of D-branes within a non-formal operator algebraic framework.

1 Introduction and Motivations

In this paper, we investigate one aspect of the problem of defining new classes of noncommutative manifolds. Currently, the main examples of noncommutative manifolds are produced by deforming the algebra of functions on torus-manifolds (i.e. on manifolds on which the Abelian Lie group $\mathbb{T}^d$ acts on). Essentially the deformation is obtained by implementing a noncommutative torus as part of the algebra via a so-called *universal deformation formula* (briefly UDF) for actions of the torus [10]. The motivation for this partly come from (open) String Theory. Indeed, a deformation of the brane in the direction of the B-field is generally understood as the (non-commutative) geometrical framework for studying interactions of strings with endpoints attached to the brane [26]. Though curved non-commutative situations have been extensively studied in the context of strict deformation theory i.e. in a purely operator algebraic framework (for a review and references, see [2]), non-commutative spaces emerging from string theory have up to now mainly been studied in the case of constant B-fields in flat (Minkowski) backgrounds.

One obstacle in attending strict deformation theory within a curved background is that one rarely disposes of a compatible action of an Abelian Lie group. However, there are interesting situations – such as WZW models with noncompact target group manifolds [3] – yielding actions of non-Abelian solvable Lie groups. This motivates us to define UDF's for solvable Lie group actions. We now precise the mathematical context.

P. Bieliavsky, P. Bonneau, and Y. Maeda: *Universal Deformation Formulae for Three-Dimensional Solvable Lie Groups*, Lect. Notes Phys. **662**, 127–141 (2005)
www.springerlink.com

Let G be a group acting on a set M. Denote by $\tau : G \times M \to M : (g,x) \mapsto \tau_g(x)$ the (left) action and by $\alpha : G \times \mathrm{Fun}(M) \to \mathrm{Fun}(M)$ the corresponding action on the space of (complex valued) functions (or formal series) on M $(\alpha_g := \tau_{g^{-1}}^*)$. Assume that on a subspace $\mathbb{A} \subset \mathrm{Fun}(G)$, one has an associative $\mathbb{C}$-algebra product $\star_{\mathbb{A}}^G : \mathbb{A} \times \mathbb{A} \to \mathbb{A}$ such that

(i) $\mathbb{A}$ is invariant under the (left) regular action of G on $\mathrm{Fun}(G)$,
(ii) the product $\star_{\mathbb{A}}^G$ is left-invariant as well i.e. for all $g \in G; a,b \in \mathbb{A}$, one has

$$(L_g^\star a) \star_{\mathbb{A}}^G (L_g^\star b) = L_g^\star (a \star_{\mathbb{A}}^G b) .$$

Given a function on M, $u \in \mathrm{Fun}(M)$, and a point $x \in M$, one denotes by $\alpha^x(u) \in \mathrm{Fun}(G)$ the function on G defined as

$$\alpha^x(u)(g) := \alpha_g(u)(x) .$$

Then one readily observes that the subspace $\mathbb{B} \subset \mathrm{Fun}(M)$ defined as

$$\mathbb{B} := \{u \in \mathrm{Fun}(M) \,|\, \forall x \in M : \alpha^x(u) \in \mathbb{A}\}$$

becomes an associative $\mathbb{C}$-algebra when endowed with the product $\star_{\mathbb{B}}^M$ given by

$$u \star_{\mathbb{B}}^M v(x) := (\alpha^x(u) \star_{\mathbb{A}}^G \alpha^x(v))(e) \tag{1}$$

(e denotes the neutral element of G). Of course, all this can be defined for right actions as well.

Definition 1 *Such a pair* $(\mathbb{A}, \star_{\mathbb{A}}^G)$ *is called a (left)* **universal deformation** *of G, while Formula (1) is called the associated* **universal deformation formula** *(briefly* **UDF**).

In the present article, we will be concerned with the case where G is a Lie group. The function space $\mathbb{A}$ will be either
- a functional subspace (or a topological completion) of $C^\infty(G, \mathbb{C})$ containing the smooth compactly supported functions in which case we will talk about **strict deformation** (following Rieffel [24]),
or,
- the space $\mathbb{A} = C^\infty(G)[[\hbar]]$ of formal power series with coefficients in the smooth functions on G in which case, we'll speak about **formal deformation**. In any case, we'll assume the product $\star_{\mathbb{A}}^G$ admits an asymptotic expansion of star-product type:

$$a \star_{\mathbb{A}}^G b \sim ab + \frac{\hbar}{2i} \mathbf{w}(du, dv) + o(\hbar^2) \qquad (a,b \in C_c^\infty(G)) ,$$

where $\mathbf{w}$ denotes some (left-invariant) Poisson bivector on G [4]. In the strict cases considered here, the product will be defined by an integral three-point kernel $K \in C^\infty(G \times G \times G)$:

$$a \star_{\mathbb{A}}^{G} b(g) := \int_{G \times G} a(g_1)\, b(g_2) K(g, g_1, g_2) \mathrm{d}g_1\, \mathrm{d}g_2 \qquad (a, b \in \mathbb{A})$$

where $\mathrm{d}g$ denotes a normalized left-invariant Haar measure on G. Moreover, our kernels will be of **WKB type** [14, 30] i.e.:

$$K = A\, e^{\frac{i}{\hbar} \Phi},$$

with A (the **amplitude**) and Φ (the **phase**) in $C^{\infty}(G \times G \times G, \mathbb{R})$ being invariant under the (diagonal) action by left-translations.

Note that in the case where the group G acts smoothly on a smooth manifold M by diffeomorphisms: $\tau : G \times M \to M : (g, x) \mapsto \tau_g(x)$, the first-order expansion term of $u \star_{\mathbb{B}}^{M} v$, $u, v \in C^{\infty}(M)$ defines a Poisson structure $\mathbf{w}^M$ on M which can be expressed in terms of a basis $\{X_i\}$ of the Lie algebra $\mathfrak{g}$ of G as:

$$\mathbf{w}^M = [\mathbf{w}_e]^{ij}\, X_i^{\star} \wedge X_j^{\star}, \tag{2}$$

where $X^{\star}$ denotes the fundamental vector field on M associated to $X \in \mathfrak{g}$.

Strict deformation theory in the WKB context was initiated by Rieffel in [25] in the cases where G is either Abelian or 1-step nilpotent. Rieffel's work has led to what is now called "Rieffel's machinery"; producing a whole class of exciting non-commutative manifolds (in Connes sense) from the data of Abelian group actions on C^*-algebras [10].

The study of formal UDF's for non-Abelian group actions in our context was initiated in [12] where the case of the group of affine transformations of the real line ('$ax + b$') was explicitly described.

In the strict (non-formal) setting, UDF's for Iwasawa subgroups of $SU(1, n)$ have been explicitly given in [9]. These were obtained by adapting a method developed by one of us in the symmetric space framework [7]. In particular when $n = 1$ one obtains strict UDF's for the group '$ax + b$' (we recall this in the appendix).

In the present work, we provide in the Hopf algebraic context (strict and formal) UDF's for every solvable three-dimensional Lie group endowed with any left-invariant Poisson structure. The method used to obtain those is based on strict quantization of solvable symmetric spaces, on existing UDF's for '$ax+b$' and on a generalization of the classical definition of smash products in the context of Hopf algebras. As an application we analyze the particular case of the dressing action of the Poisson dual Lie group of $SU(2)$ when endowed with the Lu-Weinstein Poisson structure.

2 UDF's for Three-Dimensional Solvable Lie Groups

In this section G denotes a three-dimensional solvable Lie group with Lie algebra $\mathfrak{g}$.

Definition 2 *Let* $\mathbf{w}$ *be a left-invariant Poisson bivector field on* G. *The pair* $(G, \mathbf{w})$ *is called a* **pre-symplectic** *Lie group.*

The terminology "symplectic" is after Lichnérowicz [15] who studied this type of structures.

One then observes

Proposition 1 *(i) The orthodual* $\mathfrak{s}$ *of the radical of* $\mathbf{w}_e$

$$\mathfrak{s} := (rad \ \mathbf{w}_e)^{\perp}$$

is a subalgebra of $\mathfrak{g}$.
(ii) *Every two-dimensional subalgebra of* $\mathfrak{g}$ *can be seen as the orthodual of the radical of a Poisson bivector.*
(iii) *The symplectic leaves of* $\mathbf{w}$ *are the left classes of the analytic subgroup* S *of* G *whose Lie algebra is* $\mathfrak{s}$.

We now fix a pair $(\mathfrak{g}, \mathbf{w}_e)$ to be such a pre-symplectic Lie algebra with associated symplectic Lie algebra $(\mathfrak{s}, \mathbf{w}_e|_{\mathfrak{s}^* \times \mathfrak{s}^*})$. We assume

$$\dim \mathfrak{s} = 2 \ .$$

We denote by $\eth$ the first derivative of $\mathfrak{g}$:

$$\eth := [\mathfrak{g}, \mathfrak{g}] \ .$$

Note that $\eth$ is an Abelian algebra.

2.1 Case 1: $\dim \eth = 2$

In this case $\mathfrak{g}$ can be realized as a split extension of Abelian algebras:

$$0 \to \eth \longrightarrow \mathfrak{g} \longrightarrow \mathfrak{a} \to 0 \ ,$$

with $\dim \mathfrak{a} = 1$. We denote by

$$\rho : \mathfrak{a} \to \mathrm{End}(\eth)$$

the splitting homomorphism.

2.1.1 $\mathfrak{s} \cap \eth \neq \eth$

In this case one sets

$$\mathfrak{p} := \mathfrak{s} \cap \eth \ ,$$

and one may assume

$$\mathfrak{a} \subset \mathfrak{s} \ .$$

Note that $\mathfrak{s}$ is then isomorphic to the Lie algebra of the group '$ax + b$'.

2.1.1.1 The Representation ρ is Semisimple

In this case one has $\mathfrak{q} \subset \mathfrak{d}$ such that

$$\mathfrak{q} \oplus \mathfrak{p} = \mathfrak{d} \text{ and}$$
$$[\mathfrak{a}, \mathfrak{q}] \subset \mathfrak{q} .$$

Let Q and S denote the analytic subgroups of G whose Lie algebras are $\mathfrak{q}$ and $\mathfrak{s}$ respectively. Consider the mapping

$$Q \times S \to G : (q, s) \mapsto qs . \tag{3}$$

Assume this is a global diffeomorphism. Endow the symplectic Lie group S with a left-invariant deformation quantization $\star^T$ as constructed in [9] or [8] (see also the appendix of the present paper). One then readily verifies

Theorem 1 *Let $(\mathbf{H}^S, \star^T)$ be an associative algebra of functions on S such that $\mathbf{H}^S \subset Fun(S)$ is an invariant subspace w.r.t. the left regular representation of S on $Fun(S)$ and $\star^T$ is a S-left-invariant product on $\mathbf{H}^S$. Then the function space $\mathbf{H} := \{u \in Fun(G) \mid \forall p \in Q : u(q, .) \in \mathbf{H}^S\}$ is an invariant subspace of $Fun(G)$ w.r.t. the left regular representation of G and the formula*

$$u \star v(q, s) := (u(q, .) \star^T v(q, .))(s)$$

defines a G-left-invariant associative product on $\mathbf{H}$.

2.1.1.2 The Representation ρ is not Semisimple

Since it cannot be nilpotent one can find bases $\{A\}$ of $\mathfrak{a}$ and $\{X, Y\}$ of $\mathfrak{d}$ such that

$$\text{span}\{X\} = \mathfrak{p}$$

and

$$\rho(A) = \begin{pmatrix} 1 & \lambda \\ 0 & 1 \end{pmatrix} \qquad \lambda \in \mathbb{R}_0 .$$

Lemma 1 $\mathfrak{g}$ *can be realized as a subalgebra of the transvection algebra of a four-dimensional symplectic symmetric space M.*

Proof. The symplectic triple associated to the symplectic symmetric space (cf. [5] or [7] Definition 2.5) is $(\mathfrak{G}, \sigma, \Omega)$ where

$$\mathfrak{G} = \mathfrak{K} \oplus \mathfrak{P}$$
$$\mathfrak{K} = \text{span}\{k_1, k_2\}$$
$$\mathfrak{P} = \text{span}\{e_1, f_1, e_2, f_2\} ;$$

with Lie algebra table given by

$$[f_1, k_1] = e_1$$
$$[f_1, k_2] = e_2 + \lambda e_1$$
$$[f_1, e_1] = k_1$$
$$[f_1, e_2] = k_2 + \lambda k_1 ;$$

f_2 being central. The symplectic structure on $\mathfrak{P}$ is given by

$$\Omega = \alpha e_1^* \wedge f_1^* + \beta e_2^* \wedge f_2^* \qquad (\alpha, \beta \neq 0) .$$

The subalgebra

$$\text{span} \left\{ X := \frac{1}{2}(k_1 + e_1), \, Y := \frac{1}{2}(k_2 + e_2), \, A := f_1, \right\}$$

is then isomorphic to $\mathfrak{g}$. $\square$

Note that the transvection subalgebra

$$\mathfrak{S} = \mathfrak{g} \oplus \mathbb{R} f_2$$

is transverse to the isotropy algebra $\mathfrak{K}$ in $\mathfrak{G}$. As a consequence the associated connected simply connected Lie group $\mathbf{S}$ acts acts transitively on the symmetric space M. Since $\dim M = \dim \mathbf{S} = 4$, one may identify M with the group manifold of $\mathbf{S}$. It turns out that the symplectic symmetric space M admits a strict deformation quantization which is transvection invariant [7].

Therefore one has

Theorem 2 *The symplectic Lie group* $\mathbf{S}$ *admits a left-invariant strict deformation quantization. Since G is a group direct factor of $\mathbf{S}$, the deformed product on $\mathbf{S}$ restricts to G providing a UDF for G.*

2.1.2 $\mathfrak{s} = \mathfrak{d}$

In this case, a UDF is obtained by writing Moyal's formula where one replaces the partial derivatives ∂_i's by (commuting) left-invariant vector fields $\widetilde{X_i}$ associated with generators X_i of $\mathfrak{d}$: for $a, b \in C^\infty(G)[[\hbar]]$, one sets

$$a \star_\hbar b := a . \exp\left(\mathbf{w}_e|_{\mathfrak{d}}{}^{ij} \, \overleftarrow{\widetilde{X}_i} \wedge \overrightarrow{\widetilde{X}_j} \right) .b$$

2.2 Case 2: $\dim \mathfrak{d} = 1$

In this case $\mathfrak{g}$ is isomorphic to either the Heisenberg algebra or to the direct sum of $\mathbb{R}$ with the Lie algebra of '$ax + b$'. The first case has been studied by Rieffel in [24]. Let then

$$\mathfrak{g} = \mathfrak{L} \oplus \mathbb{R}Z$$

where $\mathfrak{L} = \mathrm{span}\{A, E\}$ with $[A, E] = E$. Up to automorphism one has either $\mathfrak{s} = \mathfrak{L}$, $\mathfrak{s} = \mathrm{span}\{E, Z\}$ or $\mathfrak{s} = \mathrm{span}\{A, Z\}$. The last ones are Abelian thus UDF's in these cases are obtained the same way as in Sect. 2.1.2. The first case reduces to an action of '$ax + b$' and has therefore been treated in [9] or [8].

3 Crossed, Smash and Co-products

Every algebra, coalgebra, bialgebra, Hopf algebra and vector space is taken over the field $k = \mathbb{R}$ or $\mathbb{C}$. For classical definitions and facts on these subjects, we refer to [1, 29] or more fundamentally to [20].

To calculate with a coproduct Δ, we use the Sweedler notation [29]: $\Delta(b) = \sum_{(b)} b_{(1)} \otimes b_{(2)}$.

For the classical definitions of a B-module algebra, coalgebra and bialgebra we refer to [1].

The literature on Hopf algebras contains a large collection of what we can call generically semi-direct products, or crossed products. Let us describe some of them. The simplest example of these crossed products is usually called the smash product (see [21, 28]):

Definition 3 *Let B be a bialgebra and C a B-module algebra. The smash product $C \sharp B$ is the algebra constructed on the vector space $C \otimes B$ where the multiplication is defined as*

$$(f \otimes a) \overset{\rightarrow}{\star} (g \otimes b) = \sum_{(a)} f\,(a_{(1)} \rightharpoonup g) \otimes a_{(2)}b \,, \qquad f, g \in C \,, \quad a, b \in B \,. \tag{4}$$

Assuming B is cocommutative, we now introduce a generalization of the smash product.

Definition 4 *Let B be a cocommutative bialgebra and C a B-bimodule algebra (i.e. a B-module algebra for both, left and right, B-module structures). The **L-R-smash product** $C \natural B$ is the algebra constructed on the vector space $C \otimes B$ where the multiplication is defined by*

$$(f \otimes a) \star (g \otimes b) = \sum_{(a)(b)} (f \leftharpoonup b_{(1)})(a_{(1)} \rightharpoonup g) \otimes a_{(2)}b_{(2)} \,, \quad f, g \in C \,, \quad a, b \in B \,.$$

$$\tag{5}$$

Remark 1 The fact that the L-R-smash product $C \natural B$ is an associative algebra follows by an easy adaptation of the proof of the associativity for the smash product [21].

In the same spirit, one has

Lemma 2 *If C is a B-bimodule bialgebra, the natural tensor product coalgebra structure on $C \otimes B$ defines a bialgebra structure to $C \natural B$.*

If C and B are Hopf algebras, $C \natural B$ is a Hopf algebra as well, defining the antipode by

$$J_\star(f \otimes a) = \sum_{(a)} J_B(a_{(1)}) \rightharpoonup J_C(f) \leftharpoondown J_B(a_{(2)}) \otimes J_B(a_{(3)}) \tag{6}$$

$$= \sum_{(a)} (1_C \otimes J_B(a_{(1)})) \star (J_C(f) \otimes 1_B) \star (1_C \otimes J_B(a_{(2)})) \,.$$

Now by a careful computation, one proves

Proposition 2 *Let B be a cocommutative bialgebra, C be a B-bimodule algebra and $(C \natural B, \star)$ be their L-R-smash product.*

Let T be a linear automorphism of C (as vector space). We define:

(i)   *the product $\bullet^T$ on C by*

$$f \bullet^T g = T^{-1}(T(f).T(g)) \,;$$

(ii)   *the left and right B-module structures, $\overset{T}{\rightharpoonup}$ and $\overset{T}{\leftharpoondown}$, by*

$$a \overset{T}{\rightharpoonup} f := T^{-1}(a \rightharpoonup T(f)) \quad and \quad f \overset{T}{\leftharpoondown} a := T^{-1}(T(f) \leftharpoondown a) \,;$$

(iii) *the product, $\star^T$, on $C \otimes B$ by*

$$(f \otimes a) \star^T (g \otimes b) = \mathbf{T}^{-1}(\mathbf{T}(f \otimes a) \star \mathbf{T}(g \otimes b)) \quad where \quad \mathbf{T} := T \otimes \mathrm{Id} \,.$$

Then $(C, \bullet^T)$ is a B-bimodule algebra for $\overset{T}{\rightharpoonup}$ and $\overset{T}{\leftharpoondown}$ and $\star^T$ is the L-R-smash product defined by these structures.

Moreover, if $(C, ., \Delta_C, J_C, \rightharpoonup, \leftharpoondown)$ is a Hopf algebra and a B-module bialgebra, then

$$C_T := (C, \bullet^T, \Delta_C^T := (T^{-1} \otimes T^{-1}) \circ \Delta_C \circ T, J_C^T := T^{-1} \circ J_C \circ T, \overset{T}{\rightharpoonup}, \overset{T}{\leftharpoondown})$$

is also a Hopf algebra and a B-module bialgebra. Therefore, by Lemma 2,

$$(C_T \natural B, \star^T, \Delta^T = (23) \circ (\Delta_C^T \otimes \Delta_B), J_\star^T),$$

is a Hopf algebra, Δ^T being the natural tensor product coalgebra structure on $C_T \natural B$ (with $(23) : C \otimes C \otimes B \otimes B \to C \otimes B \otimes C \otimes B$, $c_1 \otimes c_2 \otimes b_1 \otimes b_2 \mapsto c_1 \otimes b_1 \otimes c_2 \otimes b_2$) and $J_\star^T$ being the antipode given on $C_T \natural B$ by Lemma 2. Also, one has

$$\Delta^T = (\mathbf{T}^{-1} \otimes \mathbf{T}^{-1}) \circ (23) \circ (\Delta_C \otimes \Delta_B) \circ \mathbf{T} \quad and \quad J_\star^T = \mathbf{T}^{-1} \circ J_\star \circ \mathbf{T}$$

$$with \; \mathbf{T} = T \otimes \mathrm{Id} \,.$$

4 Examples in Deformation Quantization

4.1 A Construction on $T^*(G)$

Let G be a Lie group with Lie algebra $\mathfrak{g}$ and $T^*(G)$ its cotangent bundle. We denote by $U\mathfrak{g}$, $T\mathfrak{g}$ and $S\mathfrak{g}$ respectively the enveloping, tensor and symmetric algebras of $\mathfrak{g}$. Let $\mathsf{Pol}(\mathfrak{g}^*)$ be the algebra of polynomial functions on $\mathfrak{g}^*$. We have the usual identifications:

$$\mathcal{C}^\infty(T^*G) \simeq \mathcal{C}^\infty(G \times \mathfrak{g}^*) \simeq \mathcal{C}^\infty(G) \,\hat{\otimes}\, \mathcal{C}^\infty(\mathfrak{g}^*) \supset \mathcal{C}^\infty(G) \otimes \mathsf{Pol}(\mathfrak{g}^*)$$
$$\simeq \mathcal{C}^\infty(G) \otimes S\mathfrak{g}.$$

First we deform $S\mathfrak{g}$ via the "parametrized version" $U_t\mathfrak{g}$ of $U\mathfrak{g}$ defined by

$$U_t\mathfrak{g} = \frac{T\mathfrak{g}[[t]]}{< XY - YX - t[X,Y]; X, Y \in \mathfrak{g} >}.$$

$U_t\mathfrak{g}$ is naturally a Hopf algebra with $\varDelta(X) = 1 \otimes X + X \otimes 1$, $\epsilon(X) = 0$ and $S(X) = -X$ for $X \in \mathfrak{g}$. For $X \in \mathfrak{g}$, we denote by $\widetilde{X}$ (resp. $\overline{X}$) the left- (resp. right-) invariant vector field on G such that $\widetilde{X}_e = \overline{X}_e = X$. We consider the following $k[[t]]$-bilinear actions of $B = U_t\mathfrak{g}$ on $C = \mathcal{C}^\infty(G)[[t]]$, for $f \in C$ and $\lambda \in [0,1]$:

(i) $(X \rightharpoonup f)(x) = t(\lambda - 1)\,(\widetilde{X}.f)(x)$,
(ii) $(f \leftharpoonup X)(x) = t\lambda\,(\overline{X}.f)(x)$.

One then has

Lemma 3 C is a B-bimodule algebra w.r.t. the above left and right actions (i) and (ii).

Definition 5 We denote by $\star_\lambda$ the star product on $(\mathcal{C}^\infty(G) \otimes \mathsf{Pol}(\mathfrak{g}^*))[[t]]$ given by the L-R-smash product on $\mathcal{C}^\infty(G)[[t]] \otimes U_t\mathfrak{g}$ constructed from the bimodule structure of the preceding lemma.

Proposition 3 For $G = \mathbb{R}^n$, $\star_{\frac{1}{2}}$ is the Moyal star product (Weyl ordered), $\star_0$ is the standard ordered star product and $\star_1$ the anti-standard ordered one. In general $\star_\lambda$ yields the λ-ordered quantization, within the notation of M. Pflaum [23].

Remark 2 In the general case, it would be interesting to compare our λ-ordered L-R smash product with classical constructions of star products on $T^*(G)$, with Gutt's product as one example [13].

4.2 Hopf Structures

We have discussed (see Lemma 2) the possibility of having a Hopf structure on $C \natural B$. Let us consider the particular case of $\mathcal{C}^\infty(\mathbb{R}^n)[[t]] \natural U_t \mathbb{R}^n = \mathcal{C}^\infty(\mathbb{R}^n)[[t]] \natural S\mathbb{R}^n$ ($\mathbb{R}^n$ is commutative). $S\mathbb{R}^n$ is endowed with its natural Hopf structure but we also need a Hopf structure on $\mathcal{C}^\infty(\mathbb{R}^n)[[t]] = \mathcal{C}^\infty(\mathbb{R}^n) \otimes \mathbb{R}[[t]]$. We will not use the usual one. Our alternative structure is defined as follows.

Definition 6 *We endow $\mathbb{R}[[t]]$ with the usual product, the co-product $\Delta(P)(t_1, t_2) := P(t_1 + t_2)$, the co-unit $\epsilon(P) = P(0)$ and the antipode $J(t) = -t$. We consider the Hopf algebra $(\mathcal{C}^\infty(\mathbb{R}^n), ., 1, \Delta_C, \epsilon_C, J_C)$, with pointwise multiplication, the unit $\mathbf{1}$ (the constant function of value 1), the coproduct $\Delta_C(f)(x, y) = f(x + y)$, the co-unit $\epsilon(f) = f(0)$ and the antipode $J_C(f)(x) = f(-x)$. The tensor product of these two Hopf algebras then yields a Hopf algebra denoted by*

$$(\mathcal{C}^\infty(\mathbb{R}^n)[[t]], ., \mathbf{1}, \Delta_t, \epsilon_t, J_t) .$$

Note that Δ_t and J_t are not linear in t. We then define, on the L-R smash $\mathcal{C}^\infty(\mathbb{R}^n)[[t]] \natural S\mathbb{R}^n$,

$$\Delta_\star := (23) \circ (\Delta_t \otimes \Delta_B), \quad \epsilon_\star := \epsilon_t \otimes \epsilon_B \quad and \quad J_\star \ as \ in \ Lemma \ 2 .$$

Proposition 4 $(\mathcal{C}^\infty(\mathbb{R}^n)[[t]] \natural S\mathbb{R}^n, \star_\lambda, 1 \otimes 1, \Delta_\star, \epsilon_\star, J_\star)$ *is a Hopf algebra.*

Remark 3 The case $\lambda = \frac{1}{2}$ yields the usual Hopf structure on the enveloping algebra of the Heisenberg Lie algebra.

5 The "Book" Algebra

Definition 7 *The book Lie algebra is the three-dimensional Lie algebra $\mathfrak{g}$ defined as the split extension of Abelian Lie algebras $\mathfrak{a} := \mathbb{R}.A$ and $\mathfrak{d} := \mathbb{R}^2$:*

$$0 \to \mathfrak{d} \to \mathfrak{g} \to \mathfrak{a} \to 0 ,$$

where the splitting homomorphism $\rho : \mathfrak{a} \to \mathrm{End}(\mathfrak{d})$ is defined by

$$\rho(A) := \mathrm{Id}_\mathfrak{d} .$$

The name comes from the fact that the regular co-adjoint orbits in $\mathfrak{g}^\star$ are open half planes sitting in $\mathfrak{g}^\star = \mathbb{R}^3$ ressembling the pages of an open book.

We are particularly interested in this example because the associated connected simply connected Lie group G turns out to be the solvable Lie group underlying the Poisson dual of $SU(2)$ when endowed with the Lu-Weinstein Lie-Poisson structure [17]. Explicitly, one has the following situation (see [16] or [27]). Set $\mathfrak{K} := \mathfrak{su}(2)$ and consider the Cartan decomposition of $\mathfrak{K}^\mathbb{C} = \mathfrak{sl}_2(\mathbb{C})$:

$$\mathfrak{K}^{\mathbb{C}} = \mathfrak{K} \oplus i\mathfrak{K} .$$

Fix a maximal Abelian subalgebra $\mathfrak{a}$ in $i\mathfrak{K}$, consider the corresponding root space decomposition

$$\mathfrak{K}^{\mathbb{C}} = \mathfrak{K}^{+} \oplus \mathfrak{a}^{\mathbb{C}} \oplus \mathfrak{K}^{-} ,$$

and set

$$\mathfrak{d} := \mathfrak{K}^{+} \simeq \mathbb{C} \simeq \mathbb{R}^{2} .$$

The Iwasawa part $\mathfrak{a} \oplus \mathfrak{K}^{+}$ is then a realization of the book algebra $\mathfrak{g} = \mathfrak{a} \times \mathfrak{d}$. One denotes by

$$K^{\mathbb{C}} = KG$$

the Iwasawa decomposition of $K^{\mathbb{C}}$. For an element $\gamma \in K^{\mathbb{C}}$, we have the corresponding factorization:

$$\gamma = \gamma_{K} \gamma_{G} .$$

The dressing action

$$G \times K \xrightarrow{\tau} K \qquad . \tag{7}$$

is then given by

$$\tau_{g}(k) := (gk)_{K} .$$

In the same way, denote by

$$\mathfrak{K}^{\mathbb{C}} \rightarrow \mathfrak{K} : A \mapsto A_{\mathfrak{K}}$$

the projection parallel to $\mathfrak{g}$. The infinitesimal dressing action of G on K is now given by

$$X_{k}^{\star} = -L_{k_{\star e}} \left[\left(\operatorname{Ad}(k^{-1})X \right)_{\mathfrak{K}} \right], \quad X \in \mathfrak{g} .$$

A choice of a pre-symplectic structure on G – or equivalently (up to a scalar) a choice of a two-dimensional Lie algebra $\mathfrak{s}$ in $\mathfrak{g}$ – then yields, via the dressing action (7), a Poisson structure $\mathbf{w}^{\mathfrak{s}}$ on K (cf. (2)):

$$L_{k_{\star k}^{-1}}(\mathbf{w}^{\mathfrak{s}}{}_{k}) := \left(\operatorname{Ad}(k^{-1})S_{1} \right)_{\mathfrak{K}} \wedge \left(\operatorname{Ad}(k^{-1})S_{2} \right)_{\mathfrak{K}}$$

where $\{S_{1}, S_{2}\}$ is a basis of $\mathfrak{s}$. Note that this Poisson structure is compatible with the Lu-Weinstein structure.

For each choice of symplectic Lie subalgebra $\mathfrak{s}$ of $\mathfrak{g}$, the preceding section then produces deformations of $(K, \mathbf{w}^{\mathfrak{s}})$.

Appendix: UDF's for '$ax + b$'

Let $S = $ '$ax + b$' denote the two-dimensional solvable Lie group presented as follows. As a manifold, one has

$$S = \mathbb{R}^{2} = \{(a, \ell)\} ,$$

and the multiplication law is given by

$$(a, \ell).(a', \ell') := (a + a', \, e^{-a'}\ell + \ell') \,.$$

Observe that the symplectic form on S defined by

$$\omega := da \wedge dl$$

is left-invariant.

We now define two specific real-valued functions. First, the two-point function $A \in C^\infty(S \times S)$ given by

$$A(x_1, x_2) := \cosh(a_1 - a_2)$$

where $x_i = (a_i, \ell_i)$ $i = 1, 2$. Second, the three-point function $\Phi \in C^\infty(S \times S \times S)$ given by

$$\Phi(x_0, x_1, x_2) := \oint_{0,1,2} \sinh(a_0 - a_1)\ell_2$$

where $\oint$ stands for cyclic summation. One then has

Theorem 3 *[7, 9] There exists a family of functional subspaces*

$$\{\mathcal{H}\}_{\hbar \in \mathbb{R}} \subset \mathcal{C}^\infty(S)$$

such that

(i)   *for all $\hbar \in \mathbb{R}$,*

$$C_c^\infty(S) \subset \mathcal{H}_\hbar \,;$$

(ii)   *For all $\hbar \in \mathbb{R}\backslash\{0\}$ and $u, v \in C_c^\infty(M)$, the formula:*

$$u \star_\hbar v(x_0) := \int_{M \times M} u(x_1)\, v(x_2)\, A(x_1, x_2)\, e^{\frac{i}{\hbar}\Phi(x_0, x_1, x_2)}\, dx_1\, dx_2 \qquad (8)$$

extends as an associative product on $\mathcal{H}_\hbar$ (dx denotes some normalization of the symplectic volume on (S, ω)). Moreover, (for suitable u, v and x_0) a stationary phase method yields a power series expansion of the form

$$u \star_\hbar v(x_0) \sim uv(x_0) + \frac{\hbar}{2i}\{u, v\}(x_0) + o(\hbar^2) \,; \qquad (9)$$

where $\{\,,\}$ denotes the symplectic Poisson bracket on (S, ω).

(iii)   *The pair $(\mathcal{H}_\hbar, \star_\hbar)$ is a topological (pre) Hilbert algebra on which the group S acts on the left by automorphisms.*

Now setting $\mathbb{R}^2 = \mathfrak{a} \times \mathfrak{l}$; $a \in \mathfrak{a}, \ell \in \mathfrak{l}$, one gets the linear isomorphisms

$$\mathcal{C}^\infty(S) \simeq \mathcal{C}^\infty(\mathfrak{l}) \hat{\otimes} \mathcal{C}^\infty(\mathfrak{a}) \simeq \mathcal{C}^\infty(\mathfrak{l}) \hat{\otimes} \mathcal{C}^\infty(\mathfrak{l}^*) \supset \mathcal{C}^\infty(\mathfrak{l}) \otimes Pol(\mathfrak{l}^*) \simeq \mathcal{C}^\infty(\mathfrak{l}) \otimes U\mathfrak{l} \,.$$

The quantization (8) on such a space turns out to be a L-R-smash product. Namely, one has

Proposition 5 *The formal version of the invariant quantization (8) is a L-R-smash product of the form* $\star^T$ *(cf. Proposition 2).*

Proof. Let $\mathcal{S}(\mathfrak{l})$ denote the Schwartz space on the vector space $\mathfrak{l}$. In [7], one shows that the map

$$\mathcal{S}(\mathfrak{l}) \xrightarrow{T} \mathcal{S}(\mathfrak{l})$$
$$u \mapsto T(u) := F^{-1}\phi_\hbar^\star F(u) \,,$$

where

$$\phi_\hbar : \mathfrak{l}^* \to \mathfrak{l}^* : a \mapsto \frac{2}{\hbar} \sinh\left(\frac{\hbar}{2}a\right) \,,$$

is a linear injection for all $\hbar \in \mathbb{R}$ (F denotes the partial Fourier transform w.r.t. the variable ℓ). An asymptotic expansion in a power series in $\hbar$ then yields a formal equivalence again denoted by T:

$$T := \mathtt{Id} + o(\hbar) : C^\infty(\mathfrak{l})[[\hbar]] \to C^\infty(\mathfrak{l})[[\hbar]] \,.$$

Carrying the Moyal star product on $(\mathfrak{a} \times \mathfrak{l}, \omega)$ by $\mathbf{T} := T \otimes \mathtt{Id}$ yields a star product on S which coincides with the asymptotic expansion of the left-invariant star-product (8) [7, 9]. $\square$

Subsection 4.2 then yields

Corollary 1 *The (formal) UDF (8) admits compatible co-product and antipode.*

References

1. Eiichi Abe. *Hopf algebras.* Cambridge Tracts in Mathematics, 74. Cambridge University Press, 1980.
2. V. Schomerus, *Lectures on branes in curved backgrounds*, hep-th/0209241; A.Y. Alekseev, A. Recknagel, V. Schomerus, hep-th/0104054.
3. C. Bachas, M. Petropoulos, JHEP 0102 (2001) 025, hep-th/0012234.
4. François Bayen, Moshé Flato, Christian Fronsdal, André Lichnerowicz and Daniel Sternheimer. *Deformation theory and quantization I. Deformations of symplectic structures.* Ann. Physics **111** (1978), no. 1, 61–110. *Deformation theory and quantization II. Physical applications.* Ann. Physics **111** (1978), no. 1, 111–151.
5. Pierre Bieliavsky. *Espaces symétriques symplectiques.* Ph. D. thesis, Université Libre de Bruxelles, 1995.

6. Pierre Bieliavsky. *Four-dimensional simply connected symplectic symmetric spaces.* Geom. Dedicata 69 (1998), no. 3, 291–316.

7. Pierre Bieliavsky. *Strict quantization of solvable symmetric spaces.* J. Sympl. Geom. 1 (2002), no. 2, 269–320.

8. Pierre Bieliavsky and Yoshiaki Maeda. *Convergent star product algebras on "ax + b".* Lett. Math. Phys. 62 (2002), no. 3, 233–243.

9. Pierre Bieliavsky and Marc Massar. *Oscillatory integral formulae for left-invariant star products on a class of Lie groups.* Lett. Math. Phys. 58 (2001), no. 2, 115–128.

10. Alain Connes and Giovanni Landi. *Noncommutative manifolds, the instanton algebra and isospectral deformations.* Comm. Math. Phys. 221 (2001), no. 1, 141–159.

11. Bernhard Drabant, Alfons Van Daele and Yinhuo Zhang. *Actions of multiplier Hopf algebras.* Commun. Algebra 27 (1999), no. 9, 4117–4172, .

12. Anthony Giaquinto and James J. Zhang. *Bialgebra actions, twists, and universal deformation formulas.* J. Pure Appl. Algebra 128 (1998), no. 2, 133–151.

13. Simone Gutt. *An explicit *-product on the cotangent bundle of a Lie group.* Lett. Math. Phys. 7 (1983), no. 3, 249–258.

14. Mikhail Karasev. *Formulas for non-commutative products of functions in terms of membranes and strings I.* Russian J. Math. Phys. 2 (1994), no. 4, 445–462.

15. André Lichnerowicz and Alberto Medina. *On Lie groups with left-invariant symplectic or Kählerian structures.* Lett. Math. Phys. 16 (1988), no. 3, 225–235.

16. Jiang-Hua Lu and Tudor Ratiu; *On the nonlinear convexity result of Kostant.* J. of the A.M.S. 4 (1991), no. 2, 349–363.

17. Jiang-Hua Lu and Alan Weinstein. *Poisson Lie groups, dressing transformations, and Bruhat decompositions.* J. Diff. Geom. 31 (1990), no. 2, 501–526.

18. Shahn Majid. *Physics for algebraists: Non-commutative and non-cocommutative Hopf algebras by a bicrossproduct construction.* J. Algebra 130 (1990), no. 1, 17–64.

19. Saunders Mac Lane. *Homology.* Springer-Verlag, 1975.

20. John W. Milnor and John C. Moore. *On the structure of Hopf algebras.* Ann. Math. 81 (1965), no. 2, 211–264.

21. Richard K. Molnar. *Semi-direct products of Hopf algebras.* J. Algebra 47 (1977), 29–51.

22. Gert K. Pedersen. *C^*-algebras and their automorphism groups.* London Mathematical Society Monographs, 14. Academic Press, 1979.

23. Markus J. Pflaum. *Deformation quantization on cotangent bundles.* Rep. Math. Phys. 43 (1999) no. 1-2, 291–297.

24. Marc A. Rieffel. *Deformation Quantization of Heisenberg Manifolds.* Commum. Math. Phys. 122 (1989), 531–562.

25. Marc A. Rieffel, *Deformation quantization for actions of $\mathbf{R}^d$.* Mem. Amer. Math. Soc. 106 (1993), no. 506.

26. N. Seiberg, E. Witten, JHEP 9909:032, 1999, hep-th/9908142

27. Pierre Sleewaegen. *Application moment et théorème de convexité de Kostant.* Ph.D. thesis, Brussels ULB, 1998.

28. Moss E. Sweedler. *Cohomology of algebras over Hopf algebras.* Trans. Amer. Math. Soc. 133 (1968), 205–239.

29. Moss E. Sweedler. *Hopf algebras.* W.A. Benjamin, 1969.

30. Alan Weinstein. *Traces and triangles in symmetric symplectic spaces.* Symplectic geometry and quantization (Sanda and Yokohama, 1993), Contemp. Math. 179 (1994), 261–270.

Morita Equivalence, Picard Groupoids and Noncommutative Field Theories

S. Waldmann

Fakultät für Mathematik und Physik, Physikalisches Institut, Albert-Ludwigs-Universität Freiburg, Hermann Herder Strasse 3, 79104 Freiburg, Germany
Stefan.Waldmann@physik.uni-freiburg.de

Summary. In this article we review recent developments on Morita equivalence of star products and their Picard groups. We point out the relations between noncommutative field theories and deformed vector bundles which give the Morita equivalence bimodules.

1 Introduction: Noncommutative Field Theories

Noncommutative field theory has recently become a very active field in mathematical physics, see e.g. [13, 17, 22, 31, 33] to mention just a few references. Many additional references can be found in the recent review [35] as well as in these proceedings. The basic idea is to consider field theories on a space-time which is *noncommutative* itself. Thus a considerable part of noncommutative field theory consists in defining and constructing physically plausible models for noncommutative space-times and declaring what a field theory on such a space-time should look like. Here the deformation aspect comes in very useful as one tries to *deform* the algebra of functions on a given ordinary space-time M into some noncommutative algebra. Necessarily, this is, from the mathematical framework, the same as constructing a star product for M corresponding to some Poisson structure on M. Then the spaces of noncommutative fields consist of certain modules over the deformed algebra of functions on M.

The purpose of this note is to point out some mathematical structures underlying noncommutative field theories and to clarify the relations to deformation quantization [2] and Morita equivalence of star products [6, 7, 23]. Here we focus mainly on the framework itself rather than on particular models. In this sense, all what will be said about noncommutative field theories in this note should apply to all such theories and needs more specifications in a given model.

In the commutative framework the (matter) fields are geometrically described by sections $\mathcal{E} = \Gamma^\infty(E)$ of some vector bundle $E \to M$ over the space-time manifold M. We consider here a complex vector bundle E. Since a field $\phi \in \mathcal{E}$ can be multiplied by a function $f \in C^\infty(M)$ and since clearly $(\phi f)g = \phi(fg)$ we obtain a (right) module structure of $\mathcal{E}$ over the algebra

S. Waldmann: *Morita Equivalence, Picard Groupoids and Noncommutative Field Theories*, Lect. Notes Phys. **662**, 143–155 (2005)
www.springerlink.com

of smooth complex-valued functions $C^\infty(M)$. Gauge transformations are encoded in the action of the sections of the endomorphism bundle $\Gamma^\infty(\mathsf{End}(E))$, i.e. $A \in \Gamma^\infty(\mathsf{End}(E))$ can be applied to a field ϕ by pointwise multiplication $A\phi$. In this way, $\mathcal{E}$ becomes a $\Gamma^\infty(\mathsf{End}(E))$ left module. Moreover, the action of $\Gamma^\infty(\mathsf{End}(E))$ commutes with the action of $C^\infty(M)$

$$(A\phi)f = A(\phi f) , \tag{1}$$

whence the space of fields $\mathcal{E}$ becomes a *bimodule* over the algebras $\Gamma^\infty(\mathsf{End}(E))$ and $C^\infty(M)$.

In order to formulate not only the kinematics but also the dynamics we need a Lagrange density $\mathcal{L}$ for E. Geometrically, this is a function on the first jet bundle of E. A particular important piece in $\mathcal{L}$ is the *mass term* which is encoded in a *Hermitian fibre metric* h_0 for E. Recall that a Hermitian fibre metric is a map

$$h_0 : \mathcal{E} \times \mathcal{E} \to C^\infty(M) \tag{2}$$

such that h_0 is $C^\infty(M)$-linear in the second argument, $h_0(\phi,\psi) = \overline{h_0(\psi,\phi)}$ and one has the positivity

$$h_0(\phi,\phi)(x) > 0 \quad \text{iff} \quad \phi(x) \neq 0 . \tag{3}$$

Then the mass term in $\mathcal{L}$ is just $h_0(\phi,\phi)$ and the last condition (3) is the positivity of the masses. Note that such a Hermitian fibre metric is also used to encode geometrically some polynomial interaction terms like ϕ^4. Hence it is of major importance to have a definiteness like (3).

We presented this well-known geometrical formulation, see e.g. the textbook [37], in order to motivate now the noncommutative analogs. The main idea is that at some scale (Planck, etc.) the space-time itself behaves in a noncommutative fashion. One way to encode this noncommutative nature is to consider a star product $\star$ on M which makes the algebra of functions $C^\infty(M)$ into a noncommutative algebra. Here we consider formal star products for convenience, see [2] as well as [16, 19] for recent reviews and further references.

Thus let π be a Poisson tensor on the space-time M and let $\star$ be a formal star product for π, i.e. a $\mathbb{C}[[\lambda]]$-bilinear associative multiplication for $C^\infty(M)[[\lambda]]$,

$$f \star g = \sum_{r=0}^{\infty} \lambda^r C_r(f,g) , \tag{4}$$

with some bidifferential operators C_r such that $C_0(f,g) = fg$ is the undeformed product and $C_1(f,g) - C_1(g,f) = \mathrm{i}\{f,g\}$ gives the Poisson bracket corresponding to π. Moreover, we assume $f \star 1 = f = 1 \star f$ and $\overline{f \star g} = \overline{g} \star \overline{f}$. The formal parameter λ corresponds to the scale where the noncommutativity becomes important. Two star products are called *equivalent* if there is a formal series $T = \mathrm{id} + \sum_{r=1}^{\infty} \lambda^r T_r$ of differential operators T_r such that

$T(f \star g) = Tf \star' Tg$. See [15, 18, 24, 28] for existence and [3, 14, 24, 26, 40] for the classification of such star products up to equivalence.

In order to give a geometrical framework of noncommutative field theories we want a deformed picture of the above bimodule structure. Thus we look for a right module structure $\bullet$ on the space $\Gamma^\infty(E)[[\lambda]]$ with respect to the algebra $C^\infty(M)[[\lambda]]$. Thus $\bullet$ is a $\mathbb{C}[[\lambda]]$-bilinear map

$$\phi \bullet f = \sum_{r=0}^{\infty} \lambda^r R_r(\phi, f) , \tag{5}$$

where $R_r : \Gamma^\infty(E) \times C^\infty(M) \to \Gamma^\infty(E)$ is a bidifferential operator with $R_0(\phi, f) = \phi f$ and

$$(\phi \bullet f) \bullet g = \phi \bullet (f \star g) \quad \text{and} \quad \phi \bullet 1 = \phi . \tag{6}$$

This gives the right module structure. But we also need an associative deformation $\star'$ of $\Gamma^\infty(\mathsf{End}(E))$ and a left module structure $\bullet'$ such that we have

$$(A \star' B) \bullet' \phi = A \bullet' (B \bullet' \phi), \quad \mathbb{1} \bullet' \phi = \phi \quad \text{and} \quad A \bullet' (\phi \bullet f) = (A \bullet' \phi) \bullet f . \tag{7}$$

This gives then a deformed bimodule structure on $\Gamma^\infty(E)$.

If we are interested in the analog of the Hermitian metric h_0 then we want a $\mathbb{C}[[\lambda]]$-sesquilinear deformation $h = \sum_{r=0}^{\infty} \lambda^r h_r$ of h_0 where $h_r : \Gamma^\infty(E) \times \Gamma^\infty(E) \to C^\infty(M)$ such that

$$h(\phi, \psi \bullet f) = h(\phi, \psi) \star f , \tag{8}$$

$$h(\phi, \psi) = \overline{h(\psi, \phi)} , \tag{9}$$

$$h(\phi, \phi) \text{ is positive} , \tag{10}$$

$$h(A \bullet' \phi, \psi) = h(\phi, A^* \bullet' \psi) . \tag{11}$$

The positivitiy in (10) is understood in the sense of *-algebras over ordered rings, see [6, 8] and also [32] for a treatment of various notions for positivity in *-algebras over $\mathbb{C}$. In the case of a vector bundle this just means that $h(\phi, \phi)$ can be written as a sum of squares $\sum_i \bar{f}_i \star f_i$.

Having this structure one obtains a *framework* for noncommutative field theories *beyond* the usual formulations on a flat space-time with trivial vector bundle, very much in the spirit of Connes' noncommutative geometry [12]. To formulate a physical theory one needs of course much more, like an action principle, convergence in the deformation parameter λ, a quantization of this still classical theory, etc. All these questions shall not be addressed in this work. Instead, we shall focus on the question whether and how one can prove existence, construct, and classify the structures $\bullet$, $\bullet'$, $\star'$ and h out of the given classical data and a given star product $\star$. The case of a line bundle $E = L$ plays a particularly interesting role as this corresponds exactly to (complex) scalar fields.

2 Deformation of Projective Modules

There are several different ways to construct deformed versions of a Hermitian vector bundle. We shall focus on a rather general algebraic construction before discussing the other possibilities. Fundamental is the well-known *Serre-Swan theorem* [34] in its smooth version: The $C^\infty(M)$-module of sections $\Gamma^\infty(E)$ is a finitely generated projective module. Moreover, the $C^\infty(M)$-linear module endomorphisms are just the sections of the endomorphism bundle, i.e. $\Gamma^\infty(\mathsf{End}(E)) = \mathsf{End}_{C^\infty(M)}(\Gamma^\infty(E))$. Hence one finds a projection $P_0 = P_0^2 \in M_N(C^\infty(M))$ where N is sufficiently large, such that

$$\Gamma^\infty(E) \cong P_0 C^\infty(M)^N \quad \text{and} \quad \Gamma^\infty(\mathsf{End}(E)) \cong P_0 M_N(C^\infty(M))P_0 . \quad (12)$$

If E is equipped with a Hermitian fiber metric h_0 then one can even find a Hermititan projection $P_0 = P_0^2 = P_0^*$ such that with the identification of (12) the Hermitian fiber metric becomes

$$h_0(\phi, \psi) = \sum_{i=1}^N \overline{\phi}_i \psi_i , \quad (13)$$

where $\phi = (\phi_1, \ldots, \phi_N)$, $\psi = (\psi_1, \ldots, \psi_N) \in C^\infty(M)^N$ are elements in $P_0 C^\infty(M)^N$, i.e. they satisfy $P_0\phi = \phi$, $P_0\psi = \psi$.

It is worth to look at this situation in general. Thus let $\mathcal{A}$ be an associative algebra over a ring C and let $\star$ be an associative formal deformation of $\mathcal{A}$. We denote the deformed algebra by $\boldsymbol{\mathcal{A}} = (\mathcal{A}[[\lambda]], \star)$. Now let $\mathcal{E}$ be a finitely generated projective right module over $\mathcal{A}$ and let $\mathsf{End}_\mathcal{A}(\mathcal{E})$ denote the $\mathcal{A}$-linear endomorphisms of $\mathcal{E}$. Then one has the following result, see [5]:

Theorem 1 *There exists a deformation $\bullet$ of $\mathcal{E}$ into a $\boldsymbol{\mathcal{A}}$-right module $(\mathcal{E}[[\lambda]], \bullet)$ which is unique up to equivalence such that $(\mathcal{E}[[\lambda]], \bullet)$ is finitely generated and projective over $\boldsymbol{\mathcal{A}}$ and $\mathsf{End}_{\boldsymbol{\mathcal{A}}}(\mathcal{E}[[\lambda]], \bullet)$ is isomorphic as $\mathsf{C}[[\lambda]]$-module to $\mathsf{End}_\mathcal{A}(\mathcal{E})[[\lambda]]$.*

Equivalence of two deformations $\bullet$ and $\tilde{\bullet}$ means that there is a map $T = \mathrm{id} + \sum_{r=1}^\infty \lambda^r T_r$ with $T(\phi \bullet a) = T(\phi)\tilde{\bullet}a$ for all $\phi \in \mathcal{E}[[\lambda]]$ and $a \in \mathcal{A}[[\lambda]]$.

The idea of the proof consists in first deforming the projection P_0 into a projection $\boldsymbol{P}$ with respect to the deformed product $\star$ by using the formula [18, eq. (6.1.4)]

$$\boldsymbol{P} = \frac{1}{2} + \left(P_0 - \frac{1}{2}\right) \star \frac{1}{\sqrt[*]{1 + 4(P_0 \star P_0 - P_0)}} . \quad (14)$$

Then the $\boldsymbol{\mathcal{A}}$-right module $\boldsymbol{P} \star \boldsymbol{\mathcal{A}}^N$ is obviously a finitely generated and projective $\boldsymbol{\mathcal{A}}$-module and it turns out that it is isomorphic to $\mathcal{E}[[\lambda]]$ as $\mathsf{C}[[\lambda]]$-module. Then the uniqueness of the deformation $\bullet$ up to equivalence follows from the fact the $\boldsymbol{P} \star \mathcal{A}^N$ is projective again. Indeed, let $\mathcal{E}$ be endowed with the trivial $\mathcal{A}$ right module structure given by $\phi \cdot a = \phi a_0$ for $\phi \in \mathcal{E}$ and $a = \sum_{r=0}^\infty \lambda^r a_r \in \mathcal{A}[[\lambda]]$. Then the classical limit map cl : $(\mathcal{E}[[\lambda]], \bullet) \longrightarrow (\mathcal{E}, \cdot)$ (setting $\lambda = 0$) is a module morphism for $\mathcal{A}$ right modules. The same holds for any other deformation $\tilde{\bullet}$. Since the deformation $\bullet$ is projective and since cl is obviously surjective this means we can find a module morphism $T : (\mathcal{E}[[\lambda]], \bullet) \longrightarrow (\mathcal{E}[[\lambda]], \tilde{\bullet})$ such that cl $\circ\, T$ = cl. This implies $T = \text{id} + \sum_{r=1}^\infty \lambda^r T_r$ whence we have found an equivalence, see [5] for details.

In particular, the choice of a $\mathbb{C}[[\lambda]]$-linear isomorphism between $\mathsf{End}_\mathcal{A}(\mathcal{E}[[\lambda]], \bullet)$ and $\mathsf{End}_\mathcal{A}(\mathcal{E})[[\lambda]]$ induces a new *deformed* multiplication $\star'$ for $\mathsf{End}_\mathcal{A}(\mathcal{E})[[\lambda]]$ together with a new module multiplication $\bullet'$ for $\mathcal{E}[[\lambda]]$ such that $(\mathcal{E}[[\lambda]], \bullet', \bullet)$ becomes a $(\mathsf{End}_\mathcal{A}(\mathcal{E})[[\lambda]], \star')$-$\mathcal{A}$ bimodule.

Remark 1 1. Since $(\mathcal{E}[[\lambda]], \bullet)$ is unique up to equivalence the deformation $\star'$ is unique up to *isomorphism* since $\mathsf{End}_\mathcal{A}(\mathcal{E}[[\lambda]], \bullet)$ is fixed. One can even obtain a $\star'$ which is unique up to equivalence if one imposes $\bullet'$ to be a *deformation* of the original left module structure.
Otherwise, if $\star'$ is such a deformation and Φ is an automorphism of the undeformed algebra $\mathsf{End}_\mathcal{A}(\mathcal{E})$ then

$$A \star^\Phi B := \Phi(\Phi^{-1}A \star' \Phi^{-1}B) \tag{15}$$

yields another isomorphic but not necessarily equivalent deformation of $\mathsf{End}_\mathcal{A}(\mathcal{E})$ allowing for a bimodule structure as above.
2. In general, there is an obstruction on $\star'$ to allow such a bimodule deformation $\bullet'$ for a given fixed $\star$ (and hence $\bullet$) as the algebra structure has to be isomorphic to $\mathsf{End}_\mathcal{A}(\mathcal{E}[[\lambda]], \bullet)$.
3. By analogous arguments as above one can also show the existence and uniqueness up to isometries of deformations of Hermitian fiber metrics [5].
4. In physical terms: noncommutative field theories on a classical vector bundle always exist and are even uniquely determined by the underlying deformation of the space-time, at least up to equivalence. Morally, this can be seen as the deeper reason for the existence of Seiberg-Witten maps.

Let us now mention two other constructions leading to deformed vector bundles. It is clear that the above argument has strong algebraic power but is of little use when one wants more explicit formulas as even the classical projections P_0 describing a given vector bundle $E \to M$ are typically rather in-explicit. The following two constructions provide more explicit formulas:

1. Jurčo, Schupp, and Wess [23] considered the case of a *line bundle* $L \to M$ with connection ∇^L and an arbitrary Poisson structure θ on M. Here one can use first the Kontsevich star product quantizing θ by use of a global formality map. Second, one can use the same formality map together with

the connection ∇^L to construct $\bullet$, $\star'$ and $\bullet'$ as well. The construction depends on the choice of a global formality. One also obtains a Seiberg-Witten map using the formality.

2. In [38] we considered the case of a symplectic manifold with arbitrary vector bundle $E \to M$. Given a symplectic connection ∇, Fedosov's construction yields a star product $\star$ for M, see e.g. [18]. Using a connection ∇^E for E one obtains $\bullet$, $\star'$ and $\bullet'$ depending even functorially on the inital data of the connections. Hence one obtains a very explicit and geometric construction this way.

For both approaches one can show that the resulting deformations $\bullet$, $\star'$ and $\bullet'$ can be chosen to be *local*, i.e. the deformations are formal power series in bidifferential operators acting on functions, sections and endomorphisms, respectively. Thus one can 'localize' and restrict to open subsets $U \subseteq M$. If in particular one has a good open cover $\{U_\alpha\}$ of M then $E\big|_{U_\alpha}$ becomes a *trivial* vector bundle. Since the deformation is unique up to equivalence the restricted deformation $\bullet_\alpha$ has to be equivalent to the trivial deformation of a trivial bundle. This way one arrives at a description of $\bullet$, $\bullet'$ and $\star'$ in terms of *transition matrices* $\boldsymbol{\Phi}_{\alpha\beta}$ satisfying a *deformed cocycle identity*

$$\boldsymbol{\Phi}_{\alpha\beta} \star \boldsymbol{\Phi}_{\beta\gamma} \star \boldsymbol{\Phi}_{\gamma\alpha} = \mathbb{1} \quad \text{and} \quad \boldsymbol{\Phi}_{\alpha\beta} \star \boldsymbol{\Phi}_{\beta\alpha} = \mathbb{1} \tag{16}$$

on non-trivial overlaps of U_α, U_β, and U_γ. Here $\boldsymbol{\Phi}_{\alpha\beta} = \sum_{r=0}^{\infty} \lambda^r \boldsymbol{\Phi}_{\alpha\beta}^{(r)} \in M_k(C^\infty(U_{\alpha\beta}))[[\lambda]]$ and the $\boldsymbol{\Phi}_{\alpha\beta}^{(0)}$ are the classical transition matrices. Conversely, if one finds a deformation (16) of the classical cocycle then one can construct a deformation of the vector bundle out of it. This can be seen as a *Quantum Serre-Swan Theorem*, see [39]. We conclude this section with a few further remarks:

Remark 2   1. Since the finitely generated projective modules $\mathcal{E}$ over $\mathcal{A}$ give the K_0-theory of the algebra $\mathcal{A}$ and since any such $\mathcal{E}$ can be deformed in a unique way up to equivalence and since clearly any finitely generated projective module over $\mathcal{A}$ arises this way up to isomorphism one finally obtains that the classical limit map cl induces an *isomorphism*

$$\mathrm{cl}_* : K_0(\boldsymbol{\mathcal{A}}) \xrightarrow{\cong} K_0(\mathcal{A}) . \tag{17}$$

Thus K-theory is stable under formal deformations [30].

2. If $\int : \boldsymbol{\mathcal{A}} \to \mathbb{C}[[\lambda]]$ is a trace functional, i.e.

$$\int a \star b = \int b \star a , \tag{18}$$

then $\mathrm{ind} : K_0(\boldsymbol{\mathcal{A}}) \longrightarrow \mathbb{C}[[\lambda]]$ defined by

$$[\boldsymbol{P}] \mapsto \mathrm{ind}([\boldsymbol{P}]) = \int \mathrm{tr}(\boldsymbol{P}) \tag{19}$$

gives a well-defined group morphism and for a fixed choice of $\int$ the *index* ind($\boldsymbol{P}$) depends only on the classical class $[P_0]$. In case of deformation quantization this yields the *index theorems of deformation quantization* where one has explicit formulas for ind($\boldsymbol{P}$) in terms of geometric data of E, M and the equivalence class $[\star]$ of the star product, see Fedosov's book for the symplectic case [18] as well as Nest and Tsygan [26, 27] and the work of Tamarkin and Tsygan for the Poisson case [36].

3. In the connected symplectic case the trace functional $\int$ is unique up to normalization [26] and given by a deformation of the integration over M with respect to the Liouville measure. In the Poisson case one may have many different trace functionals, see e.g. [4].

4. Physically, such trace functionals are needed for the formulation of gauge invariant action functionals which are used to define dynamics for the noncommutative field theories. Recall that the structure of a deformed vector bundle is only the kinematical framework.

3 Morita Equivalence

Let us now discuss how Morita theory enters the picture of deformed vector bundles. Vector bundles do not only correspond to projective modules but the projections $P_0 \in M_N(C^\infty(M))$ are always *full projections* which means that the ideal in $C^\infty(M)$ generated by the components $(P_0)_{ij}$ is the whole algebra $C^\infty(M)$. We exclude the trivial case $P_0 = 0$ from our discussion in order to avoid trivialities. Then the following statement is implied by general Morita theory, see e.g. [25] as well as [5].

Theorem 2 *The bimodule $\mathcal{E} = \Gamma^\infty(E)$ is actually a Morita equivalence bimodule for the algebras $C^\infty(M)$ and $\Gamma^\infty(\mathsf{End}(E))$. In particular, these algebras are Morita equivalent.*

In the general algebraic case, it is easy to check that the deformation $\boldsymbol{P}$ of a full projection P_0 is again full whence we conclude that $(\mathcal{E}[[\lambda]], \bullet, \bullet')$ is a Morita equivalence bimodule for the algebras $(\mathcal{A}[[\lambda]], \star)$ and $(\mathsf{End}_\mathcal{A}(\mathcal{E})[[\lambda]], \star')$ and the later two algebras are Morita equivalent. Moreover, any Morita equivalence bimodule between the deformed algebras arises as such a deformation of a classical Morita equivalence bimodule up to isomorphism, see e.g. [9] for a detailed discussion. Since the deformation $\star'$ was already fixed up to isomorphism by the classical right module structure of $\mathcal{E}$, one has to expect obstructions that an a priori given deformation $\tilde{\star}$ of $\mathsf{End}_\mathcal{A}(\mathcal{E})$ is Morita equivalent to the deformation $\star$ of $\mathcal{A}$. These obstructions make the classification of the Morita equivalent deformations difficult in the general framework. We shall come back to this effect when considering the Picard groupoid.

However, for symplectic star products one has the following explicit classification of Morita equivalent star products [7], see also [23] for a related statement in the Poisson case. Note that for star products $\star$ and $\star'$ we want the endomorphisms $\Gamma^\infty(\mathsf{End}(E))$ classically to be isomorphic to the functions $C^\infty(M)$ whence the Morita equivalence bimodules arise as deformations of *line bundles*.

Theorem 3 *Let (M, ω) be symplectic. Then two star products $\star$ and $\star'$ are Morita equivalent if and only if there exists a symplectic diffeomorphism $\psi :$ $M \to M$ such that*

$$\psi^* c(\star') - c(\star) \in 2\pi \mathrm{i} H^2_{\mathrm{dR}}(M, \mathbb{Z}) , \tag{20}$$

where $c(\star) \in \frac{[\omega]}{\mathrm{i}\lambda} + H^2_{\mathrm{dR}}(M, \mathbb{C})[[\lambda]]$ is the characteristic class of $\star$. The equivalence bimodule can be obtained by deforming a line bundle $L \to M$ whose Chern class $c_1(L)$ is given by the above integer class.

The most suitable definition of the characteristic class of a symplectic star product which is used in this theorem is the Čech cohomological description as it can be found in [20]. Then the first proof in [7] consists in examining the deformed transition functions (16). In the approach of [38] using Fedosov's construction there is an almost trivial proof for the above theorem as the Chern class of the line bundle L can be build into the Fedosov construction as a curvature term of a connection ∇^L on L directly.

The additional diffeomorphism ψ is necessary as $\star'$ is only determined by L up to isomorphism and not up to equivalence as this is encoded in the characteristic class.

Remark 3 There is even a stronger result: For *-algebras one has a notion of *strong Morita equivalence* [6] which is a generalization of Rieffel's notion of strong Morita equivalence for C^*-algebras [29]. Applying this for Hermitian star products, i.e. those with $\overline{f \star g} = \bar{g} \star \bar{f}$, one has the statement that two Hermitian star products are strongly Morita equivalent if and only if they are Morita equivalent [7, Thm. 2]. One uses a deformed Hermitian fiber metric in order to get this stronger result. Physically, this is the relevant notion of Morita equivalence as one also needs to keep track of the *-involutions and positivity requirements as we have discussed above. Thanks to [7, Thm. 2], we can focus on the purely ring-theoretical Morita theory without restriction.

4 The Picard Groupoid

In this last section we shall consider the question in 'how many ways' two Morita equivalent algebras can actually be Morita equivalent. In particular, we want to investigate how Morita equivalence bimodules behave under formal deformations.

First we note that this is physically an important questions since we have already seen that the algebra $C^\infty(M)$ and the algebra $\Gamma^\infty(\mathsf{End}(E))$, which encodes the gauge transformations, are Morita equivalent via the sections $\Gamma^\infty(E)$ of the vector bundle E. Thus the above question wants to answer how many 'different' vector bundles, i.e. field theories, one can find which allow for such a bimodule structure for the same algebra of gauge transformations.

To formulate these questions one uses the following definitions for unital algebras $\mathcal{A}$, $\mathcal{B}$, ... over some ring C:

Definition 1 Let $\underline{\mathrm{Pic}}(\mathcal{A}, \mathcal{B})$ denote the category of $\mathcal{A}$-$\mathcal{B}$ Morita equivalence bimodules with bimodule homomorphisms as morphisms. The set of isomorphism classes of bimodules in $\underline{\mathrm{Pic}}(\mathcal{A}, \mathcal{B})$ is denoted by $\mathrm{Pic}(\mathcal{A}, \mathcal{B})$.

From Morita theory we know that $\mathcal{E} \in \underline{\mathrm{Pic}}(\mathcal{A}, \mathcal{B})$ is a finitely generated and projective module whence the isomorphism classes are a set indeed.

It is a well-known fact that tensoring equivalence bimodules gives again an equivalence bimodule. Hence if $\mathcal{E} \in \underline{\mathrm{Pic}}(\mathcal{A}, \mathcal{B})$ and $\mathcal{F} \in \underline{\mathrm{Pic}}(\mathcal{B}, \mathcal{C})$ then $\mathcal{E} \otimes_{\mathcal{B}} \mathcal{F} \in \underline{\mathrm{Pic}}(\mathcal{A}, \mathcal{C})$. Moreover, it is clear that this tensor product is compatible with the notion of isomorphisms of equivalence bimodules. Thus this gives a composition law

$$\otimes : \mathrm{Pic}(\mathcal{A}, \mathcal{B}) \times \mathrm{Pic}(\mathcal{B}, \mathcal{C}) \longrightarrow \mathrm{Pic}(\mathcal{A}, \mathcal{C}) . \tag{21}$$

Then the tensor product is associative on the level of isomorphism classes, whenever the composition is defined. We also note that the trivial self-equivalence bimodule $\mathcal{A}$ behaves like a unit with respect to $\otimes$ at least on the level of isomorphism classes. Finally, the dual module to $\mathcal{E}$ gives an inverse whence we end up with a *groupoid structure*, called the *Picard groupoid* $\mathrm{Pic}(\cdot, \cdot)$. The units are just trivial self-equivalence bimodules and the spaces of arrows are just the $\mathrm{Pic}(\mathcal{A}, \mathcal{B})$. The isotropy groups of this groupoid are the *Picard groups* $\mathrm{Pic}(\mathcal{A}) = \mathrm{Pic}(\mathcal{A}, \mathcal{A})$, see e.g. [25, 1].

After this excursion let us now focus again on the deformation problem. Assume $\boldsymbol{\mathcal{A}} = (\mathcal{A}[[\lambda]], \star)$ and $\boldsymbol{\mathcal{B}} = (\mathcal{B}[[\lambda]], \star')$ are associative deformations such that the resulting algebras are Morita equivalent. Then $\underline{\mathrm{Pic}}(\boldsymbol{\mathcal{A}}, \boldsymbol{\mathcal{B}})$ is non-empty and any $\boldsymbol{\mathcal{E}}$ is isomorphic to a deformation $(\mathcal{E}[[\lambda]], \bullet, \bullet')$ of an equivalence bimodule $\mathcal{E}$ of the undeformed algebras $\mathcal{A}$, $\mathcal{B}$. In particular, $\mathcal{A}$ and $\mathcal{B}$ have to be Morita equivalent, too. Moreover, $\mathcal{E}$ is uniquely determined up to isomorphism whence one obtains a well-defined *classical limit map*

$$\mathrm{cl}_* : \mathrm{Pic}(\boldsymbol{\mathcal{A}}, \boldsymbol{\mathcal{B}}) \longrightarrow \mathrm{Pic}(\mathcal{A}, \mathcal{B}) . \tag{22}$$

It is easy to see that the classical limit map behaves well with respect to tensor products of bimodules whence on the level of isomorphism classes we obtain a groupoid morphism, see [9] where the case of the group morphism is discussed:

Proposition 1 *The classical limit map* cl_* *is a groupoid morphism. In particular,*

$$\mathrm{cl}_* : \mathrm{Pic}(\mathcal{A}) \longrightarrow \mathrm{Pic}(A) \tag{23}$$

is a group morphism.

Note that this is a very similar situation as for the K-theory (17). However, here cl_* is far from being an isomorphism in general. Thus we would like to find a description of the kernel and the image of the map cl_*, at least for the cases where $\mathcal{A}$ is commutative.

For the kernel one obtains the following characterization. Let

$$\mathrm{Equiv}(\mathcal{A}) = \left\{ T = \mathrm{id} + \sum_{r=1}^{\infty} \lambda T_r \mid T \in \mathrm{Aut}(\mathcal{A}) \right\} \tag{24}$$

denote the *self-equivalences* of the deformed algebra. Since we assume that the undeformed algebra $\mathcal{A}$ is commutative, the inner automorphisms of $\mathcal{A}$ are necessarily self-equivalences. Thus one can define the group of outer self-equivalences

$$\mathrm{OutEquiv}(\mathcal{A}) = \frac{\mathrm{Equiv}(\mathcal{A})}{\mathrm{InnAut}(\mathcal{A})} . \tag{25}$$

Then one has

$$\ker \mathrm{cl}_* \cong \mathrm{OutEquiv}(\mathcal{A}) \tag{26}$$

as groups [9, Cor. 3.11].

In the case of star products one can describe $\ker \mathrm{cl}_*$ even more explicitly. Assume that $\star$ is a star product on (M, π) with the property that any π-central function can be deformed into a $\star$-central function and any π-derivation can be deformed into a $\star$-derivation. There are many star products which actually have this property, e.g. all symplectic star product, the Kontsevich star product for a formal Poisson structure which is equal to the classical one and the star products constructed in [10, 11]. Under these assumptions one has [9, Thm. 7.1]

$$\mathrm{OutEquiv}(\star) \cong \frac{H^1_\pi(M, \mathbb{C})}{2\pi \mathrm{i} H^1_\pi(M, \mathbb{Z})} + \lambda H^1_\pi(M, \mathbb{C})[[\lambda]] \tag{27}$$

as *sets*, where $H^1_\pi(M, \mathbb{C})$ denotes the first complex Poisson cohomology of (M, π) and $H^1_\pi(M, \mathbb{Z})$ the first integral Poisson cohomology, i.e. the image of the integral deRham classes under the natural map $H^1_{\mathrm{dR}}(M, \mathbb{Z}) \longrightarrow H^1_\pi(M, \mathbb{C})$.

The identification above is even a *group isomorphism* for symplectic star products where the right hand side is endowed with its canonical abelian group structure. However, in the general Poisson case the group structure on the left hand side is nonabelian.

The situation for the image of the classical limit map cl_* is more mysterious [9]: From the condition (20) one obtains that the torsion line bundles are always in the image in the case of symplectic star products. However, there are examples where the image contains also non-torsion elements and it seems to depend strongly on the example how big the image actually can be. In the Poisson case even less is known.

Acknowledgement

I would like to thank the organizers of the workshop for their excellent working conditions and warm hospitality. Moreover, I would like to thank the participants for their remarks, ideas and suggestions on the topic as well as Henrique Bursztyn and Stefan Jansen for comments on the manuscript.

References

1. Bass, H.: *Algebraic K-theory*. W. A. Benjamin, Inc., New York, Amsterdam, 1968.
2. Bayen, F., Flato, M., Frønsdal, C., Lichnerowicz, A., Sternheimer, D.: *Deformation Theory and Quantization*. Ann. Phys. **111** (1978), 61–151.
3. Bertelson, M., Cahen, M., Gutt, S.: *Equivalence of Star Products*. Class. Quant. Grav. **14** (1997), A93–A107.
4. Bieliavsky, P., Bordemann, M., Gutt, S., Waldmann, S.: *Traces for star products on the dual of a Lie algebra*. Rev. Math. Phys. **15** (2003), 425–445.
5. Bursztyn, H., Waldmann, S.: *Deformation Quantization of Hermitian Vector Bundles*. Lett. Math. Phys. **53** (2000), 349–365.
6. Bursztyn, H., Waldmann, S.: *Algebraic Rieffel Induction, Formal Morita Equivalence and Applications to Deformation Quantization*. J. Geom. Phys. **37** (2001), 307–364.
7. Bursztyn, H., Waldmann, S.: *The characteristic classes of Morita equivalent star products on symplectic manifolds*. Commun. Math. Phys. **228** (2002), 103–121.
8. Bursztyn, H., Waldmann, S.: *Completely positive inner products and strong Morita equivalence*. Preprint (FR-THEP 2003/12) **math.QA/0309402** (September 2003), 36 pages. To appear in Pacific J. Math.
9. Bursztyn, H., Waldmann, S.: *Bimodule deformations, Picard groups and contravariant connections*. K-Theory **31** (2004), 1–37.
10. Cattaneo, A. S., Felder, G., Tomassini, L.: *Fedosov connections on jet bundles and deformation quantization*. In: Halbout, G. (eds.): *Deformation quantization*. [21], 191–202.
11. Cattaneo, A. S., Felder, G., Tomassini, L.: *From local to global deformation quantization of Poisson manifolds*. Duke Math. J. **115**.2 (2002), 329–352.
12. Connes, A.: *Noncommutative Geometry*. Academic Press, San Diego, New York, London, 1994.
13. Connes, A., Douglas, M. R., Schwarz, A.: *Noncommutative geometry and matrix theory: compactification on tori*. J. High Energy Phys. **02** (1998), 003.
14. Deligne, P.: *Déformations de l'Algèbre des Fonctions d'une Variété Symplectique: Comparaison entre Fedosov et DeWilde, Lecomte*. Sel. Math. New Series **1**.4 (1995), 667–697.

15. DeWilde, M., Lecomte, P.B.A.: *Existence of Star-Products and of Formal Deformations of the Poisson Lie Algebra of Arbitrary Symplectic Manifolds.* Lett. Math. Phys. **7** (1983), 487–496.

16. Dito, G., Sternheimer, D.: *Deformation quantization: genesis, developments and metamorphoses.* In: Halbout, G. (eds.): *Deformation quantization.* [21], 9–54.

17. Doplicher, S., Fredenhagen, K., Roberts, J.E.: *The Quantum Structure of Spacetime at the Planck Scale and Quantum Fields.* Commun. Math. Phys. **172** (1995), 187–220.

18. Fedosov, B. V.: *Deformation Quantization and Index Theory.* Akademie Verlag, Berlin, 1996.

19. Gutt, S.: *Variations on deformation quantization.* In: Dito, G., Sternheimer, D. (eds.): *Conférence Moshé Flato 1999. Quantization, Deformations, and Symmetries, Mathematical Physics Studies* no. **21**, 217–254. Kluwer Academic Publishers, Dordrecht, Boston, London, 2000.

20. Gutt, S., Rawnsley, J.: *Equivalence of star products on a symplectic manifold; an introduction to Deligne's Čech cohomology classes.* J. Geom. Phys. **29** (1999), 347–392.

21. Halbout, G. (eds.): *Deformation Quantization,* vol. 1 in *IRMA Lectures in Mathematics and Theoretical Physics.* Walter de Gruyter, Berlin, New York, 2002.

22. Jurčo, B., Schupp, P., Wess, J.: *Noncommutative gauge theory for Poisson manifolds.* Nucl. Phys. **B584** (2000), 784–794.

23. Jurčo, B., Schupp, P., Wess, J.: *Noncommutative Line Bundles and Morita Equivalence.* Lett. Math. Phys. **61** (2002), 171–186.

24. Kontsevich, M.: *Deformation Quantization of Poisson Manifolds, I.* Preprint **q-alg/9709040** (September 1997).

25. Lam, T. Y.: *Lectures on Modules and Rings,* vol. 189 in *Graduate Texts in Mathematics.* Springer-Verlag, Berlin, Heidelberg, New York, 1999.

26. Nest, R., Tsygan, B.: *Algebraic Index Theorem.* Commun. Math. Phys. **172** (1995), 223–262.

27. Nest, R., Tsygan, B.: *Algebraic Index Theorem for Families.* Adv. Math. **113** (1995), 151–205.

28. Omori, H., Maeda, Y., Yoshioka, A.: *Weyl Manifolds and Deformation Quantization.* Adv. Math. **85** (1991), 224–255.

29. Rieffel, M.A.: *Morita equivalence for C^*-algebras and W^*-algebras.* J. Pure. Appl. Math. **5** (1974), 51–96.

30. Rosenberg, J.: *Rigidity of K-theory under deformation quantization.* Preprint **q-alg/9607021** (July 1996).

31. Schomerus, V.: *D-branes and deformation quantization.* J. High. Energy. Phys. **06** (1999), 030.

32. Schmüdgen, K.: *Unbounded Operator Algebras and Representation Theory,* Vol. 37 of *Operator Theory: Advances and Applications.* Birkhäuser Verlag, Basel, Boston, Berlin, 1990.

33. Seiberg, N., Witten, E.: *String Theory and Noncommutative Geometry.* J. High. Energy Phys. **09** (1999), 032.

34. Swan, R.G.: *Vector bundles and projective modules.* Trans. Amer. Math. Soc. **105** (1962), 264–277.

35. Szabo, R.: *Quantum Field Theory on Noncommutative Spaces.* Preprint **hep-th/0109162** (September 2001), 111 pages. To appear in Phys. Rep.

36. Tamarkin, D., Tsygan, B.: *Cyclic Formality and Index Theorems*. Lett. Math. Phys. **56** (2001), 85–97.
37. Thirring, W.: *Klassische Feldtheorie*, vol. 2 in *Lehrbuch der Mathematischen Physik*. Springer-Verlag, Wien, New York, 2. edition, 1990.
38. Waldmann, S.: *Morita equivalence of Fedosov star products and deformed Hermitian vector bundles*. Lett. Math. Phys. **60** (2002), 157–170.
39. Waldmann, S.: *On the representation theory of deformation quantization*. In: Halbout, G. (eds.): *Deformation quantization*. [21], 107–133.
40. Weinstein, A., Xu, P.: *Hochschild cohomology and characteristic classes for star-products*. In: Khovanskij, A., Varchenko, A., Vassiliev, V. (eds.): *Geometry of differential equations. Dedicated to V. I. Arnold on the occasion of his 60th birthday*, 177–194. American Mathematical Society, Providence, 1998.

Secondary Characteristic Classes
of Lie Algebroids

M. Crainic[1*] and R.L. Fernandes[2**]

[1] Depart. of Math., Utrecht University 3508 TA Utrecht, The Netherlands
`crainic@math.uu.nl`
[2] Depart. de Matem., Instituto Superior Técnico 1049-001 Lisbon, Portugal
`rfern@math.ist.utl.pt`

Summary. We show how the intrinsic characteristic classes of Lie algebroids can be seen as characteristic classes of representations. We present two alternative ways: The first one consists of thinking of the adjoint representation as a connection up to homotopy. The second one is by viewing the adjoint representation as a honest representation on the first jet bundle of a Lie algebroid.

1 Introduction

Lie algebroids are geometric versions of vector bundles which are useful to describe various geometric setups. As important classes of examples one can mention foliated geometry, equivariant geometry or Poisson geometry. We refer the reader to [4] for an introduction to the subject and its relation to the noncommutative world.

The classical theory of characteristic classes, such as Pontriagin classes or Chern classes, extends to Lie algebroids. The reason is that the usual Chern-Weil construction can be defined in the general context of Lie algebroids, as was explained in [8]. However, the classes one obtains in this way, are the image by the anchor map of the usual characteristic classes. Much more interesting are the secondary characteristic classes of Lie algebroids, introduced in [5, 8] in two distinct disguises.

On one hand, Lie algebroids have representations, which are flat Lie algebroid connections generalizing the flat vector bundles of ordinary geometry. It was shown in [5] that one can define secondary characteristic classes of representations of Lie algebroids, much like the characteristic classes of ordinary flat bundles.

On the other hand, every Lie algebroid has an underlying characteristic foliation, which will be singular in general. Again, similar to the theory of foliations (see e.g. [3, 11]), it was shown in [8] that one can define intrinsic characteristic classes of the Lie algebroid.

---

* Supported in part by NWO and a Miller Research Fellowship.
** Supported in part by FCT through program POCTI-Research Units Pluriannual Funding Program and grant POCTI/1999/MAT/33081.

In the theory of foliations, one can describe the intrinsic secondary characteristic classes as the characteristic classes of a special representation. Namely, the normal bundle to the foliation carries a canonical flat connection, the Bott connection [3], which plays the role of the "adjoint representation" of the foliation. The purpose of this work is to give a similar relation between the intrinsic secondary characteristic classes of a Lie algebroid and the characteristic classes of a representation. The additional complication in this case is that, in general, a Lie algebroid does not carry an adjoint representation.

We will give two distinct, alternative, solutions to this problem. Both solutions consist in giving an appropriate meaning to the notion of an "adjoint representation" of a Lie algebroid.

In the first solution to our problem, one enlarges the notion of representation, allowing *representations up to homotopy*. This was first proposed by Evans, Lu and Weinstein in [7], where they view the adjoint representation as a representation up to homotopy, and they use it to construct the most simple example of a secondary characteristic class, namely, the modular class. Here, we will see that one can define characteristic classes of representations up to homotopy, and that for the adjoint representation (up to homotopy) one obtains the intrinsic secondary characteristic classes of the Lie algebroid.

For the second solution to our problem, we observe that the first jet bundle of a Lie algebroid has a natural prolonged Lie algebroid structure. Moreover, this jet Lie algebroid carries a natural, honest, representation, which one can also view as the "adjoint representation" of the original Lie algebroid. By a straightforward application of the theory of characteristic classes of representations, we obtain classes in the Lie algebroid cohomology of the jet bundle. We then check that these classes are the pull-back of the intrinsic characteristic classes of the original Lie algebroid.

The remainder of the paper is organized into three sections. In Sect. 2, we recall the constructions of the intrinsic characteristic classes and of the characteristic classes of representations. In Sect. 3, we clarify the relevance of connections up to homotopy to the theory of characteristic classes, and we recover the intrinsic characteristic classes from the adjoint representation up to homotopy. In Sect. 4, we discuss the prolonged Lie algebroid structure on the jet bundle of a Lie algebroid, and we construct the intrinsic characteristic classes via the jet adjoint representation.

2 Secondary Characteristic Classes of Lie Algebroids

In this work we will denote by A a **Lie algebroid** $\pi : A \to M$, with **anchor** $\# : A \to TM$, and **Lie bracket** $[\,,\,] : \Gamma(A) \times \Gamma(A) \to \Gamma(A)$. Underlying the Lie algebroid we have a (singular) foliation $\mathcal{F}$, which integrates the (singular) involutive distribution $\operatorname{Im} \#$. We recall that the space of A-**forms** $\Omega^\bullet(A)$ is formed by the sections of the exterior bundles $\Gamma(\wedge^\bullet(A^*))$, and that the A-**differential**

$$d : \Omega^\bullet(A) \to \Omega^{\bullet+1}(A)$$

is given by the usual Cartan formula:

$$d\omega(\alpha_0, \dots, \alpha_k) = \sum_{i=0}^{k+1} (-1)^i \#\alpha_i(\omega(\alpha_0, \dots, \widehat{\alpha}_i, \dots, \alpha_k))$$

$$+ \sum_{i<j} (-1)^{i+j+1} \omega([\alpha_i, \alpha_j], \alpha_0, \dots, \widehat{\alpha}_i, \dots, \widehat{\alpha}_j, \dots, \alpha_k) . \quad (1)$$

The cohomology of the complex $(\Omega^\bullet(A), d)$ is the **Lie algebroid cohomology** of A (with trivial coefficients), and is denoted $H^\bullet(A)$.

2.1 The Chern-Weil Construction

Let us recall briefly the Chern-Weil construction for a Lie algebroid A (see [8, 5]). In the case $A = TM$ we recover the usual construction.

Given a vector bundle $E \to M$ we will consider the E-valued A-forms:

$$\Omega^\bullet(A; E) = \Omega^\bullet(A) \otimes \Gamma(E) .$$

An **A-connection** on E is a linear operator $\nabla : \Omega^0(A; E) \to \Omega^1(A; E)$, satisfying the Leibniz identity

$$\nabla_\alpha(fs) = f\nabla_\alpha s + \#\alpha(f)s .$$

It has a unique extension to an operator

$$d_\nabla : \Omega^\bullet(A; E) \to \Omega^{\bullet+1}(A; E) ,$$

also satisfying the Leibniz identity. Explicitly

$$d_\nabla\omega(\alpha_0, \dots, \alpha_k) = \sum_{i=0}^{k+1} (-1)^i \nabla_{\alpha_i}(\omega(\alpha_0, \dots, \widehat{\alpha}_i, \dots, \alpha_k))$$

$$+ \sum_{i<j} (-1)^{i+j+1} \omega([\alpha_i, \alpha_j], \alpha_0, \dots, \widehat{\alpha}_i, \dots, \widehat{\alpha}_j, \dots, \alpha_k) . \quad (2)$$

This will be a differential provided the curvature

$$R_\nabla(\alpha, \beta) = \nabla_\alpha \nabla_\beta - \nabla_\beta \nabla_\alpha - \nabla_{[\alpha,\beta]} ,$$

vanishes. If such is the case, we obtain the **Lie algebroid cohomology with coefficients** in E, denoted $H^\bullet(A; E)$. In general, the curvature will not vanish, but it will satisfy the Bianchi identity

$$d_\nabla R_\nabla = 0 , \quad (3)$$

where on $\mathrm{End}(E)$ we take the induced connection from E.

Now the usual trace on $\mathrm{End}(E)$ induces a **trace**

$$\mathrm{Tr} : (\Omega^\bullet(A; \mathrm{End}(E)), d_\nabla) \to (\Omega^\bullet(A), d) ,$$

which satisfies $d\,\mathrm{Tr} = \mathrm{Tr}\,d_\nabla$. Hence, we can define the **Chern characters** by setting

$$\mathrm{ch}_k(\nabla) = \mathrm{Tr}(R_\nabla^k) \in \Omega^{2k}(A) . \tag{4}$$

and we have:

Lemma 1. *The Chern characters* $\mathrm{ch}_k(\nabla)$ *are closed A-forms.*

A basic fact is that the cohomology class $[\mathrm{ch}_k(\nabla)] \in H^{2k}(A)$ does not depend on the connection. This can be seen through the Chern-Simons construction, which we also recall briefly in the context of Lie algebroids.

Let $\nabla_0, \dots, \nabla_l$ be A-connections on E. Also let

$$\Delta^l = \left\{ (t_0, \dots, t_l) : t_i \geq 0, \sum_{i=0}^l t_i = 1 \right\},$$

be the standard l-simplex, and denote by $p : M \times \Delta^l \to M$ the projection on the first factor. Then both E and A can be pull-backed to $M \times \Delta^l$ and $p^* A$ had a natural Lie algebroid structure. We can define a $p^* A$-connection on $p^* E$ by forming the affine combination:

$$\nabla^{\mathrm{aff}} = \sum_{i=0}^l t_i \nabla_i .$$

The classical integration along the fibers has also an analogue:

$$\int_{\Delta^l} : \Omega^\bullet(p^* A) \to \Omega^{\bullet-l}(A),$$

which is given explicitly by the formula:

$$\left(\int_{\Delta^l} \omega \right)(\alpha_1, \dots, \alpha_{n-l}) = \int_{\Delta^l} \omega \left(\frac{\partial}{\partial t_1}, \dots, \frac{\partial}{\partial t_l}, \alpha_1, \dots, \alpha_{n-l} \right) dt_1 \dots dt_l .$$

We can now define the **Chern-Simons** transgression by

$$\mathrm{cs}_k(\nabla_0, \dots, \nabla_l) = \int_{\Delta^l} \mathrm{ch}_k(\nabla^{\mathrm{aff}}) . \tag{5}$$

With the convention that for $l = 0$ we set $\mathrm{cs}_k(\nabla) = \mathrm{ch}_k(\nabla)$, we obtain the following lemma:

Lemma 2. *The Chern-Simons transgressions satisfy:*

$$d\,\mathrm{cs}_k(\nabla_0, \dots, \nabla_l) = \sum_{i=0}^l (-1)^i \mathrm{cs}_k(\nabla_0, \dots, \widehat{\nabla}_i, \dots, \nabla_l) . \tag{6}$$

The proof is a simple application of integration by parts. We conclude that for any vector bundle E we have a well defined **Chern character** $\mathrm{ch}(A; E)$, with components the cohomology classes of $\mathrm{ch}_k(\nabla)$, for any choice of A-connection ∇ on E. Whenever there is no confusion, we shall abbreviate the Chern character to $\mathrm{ch}(E)$.

2.2 Characteristic Classes of a Representation

Let E be a representation of a Lie algebroid A. This just means that E is a flat vector bundle for an A-connection ∇. From the previous paragraph we conclude immediately that:

Corollary 1. *For a representation E of A we have* $\mathrm{ch}(E) = 0$.

The vanishing of the Chern character of a representation is the origin of new *secondary characteristic classes*. These characteristic classes of a representation where first introduced in [5].

From now on, unless otherwise stated, we assume that E is a *complex* vector bundle. Let us choose some metric g on E. The connection ∇ induces an adjoint connection ∇^g on E, which is defined in the usual way:

$$\#\alpha(g(s_1, s_2)) = g(\nabla_\alpha s_1, s_2) + g(s_1, \nabla_\alpha s_2) .$$

We leave it to reader the easy check that:

Lemma 3. *Let* ∇, ∇_0, ∇_1 *be connections on E. For any metric g:*

$$\mathrm{ch}_k(\nabla^g) = (-1)^k \overline{\mathrm{ch}_k(\nabla)} ,$$
$$\mathrm{cs}_k(\nabla_0^g, \nabla_1^g) = (-1)^k \overline{\mathrm{cs}_k(\nabla_0, \nabla_1)}.$$

For any representation E we fix a metric g on E and we define elements

$$u_{2k-1}(E, \nabla) = i^{k+1} \, \mathrm{cs}_k(\nabla, \nabla^g) \in \Omega^{2k-1}(A) ,$$

where i denotes the imaginary unit.

Proposition 1. *The A-forms u_{2k-1} are real, closed, and their cohomology class is independent of the metric.*

Proof. By (6) and the previous lemma, we find

$$d \, \mathrm{cs}_k(\nabla, \nabla^g) = \mathrm{ch}_k(\nabla) - \mathrm{ch}_k(\nabla^g)$$
$$= \mathrm{ch}_k(\nabla) - (-1)^k \overline{\mathrm{ch}_k(\nabla)} .$$

Sine ∇ is flat, both terms vanish. Therefore, the u_{2k-1} are closed, and they are also real by the previous lemma.

If g and g' are metrics on E, formula (6) gives:

$$\mathrm{cs}_k(\nabla, \nabla^g) - \mathrm{cs}_k(\nabla, \nabla^{g'}) = \mathrm{cs}_k(\nabla^g, \nabla^{g'}) - \mathrm{d\,cs}_k(\nabla, \nabla^g, \nabla^{g'}) \,.$$

Therefore, u_{2k-1} will be independent of the metric provided the A-form $\mathrm{cs}_k(\nabla^g, \nabla^{g'})$ is exact. For that we choose a family g_t of metrics joining $g = g_0$ to $g' = g_1$, and define $u_t \in \mathrm{End}(E)$ by $g_t(e_1, e_2) = g(u_t(e_1), e_2)$. Then

$$\mathrm{cs}_k(\nabla^g, \nabla^{g_t}) = \mathrm{Tr}(\omega_t^{2k-1}),$$

and all we need to show is that $\frac{\partial}{\partial t}\mathrm{Tr}(\omega_t^{2k-1})$ is an exact form. A simple computation shows that

$$\frac{\partial \omega_t}{\partial t} = \mathrm{d}_{\nabla^g}(v_t) + [\omega_t, v_t],$$

where $v_t = u_t^{-1}\frac{\partial u_t}{\partial t}$. Since $\mathrm{d}_{\nabla^g}(\omega_t^2) = 0$, this implies that:

$$\frac{\partial \omega_t}{\partial t}\omega_t^{2k-2} = \mathrm{d}_{\nabla^g}(v_t\omega_t^{2k-2}) + [\omega_t, v_t\omega_t^{2k-2}] \,.$$

Now, by the properties of the trace, it follows that

$$\frac{\partial}{\partial t}\mathrm{Tr}(\omega_t^{2k-1}) = \mathrm{d\,Tr}(v_t\omega_t^{2k-2}) \,,$$

as desired.   □

We can now introduce:

Definition 1. *The **characteristic classes of a representation** E are the cohomology classes*

$$u_{2k-1}(E) = [u_{2k-1}(E, \nabla)] \quad (k = 1, \dots, r),$$

where r is the rank of E.

Notice that, if E admits an invariant metric g, then these classes vanish, so they can be seen as obstructions to the existence of an invariant metric. The main properties of these classes are:

(i) $u_{2k-1}(E \oplus F) = u_{2k-1}(E) + u_{2k-1}(F)$;
(ii) $u_{2k-1}(E \otimes F) = \mathrm{rank}(E)u_{2k-1}(F) + \mathrm{rank}(F)u_{2k-1}(E)$;
(iii) $u_{2k-1}(E^*) = -u_{2k-1}(E)$.

We refer to [5] for a proof of these facts. They can also be summarized by saying that the map $\mathrm{Rep}(A) \to \mathbb{Z} \times H^{\mathrm{odd}}(A)$ defined by

$$E \mapsto (\mathrm{rank}(E), u_1(E), \dots, u_{2r-1}(E)) \,,$$

is a morphism of *-semi-rings.

Let us consider now the case of a *real* vector bundle. For these we have:

Proposition 2. *Assume that (E, ∇) is a real representation of A. If k is even, then $u_{2k-1}(E, \nabla) = 0$. If k is odd, then for any metric connection ∇_m, the differential form*

$$(-1)^{\frac{k+1}{2}} \operatorname{cs}_k(\nabla_0, \nabla_m) \in \Omega^{2k-1}(A)$$

is closed, and its cohomology class equals $\frac{1}{2} u_{2k-1}(E, \nabla)$.

Proof. Let ∇_m be a metric connection for some metric g, so that $\nabla_m^g = \nabla_m$. From Lemma 3, we find

$$\operatorname{cs}_k(\nabla_m, \nabla^g) = (-1)^k \operatorname{cs}_k(\nabla_m^g, \nabla) = (-1)^{k+1} \operatorname{cs}_k(\nabla, \nabla_m) .$$

This, combined with the transgression formula (6), implies

$$\begin{aligned}
d \operatorname{cs}_k(\nabla, \nabla_m, \nabla^g) &= \operatorname{cs}_k(\nabla_m, \nabla^g) - \operatorname{cs}_k(\nabla, \nabla^g) + \operatorname{cs}_k(\nabla, \nabla_m) \\
&= (1 + (-1)^{k+1}) \operatorname{cs}_k(\nabla, \nabla_m) - \operatorname{cs}_k(\nabla, \nabla^g) ,
\end{aligned}$$

which proves the proposition. $\square$

2.3 Intrinsic Secondary Characteristic Classes

In order to motivate the introduction of these characteristic classes let us start by looking at the special case of regular Lie algebroids.

Let A be a regular Lie algebroid so that the characteristic foliation $\mathcal{F}$ integrating Im $\#$ is non-singular. Denote the normal bundle by $\nu = TM/T\mathcal{F}$, and the kernel of the anchor by $K = \operatorname{Ker} \#$. These are both vector bundles, since A is regular, and they carry canonical flat connections, namely the *Bott connections*:

$$\nabla_\alpha \beta = [\alpha, \beta], \quad \beta \in \Gamma(K) , \tag{7}$$

$$\nabla_\alpha \overline{X} = \overline{\mathcal{L}_{\#\alpha} X}, \quad \overline{X} \in \Gamma(\nu) . \tag{8}$$

This means that we can define intrinsic secondary characteristic classes of A, by letting:

$$u_{2k-1}(A) = u_{2k-1}(K) - u_{2k-1}(\nu) .$$

Notice that the origin of these secondary classes is the vanishing of the Chern character of the formal difference $K - \nu$. Now observe that we have the following short exact sequences of vector bundles:

$$0 \longrightarrow K \longrightarrow A \longrightarrow T\mathcal{F} \longrightarrow 0 ,$$

$$0 \longrightarrow T\mathcal{F} \longrightarrow TM \longrightarrow \nu \longrightarrow 0 .$$

Hence, the difference $K - \nu$ equals the difference $A - TM$, and we have the following:

Corollary 2. *For any regular Lie algebroid,* $\mathrm{ch}(A - TM) = 0$.

Let us now turn to the non-regular case. While the difference $K - \nu$ only makes sense for regular Lie algebroids, the difference $A - TM$ always makes sense. Also, we can introduce A-connections on A and TM, which are not flat, but which give rise to a flat connection on the formal difference. These naturally extend Bott's basic connections for foliations (see [3]).

Definition 2. *A connection* $(\hat{\nabla}, \check{\nabla})$ *on* $A \oplus T^*M$ *is called a* basic connection *if* $\#\hat{\nabla} = \check{\nabla}\#$ *and if they restrict to the Bott connections on each leaf L of the characteristic foliation* $\mathcal{F}$.

Notice that what is left in the formal difference $A - TM$ is precisely the Bott part of the basic connection. The vanishing of the corollary above is now replaced by the following result (see [8]):

Lemma 4. *The curvature R of a basic connection vanishes along* $K \oplus (T\mathcal{F})^0$.

A simple procedure to obtain basic connections is as follows. One chooses an ordinary connection $\bar{\nabla}$ on A, and defines A-connections on A and on TM by the formulas:

$$\hat{\nabla}_\alpha \beta = \bar{\nabla}_{\#\beta}\alpha + [\alpha, \beta], \quad \check{\nabla}_\alpha X = \bar{\nabla}_X \alpha + [\#\alpha, X] . \tag{9}$$

One checks readily that the pair $\nabla = (\hat{\nabla}, \check{\nabla})$ is a basic connection.

Now we can define our characteristic classes. We pick a basic connection ∇ and a metric connection ∇_{m} (i.e., ∇_{m} preserves some metric g on $A \oplus T^*M$).

Definition 3. *The* **intrinsic characteristic classes** *of A are the cohomology classes*

$$u_{2k-1}(\mathrm{Ad}\, A) = 2 \left[(-1)^{\frac{k+1}{2}} \mathrm{cs}_k(\nabla, \nabla_m) \right] \in H^{2k-1}(A) ,$$

where $1 \leq 2k - 1 \leq 2r - 1$, *and k is an odd integer.*

The fact that these classes are well-defined and independent of any choices, is similar to the proof of the same fact for the characteristic classes of representations, given in the previous paragraph. We refer to [8] for details.

The notation $u_{2k-1}(\mathrm{Ad}\, A)$ suggests that these are the characteristic classes of the adjoint representation of A. To which extend this is true, is the main subject of this paper, and will be discussed is the next sections. Before we do that, we look at the intrinsic characteristic class of lowest degree.

Example 1. The modular class of a Lie algebroid was introduced in [14], and further discussed in [7, 9, 15]. We recall here the construction given in [7]. Consider the line bundle $Q_A = \wedge^r A \otimes \wedge^m T^*M$. On this line bundle we have a flat A-connection ∇ defined by:

$$\nabla_\alpha(\alpha^1 \wedge \cdots \wedge \alpha^r \otimes \mu) = \sum_{j=1}^{r} \alpha^1 \wedge \cdots \wedge [\alpha, \alpha^j] \wedge \cdots \wedge \alpha^r \otimes \mu +$$

$$\alpha^1 \wedge \cdots \wedge \alpha^r \otimes \mathcal{L}_{\#\alpha}\mu , \quad (10)$$

whenever $\alpha, \alpha^1, \ldots, \alpha^r \in \Gamma(A)$ and $\mu \in \Gamma(\wedge^m T^*M)$.

Assume first that Q_A is trivial. Then we have a global section $s \in \Gamma(Q_A)$ so that

$$\nabla_\alpha s = \theta_s(\alpha)s, \qquad \forall \alpha \in \Gamma(A) .$$

Since ∇ is flat, we see that θ_s defines a section of $\Gamma(A^*)$ which is closed: $d\theta_s = 0$. If s' is another global section in $\Gamma(Q_A)$, we have $s' = fs$ for some non-vanishing smooth function f on M, and we find

$$\theta_{s'} = \theta_s + d\log|f| .$$

Therefore, we have a well defined cohomology class

$$\mathrm{mod}(A) \equiv [\theta_s] \in H^1(A)$$

which is independent of the section s. If the line bundle Q_A is not trivial one considers the square $L = Q_A \otimes Q_A$, which is trivial, and defines

$$\mathrm{mod}(A) = \frac{1}{2}[\theta_s] ,$$

for some global section $s \in \Gamma(L)$. The class $\mathrm{mod}(A)$ is called the **modular class** of the Lie algebroid A.

Now we have proved in [5, 8] that

$$u_1(\mathrm{Ad}\, A) = \frac{1}{2\pi} \mathrm{mod}(A) . \quad (11)$$

This gives a geometric interpretation of $u_1(\mathrm{Ad}\, A)$ as an obstruction class. In fact, as was argued in [7] one can think of global sections of Q_A (or $Q_A \otimes Q_A$) as "transverse measures" to A. The modular class is trivial iff there exists a transverse measure which is invariant under the flows of every section $\alpha \in \Gamma(A)$. Therefore, the modular class (i.e., the class $u_1(\mathrm{Ad}\, A)$) is an obstruction lying in the first Lie algebroid cohomology group $H^1(A)$ to the existence of a transverse invariant measure to A.

3 Characteristic Classes and Connections up to Homotopy

3.1 Non-Linear Connections

As we have mentioned before, for a general Lie algebroid, there is no adjoint representation. One way around this difficulty is to relax the notion of connection and allow for more general connections.

Let $\pi : A \to M$ be a Lie algebroid and let $E = E^1 \oplus E^0$ be a super-vector bundle over M. We consider $\mathbb{R}$-linear operators

$$\Gamma(A) \otimes \Gamma(E) \to \Gamma(E) , \quad (\alpha, s) \mapsto \nabla_\alpha s ,$$

which satisfy the identity

$$\nabla_\alpha(f\mathbf{s}) = f\nabla_\alpha s + \#\alpha(f)s ,$$

for all $f \in C^\infty(M)$, and preserve the grading. We will say that ∇ is a **non-linear connection** if ∇_α is local in α. This is a relaxation of the $C^\infty(M)$-linearity that one usually requires.

A **non-linear differential form** is an anti-symmetric, $\mathbb{R}$-multilinear map

$$\omega : \Gamma(A) \times \cdots \times \Gamma(A) \to C^\infty(M) ,$$

which is local. Many of the usual operations on forms don't use $C^\infty(M)$-linearity. For example, we have a de Rham operator $d : \Omega_{\mathrm{nl}}^\bullet(A) \to \Omega_{\mathrm{nl}}^\bullet + 1(A)$. We can also consider E-values non-linear forms, which we denote by $\Omega_{\mathrm{nl}}^\bullet(A; E)$.

The Chern-Weil and all other constructions of Sect. 2.1 immediately generalize to non-linear connections provided we use non-linear forms. For example, the usual super-trace on $\mathrm{End}(E)$ induces a **super-trace**

$$\mathrm{Tr} : (\Omega_{\mathrm{nl}}^\bullet(A; \mathrm{End}(E)), d_\nabla) \to (\Omega_{\mathrm{nl}}^\bullet(A), d) ,$$

and we obtain the **Chern characters** of the non-linear connection

$$\mathrm{ch}_k(\nabla) = \mathrm{Tr}(R_\nabla^k) \in \Omega^{2k}(A) .$$

As before, the Chern characters $\mathrm{ch}_k(\nabla)$ are closed, non-linear A-forms, and up to a boundary these classes do not depend on the connection. This, of course, is because the Chern-Simons construction also generalizes to this setting, giving a non-linear version of the **Chern-Simons transgressions** $\mathrm{cs}_k(\nabla^0, \ldots, \nabla^l)$ which still satisfy equation (5), which is now a equality between non-linear forms.

From now on, we let (E, ∂) be a super-complex of vector bundles over the manifold M,

$$(E, \partial) : \quad E^0 \underset{\partial}{\overset{\partial}{\rightleftarrows}} E^1 . \tag{12}$$

We consider also a non-linear connection ∇ on E such that $\nabla_\alpha \partial = \partial \nabla_\alpha$ for all $\alpha \in \Gamma(A)$. The notion of **connection up to homotopy** [7] on (E, ∂) is obtained by requiring linearity up to homotopy. In other words we require that

$$\nabla_{f\alpha} s = f\nabla_\alpha s + [H_\nabla(f, \alpha), \partial] ,$$

where $H_\nabla(f, \alpha) \in \Gamma(\mathrm{End}(E))$ are odd elements which are $\mathbb{R}$-linear and local in α and f.

We say that two non-linear connections ∇ and ∇' are **equivalent** (or homotopic) if, for all $\alpha \in \Gamma(A)$, we have

$$\nabla'_\alpha = \nabla_\alpha + [\theta(\alpha), \partial] \, ,$$

for some $\theta \in \Omega^1_{\mathrm{nl}}(A; \mathrm{End}(E))$ of odd degree. In this case, we write $\nabla \sim \nabla'$. There are two basic properties of this equivalence relation which we state as our next two lemmas.

Lemma 5. *A non-linear connection is a connection up to homotopy if and only if it is equivalent to a (linear) connection.*

Proof. Assume that ∇ is a connection up to homotopy. Let U_a be the domain of local coordinates x^k for M over which the bundle A trivializes, and denote by $\{e_1, \ldots, e_r\}$ a basis of local sections. We define a local linear connection

$$\nabla^a_X = \nabla_X + [u^a(X), \partial] \, ,$$

where $u_a \in \Omega^1_{\mathrm{nl}}(A|_{U_a}; \mathrm{End}(E))$ is given by

$$u_a \left(\sum_k f_k e_k \right) = - \sum_k H_\nabla(f_k, e_k) \, ,$$

for all $f_k \in C^\infty(U_a)$. Next we take $\{\phi_a\}$ to be a partition of unity subordinate to an open cover $\{U_a\}$ by such coordinate domains and set

$$\nabla'_\alpha = \sum_a \phi_a \nabla^a_\alpha \, , \quad u(\alpha) = \sum_a \phi_a u^a(\alpha) \, .$$

Then $\nabla' = \nabla + [u, \partial]$ is a connection equivalent to ∇. $\square$

Lemma 6. *If ∇^0 and ∇^1 are equivalent connections, then $\mathrm{ch}(\nabla^0) = \mathrm{ch}(\nabla^1)$.*

Proof. Let ∇^0 and ∇^1 be connections such that $\nabla^1 = \nabla^0 + [\theta, \partial]$. A simple computation shows that

$$R_{\nabla^1} = R_{\nabla^0} + [d_{\nabla^0}\theta + Q, \partial] \, , \tag{13}$$

where $Q(\alpha, \beta) = [\theta(\alpha), [\theta(\beta), \partial]]$. Let us denote by $Z \subset \Omega^\bullet_{\mathrm{nl}}(A; \mathrm{End}(E))$ the space of non-linear forms ω with the property that $[\omega, \partial] = 0$, and by $B \subset Z$ the subspace consisting of element of the form $[\eta, \partial]$ for some non-linear form η. Since we have

$$[\partial, \omega\eta] = [\partial, \omega]\eta + (-1)^{|\omega|}\omega[\partial, \eta] \, ,$$

we see that $ZB \subset B$, hence (13) implies that $R^k_{\nabla^1} \equiv R^k_{\nabla^0}$ modulo B. The desired equality follows now from the fact that Tr vanishes on B. $\square$

Observe now that if ∇ is a *linear* connections on (E, ∂), then the Chern characters $\mathrm{ch}_k(\nabla)$ are *linear* differential forms, whose cohomology classes are the components of the Chern character $\mathrm{ch}(E) = \mathrm{ch}(E^0) - \mathrm{ch}(E^1)$. Hence, an immediate consequence of the previous two lemmas is the following:

Proposition 3. *If ∇ is a connection up to homotopy on (E, ∂), then*

$$\mathrm{ch}_k(\nabla) = \mathrm{Tr}(R_\nabla^k),$$

are closed differential forms whose cohomology classes are the components of the Chern character $\mathrm{ch}(E)$.

The Chern-Simons transgressions forms of non-linear connections are non-linear forms, and they satisfy the obvious properties. We list here the relevant properties:

Lemma 7. *Let ∇, ∇_0, ∇_1 be non-linear connections. Then:*

(i) *If ∇_0 and ∇_1 are connections up to homotopy then $\mathrm{cs}_k(\nabla_0, \nabla_1)$ are linear differential forms;*
(ii) *If $\nabla_0 \sim \nabla_1$, then $\mathrm{cs}_k(\nabla_0, \nabla_1) = 0$;*
(iii) *For any metric g:*

$$\mathrm{ch}_k(\nabla^g) = (-1)^k \overline{\mathrm{ch}_k(\nabla)} \quad and \quad \mathrm{cs}_k(\nabla_0^g, \nabla_1^g) = (-1)^k \overline{\mathrm{cs}_k(\nabla_0, \nabla_1)}\,.$$

Proof. Part (i) follows from the fact that Chern characters of connections up to homotopy are differential forms.

For part (ii) we observe that the affine combination ∇^{aff} used in the definition of $\mathrm{cs}_k(\nabla_0, \nabla_1)$ is equivalent to the pull-back $\tilde{\nabla}_0$ of ∇_0 to $M \times \Delta^1$, since $\nabla^{\mathrm{aff}} = \tilde{\nabla}_0 + t[\theta, \partial]$. But $\mathrm{ch}_k(\tilde{\nabla}_0)$ vanishes so, by Lemma 6, we conclude that $\mathrm{cs}_k(\nabla_0, \nabla_1) = \mathrm{ch}_k(\nabla^{\mathrm{aff}}) = 0$.

Finally, if g is a metric on E, a simple computation shows that $R_{\nabla^g} = -R_\nabla^*$, where $*$ denotes the adjoint (with respect to g). Then (iii) follows from the equality $\mathrm{Tr}(C^*) = \overline{\mathrm{Tr}(C)}$, for any matrix C. $\square$

3.2 Characteristic Classes of Representations up to Homotopy

A **representation up to homotopy** is a super-vector bundle (E, ∂) with a connection up to homotopy which is flat. We are now ready to extend the construction of the characteristic classes of representations to representations up to homotopy. Again, the origin of these classes is the following vanishing result, which is an immediate consequence of Proposition 3.

Corollary 3. *If (E, ∂) is a representation up to homotopy, then $\mathrm{ch}(E) = 0$.*

For any representation up to homotopy (E, ∂, ∇), we choose a metric g on E and we consider the forms:

$$u_{2k-1}(E, \partial, \nabla) = i^{k+1} \, \mathrm{cs}(\nabla, \nabla^g) \in \Omega^{2k-1}(A)\,.$$

Proposition 4. *Let* (E, ∂, ∇) *be a flat representation up to homotopy. Then:*

(i) *The differential forms* $u_{2k-1}(E, \partial, \nabla)$ *are real and closed, and the induced cohomology classes do not depend on the choice of the metric.*

(ii) *If* $\nabla \sim \nabla'$*, then* $u_{2k-1}(E, \partial, \nabla) = u_{2k-1}(E, \partial, \nabla')$*.*

(iii) *If* ∇ *is equivalent to a metric connection (i.e., a connection which is compatible with a metric), then all the classes* $u_{2k-1}(E, \partial, \nabla)$ *vanish.*

Proof. If we use Proposition 3 and Lemma 7, the proof of (i) is analogous to the proof of Proposition 1, and so we omit it. To prove (ii), we observe that the non-linear version of the transgression formula (5) gives, for any connection ∇_0,

$$\operatorname{d} \operatorname{cs}_k(\nabla, \nabla_0, \nabla^g) = \operatorname{cs}_k(\nabla_0, \nabla^g) - \operatorname{cs}_k(\nabla, \nabla^g) + \operatorname{cs}_k(\nabla, \nabla_0)$$
$$\operatorname{d} \operatorname{cs}_k(\nabla^g, \nabla_0, \nabla_0^g) = \operatorname{cs}_k(\nabla_0, \nabla_0^g) - \operatorname{cs}_k(\nabla^g, \nabla_0^g) + \operatorname{cs}_k(\nabla^g, \nabla_0) \,.$$

Adding up these two relations, we conclude that the class $u_{2k-1}(E, \partial, \nabla)$ equals the cohomology class of

$$\operatorname{i}^k \left(\operatorname{cs}_k(\nabla, \nabla_0) + \operatorname{cs}_k(\nabla_0, \nabla_0^g) + \operatorname{cs}_k(\nabla_0^g, \nabla) \right),$$

for any connection ∇_0. On the other hand, if ∇' is equivalent to ∇, Lemma 7 (ii) gives:

$$\operatorname{cs}_k(\nabla, \nabla_0) - \operatorname{cs}_k(\nabla', \nabla_0) = \operatorname{d} \operatorname{cs}_k(\nabla, \nabla', \nabla_0) \,.$$

Hence, we conclude that

$$\begin{aligned} u_{2k-1}(E, \partial, \nabla) &= \operatorname{i}^k \left[\operatorname{cs}_k(\nabla, \nabla_0) + \operatorname{cs}_k(\nabla_0, \nabla_0^g) + \operatorname{cs}_k(\nabla_0^g, \nabla) \right] \\ &= \operatorname{i}^k \left[\operatorname{cs}_k(\nabla', \nabla_0) + \operatorname{cs}_k(\nabla_0, \nabla_0^g) + \operatorname{cs}_k(\nabla_0^g, \nabla') \right] \\ &= u_{2k-1}(E, \partial, \nabla') \,, \end{aligned}$$

which proves (ii).

Finally, (iii) follows from (i) and (ii). $\quad\square$

In the real case, we obtain the analogue of Proposition 2. The proof is entirely similar.

Proposition 5. *Assume that* E *is a real vector bundle. If* k *is even then* $u_{2k-1}(E, \partial, \nabla) = 0$*. If* k *is odd, then for any connection* ∇_0 *equivalent to* ∇*, and any metric connection* ∇_m*,*

$$(-1)^{\frac{k+1}{2}} \operatorname{cs}_k(\nabla_0, \nabla_m) \in \Omega^{2k-1}(A)$$

is a closed differential form whose cohomology class equals $\frac{1}{2} u_{2k-1}(E, \partial, \nabla)$*.*

In this way we have extended the theory of secondary characteristic classes of representations to representations up to homotopy. Note that the construction presented here actually works for connections which are *flat up to homotopy*, i.e., whose curvature forms are of the type $[-, \partial]$. Moreover, this notion is stable under equivalence, and the characteristic classes will only depend on the equivalence class of ∇ (cf. Proposition 4 (ii)).

Note also that, as in [6] (and following [2]), there is a version of our discussion for super-connections ([13]) up to homotopy. Some of our constructions can then be interpreted in terms of the super-connection $\partial + \nabla$.

If E is regular in the sense that $\operatorname{Ker} \partial$ and $\operatorname{Im} \partial$ are vector bundles, then so is the cohomology $H^\bullet(E, \partial) = \operatorname{Ker} \partial / \operatorname{Im} \partial$, and any connection up to homotopy ∇ on (E, ∂) defines a linear connection on $H^\bullet(E, \partial)$. Moreover, this connection is flat if ∇ is, and the characteristic classes $u_{2k-1}(E, \partial, \nabla)$ coincide with the classical characteristic classes of the flat vector bundle $H^\bullet(E, \partial)$ (see [2, 10]). In general, the classes $u_{2k-1}(E, \partial, \nabla)$ should be viewed as invariants of $H^\bullet(E, \partial)$ constructed in such a way that no regularity assumption is required.

3.3 Intrinsic Classes via Representations up to Homotopy

Let us turn now to the adjoint representation of a Lie algebroid A. The case of a regular Lie algebroid, considered at the start of Sect. 2.3, suggests that one should look at the formal difference $A - TM$. This can be made precise, by working up to homotopy. We consider the super-vector bundle

$$\operatorname{Ad}(A): \ A \underset{\#}{\overset{0}{\rightleftarrows}} TM \ , \tag{14}$$

with the flat connection up to homotopy ∇^{ad} given by:

$$\nabla_\alpha^{\mathrm{ad}} \beta = [\alpha, \beta] \ , \qquad \nabla_\alpha^{\mathrm{ad}} X = [\#\alpha, X] \ ,$$

for which the homotopies are $H(f, \alpha)(\beta) = 0$ and $H(f, \alpha)(X) = X(f)\alpha$, for all $\alpha, \beta \in \Gamma(A)$, $X \in \mathfrak{X}(M)$, $f \in C^\infty(M)$.

The following result shows that the characteristic classes of the adjoint representation up to homotopy, as in the previous paragraph, coincide with the intrinsic characteristic classes we have discussed in Sect. 2.3.

Theorem 1. *For any Lie algebroid A and any k:*

$$u_{2k-1}(\operatorname{Ad} A) = u_{2k-1}(A, \partial, \nabla^{\mathrm{ad}}) \ .$$

Proof. The clue is the following proposition relating the basic connections we have discussed in Sect. 2.3, to the adjoint connection.

Proposition 6. *If a linear connection ∇ on A is equivalent to the adjoint connection ∇^{ad} then it is a basic connection.*

Assuming the proposition holds, by Lemma 5 there exists a basic connection ∇ equivalent to ∇^{ad}. Fixing also a metric connection ∇_m, we find:

$$u_{2k-1}(\operatorname{Ad} A) = 2[(-1)^{\frac{k+1}{2}} \operatorname{cs}_k(\nabla, \nabla_m)] \qquad \text{(by Definition 3)},$$

$$= u_{2k-1}(A, \partial, \nabla^{\mathrm{ad}}) \qquad \text{(by Proposition 5)},$$

which proves the theorem.

Proof of Proposition 6. The condition that ∇ is equivalent to ∇^{ad} means that there exists some $\theta \in \Omega^1_{\mathrm{nl}}(A; \operatorname{End}(A \oplus TM))$ of odd degree such that:

$$\nabla_\alpha = \nabla^{\mathrm{ad}}_\alpha + [\theta(\alpha), \partial]$$

for all $\alpha \in \Gamma(A)$. Writing $\nabla = (\hat{\nabla}, \check{\nabla})$, this condition translates into:

$$\hat{\nabla}_\alpha \beta = [\alpha, \beta] + \theta(\alpha)\#\beta,$$
$$\check{\nabla}_\alpha X = [\#\alpha, X] + \#\theta(\alpha)X.$$

If we restrict, over a leaf L of A, the first connection to $\operatorname{Ker} \#|_L$ and the second connection to $\nu(L)$, the terms involving θ vanish. Therefore, both $\hat{\nabla}$ and $\check{\nabla}$ restrict over a leaf to the Bott connections. On the other hand, we compute:

$$\#\hat{\nabla}_\alpha \beta = \#[\alpha, \beta] + \#\theta(\alpha)\#\beta$$
$$= [\#\alpha, \#\beta] + \#\theta(\alpha)\#\beta = \check{\nabla}_\alpha \#\beta.$$

Hence, $\nabla = (\hat{\nabla}, \check{\nabla})$ is a basic connection. $\square$

Notice that Proposition 6 gives some further geometric insight to the notion of a basic connection. Moreover, in the regular case, it is easy to check that a linear connection is basic iff it is equivalent to the adjoint connection.

4 Jets and Characteristic Classes

In the previous section we saw that the adjoint representation is a representation only up to homotopy, and we used this fact to show how the intrinsic classed can be seen as classes of representations. In this section, we consider a different interpretation of the adjoint representation, as a honest representation. The price to pay is that we have to work on the jet Lie algebroid.

4.1 The Jet of a Lie Algebroid

Let us explain that for any Lie algebroid A, each jet bundle $J^k A$ inherits a natural Lie algebroid structure. This construction of the *jet Lie algebroid* can be traced back to the works of Kumpera, Libermann and Spencer (see [1] and references thereof).

Let $\pi : E \to M$ be a vector bundle. For each non-negative integer k, we will denote by $\pi^k : J^k E \to M$ the vector bundle of k-order jets of sections of E. If $l < k$, we denote by $\pi^k_l : J^k E \to J^l E$ the canonical projection. Since $J^0 E = E$ we have $\pi^k_0 = \pi^k$. If $\alpha \in \Gamma(E)$ is a section of E, we will denote by $j^k \alpha$ the induced section of $J^k E$. For basic facts on jet bundles we refer the reader to [12]. Although we will assume that $0 \leq k < \infty$, many constructions below hold if $k = \infty$.

Our first observation is the following:

Proposition 7. *If A is a Lie algebroid, there exists a unique Lie algebroid structure on $J^k A$ such that:*

(i) For any section $\alpha \in \Gamma(A)$ the anchors are related by:

$$\# j^k \alpha = \# \alpha \ .$$

(ii) For any sections $\alpha, \beta \in \Gamma(A)$ the Lie brackets are related by

$$[j^k \alpha, j^k \alpha] = j^k ([\alpha, \beta]) \ .$$

Proof. First we prove uniqueness. Property (i) clearly defines (uniquely) the anchor as the composition $\# \circ \pi_k$, where $\pi_k : J^k A \to A$ is the canonical projection. So let $[\ ,\ ]$ be a Lie algebroid bracket on $J^k A$ with anchor $\# = \# \circ \pi_k$. The sections of $J^k A$ are generated over $C^\infty(M)$ by sections of the form $j^k \alpha$, where $\alpha \in \Gamma(A)$. Using the Leibniz identity, we obtain:

$$[g_1 j^k \alpha_1, g_2 j^k \alpha_2] = g_1 g_2 [j^k \alpha_1, j^k \alpha_2] +$$
$$+ g_1 \# \alpha_1 (g_2) j^k \alpha_2 - g_2 \# \alpha_2 (g_1) j^k \alpha_1 \ . \quad (15)$$

This shows that, if (ii) also holds, then $[\ ,\ ]$ is uniquely determined.

Since uniqueness holds, it remains to show every point $x \in M$ has a neighborhood U where such a Lie bracket exists. So let $(x^1, \ldots, x^m)$ be local coordinates on a open set $U \subset M$, over which the bundle A trivializes. Let $\{e_1, \ldots, e_r\}$ be a basis of sections of $A|_U$. For a multi-index $I = (i_1, \ldots, i_m)$ of non-negative integers, we set $|I| = i_1 + \cdots + i_m$ and denote by x^I the monomial $(x^1)^{i_1} \cdots (x^m)^{i_r}$. The sections defined by

$$e^I_a = \frac{1}{|I|!} j^k (x^I e_a) \ , \quad 1 \leq a \leq m, |I| \leq k \ ,$$

form a generating set of sections for $J^k A|_U$. A basis can be obtained by considering, for example, multi-indices I with $i_1 \leq i_2 \leq \ldots i_m$. Now we can define a Lie bracket satisfying (ii), by setting

$$[e^I_a, e^J_b] = \frac{1}{|I|! |J|!} j^k ([x^I e_a, x^J e_b]),$$

and requiring the Leibniz identity to hold. $\square$

For the jet adjoint representation, to be introduced in the next section, we will be interested in the case of $J^1 A$. So let us give the local expression for the structure constants of the jet Lie algebroid $J^1 A$. Let $(x^1, \ldots, x^m)$ be local coordinates on a open set $U \subset M$, over which the bundle A trivializes. Let $\{e_1, \ldots, e_r\}$ be a basis of sections of $A|_U$. The Lie algebroid A has structure functions B_a^i and C_{ab}^c defined by

$$\#e_a = B_a^i \frac{\partial}{\partial x^i}, \qquad [e_a, e_b] = C_{ab}^c e_c,$$

where we have used the convention of summing over repeated indices. Now, we have an induced basis $\{e_a, e_a^i\}$ of $J^1 A$ so that, for every local section $s \in \Gamma(A)$,

$$j^1 s(x) = s^a(x) e_a + \frac{\partial s^a}{\partial x^i}(x) e_a^i.$$

Explicitly, the section e_a^i is given by:

$$y \mapsto j^1 ((x^i - y^i) e_a)|_{x=y}.$$

A straightforward computation using properties (i) and (ii) of Proposition 7, gives the structure functions for $J^1 A$:

$$[e_a, e_b] = C_{ab}^c e_c + \frac{\partial C_{ab}^c}{\partial x^i} e_a^i, \tag{16}$$

$$[e_a, e_b^i] = C_{ab}^c e_c^i + \frac{\partial B_a^i}{\partial x^j} e_b^j, \tag{17}$$

$$[e_a^i, e_b^j] = B_a^j e_b^i - B_b^i e_a^j. \tag{18}$$

There are similar formulas for the higher jet Lie algebroids $J^k A$.

Note that the characteristic foliations of A and $J^k A$ coincide. Also, it is easy to check that the Lie algebroid structure on $J^k A$ makes the projections $\pi_l^k : J^k A \to J^l A$ into Lie algebroid homomorphisms.

The operation of taking jets is functorial: if $\phi : A_1 \to A_2$ is a Lie algebroid homomorphism then $j^k \phi : J^k A_1 \to J^k A_2$ is also a homomorphism of Lie algebroids, and we have a commutative diagram:

$$
\begin{array}{ccc}
J^k A_1 & \xrightarrow{j^k \phi} & J^k A_2 \\
\pi_l^k \downarrow & & \downarrow \pi_l^k \\
J^l A_1 & \xrightarrow{j^l \phi} & J^l A_2
\end{array}
$$

If a Lie algebroid $A \to M$ integrates to a Lie groupoid $\mathcal{G} \rightrightarrows M$, then the jet Lie algebroid $J^k A$ integrates to the *jet Lie groupoid* $J^k \mathcal{G} \rightrightarrows M$: the arrows of $J^k \mathcal{G}$ are the k-order jets of bisections of $\mathcal{G}$, and the operations are the obvious ones. This groupoid structure, makes the natural projection $\pi_l^k :$

$J^k\mathcal{G} \to J^l\mathcal{G}$ into a Lie groupoid homomorphism, and the map of bisections $j^k : \mathcal{B}(\mathcal{G}) \to \mathcal{B}(J^k\mathcal{G})$ into a group homomorphism. Also, if $\Phi : \mathcal{G}_1 \to \mathcal{G}_2$ is a Lie groupoid homomorphism, then it induces a Lie groupoid homomorphism $j^k\Phi : J^k\mathcal{G}_1 \to J^k\mathcal{G}_2$ of the associated jet groupoids and, for each pair of indices, we have a commutative diagram:

$$
\begin{array}{ccc}
J^k\mathcal{G}_1 & \xrightarrow{\;j^k\Phi\;} & J^k\mathcal{G}_2 \\
\pi_l^k \downarrow & & \downarrow \pi_l^k \\
J^l\mathcal{G}_1 & \xrightarrow{\;j^l\Phi\;} & J^l\mathcal{G}_2
\end{array}
$$

4.2 The Jet Adjoint Representation

We saw above that the adjoint representation of a Lie algebroid A is a representation only up to homotopy. It turns out that we can also view the adjoint representation A as a honest representation of J^1A.

Proposition 8. *There is a unique representation ∇ of J^1A on the bundle A, such that for any sections $\alpha, \beta \in \Gamma(A)$:*

$$
\nabla_{j^1\alpha}\beta = [\alpha, \beta] \ . \tag{19}
$$

Proof. First we observe that there exists at most one connection satisfying (19). In fact, the sections of J^1A are generated over $C^\infty(M)$ by sections of the form $j^1\alpha$, where $\alpha \in \Gamma(A)$. Hence, any J^1A-connection ∇ is determined by its values on sections of this form. In particular, if ∇ satisfies (19), we find:

$$
\nabla_{gj^1\alpha}\beta = g[\alpha, \beta] \ ,
$$

so uniqueness holds.

Since uniqueness holds, existence will follow if we show that every point $x \in M$ has a neighborhood U where there exists a connection satisfying (19). Again, we let $(x^1, \ldots, x^m)$ be local coordinates on a open set $U \subset M$, over which the bundle A trivializes, and we let $\{e_1, \ldots, e_r\}$ be a basis of sections of $A|_U$. Using the notation above, we define a connection in $A|_U$ by

$$
\nabla_{e_a}e_b = C_{ab}^c e_c \ , \quad \nabla_{e_a^i}e_b = -B_b^i e_a \ .
$$

This connection clearly satisfies (19).

Finally, this connection is flat, since we have:

$$
R(j^1\alpha, j^1\beta)\gamma = [\alpha, [\beta, \gamma]] - [\beta, [\alpha, \gamma]] - [[\alpha, \beta], \gamma] = 0 \ ,
$$

which implies that $R \equiv 0$. $\square$

4.3 Intrinsic Classes via Jet Representations

For the jet adjoint representation ∇^{j^1}, which is a representation of the Lie algebroid J^1A on A, we can take its characteristic classes (see Sect. 2.2), which we denote by

$$u_{2k-1}(A, \nabla^{j^1}) \in \Omega^{2k-1}(J^1A) .$$

Now, the Lie algebroid morphism $\pi^1 : J^1A \to A$ determines a pull-back map $(\pi^1)^* : \Omega^\bullet(A) \to \Omega^\bullet(J^1A)$, which preserves differentials. Hence, we also have a map at the level of cohomology:

$$(\pi^1)^* : H^\bullet(A) \to H^\bullet(J^1A) .$$

We have:

Theorem 2. *The intrinsic characteristic classes of A pull-back to the characteristic classes of the jet adjoint representation:*

$$u_{2k-1}(A, \nabla^{j^1}) = (\pi^1)^* u_{2k-1}(\operatorname{Ad} A) .$$

Proof. For a section α of J^1A we will denote by $\pi^1_*\alpha$ the induced section of A. If ∇ is a A-connection on a vector bundle E, then we have a pull-back J^1A-connection on E, denoted $(\pi^1)^*\nabla$, and which is defined by the formula

$$(\pi^1)^*\nabla_\alpha s = \nabla_{\pi^1_*\alpha} s ,$$

for any sections $\alpha \in \Gamma(J^1A)$ and $s \in \Gamma(E)$. If we twist the connection ∇ by a metric g in E, then its pull-back is the twisted pull-back connection:

$$(\pi^1)^*\nabla^g = ((\pi^1)^*\nabla)^g .$$

Also, it follows from the definitions of the Chern-Simons transgressions, that we have:

$$\operatorname{cs}_k((\pi^1)^*\nabla^0, \ldots, (\pi^1)^*\nabla^l) = (\pi^1)^* \operatorname{cs}_k(\nabla^0, \ldots, \nabla^l) .$$

Now, all this is still true even for non-linear connections. For example, the adjoint connection up to homotopy ∇^{ad} pulls-back to the jet adjoint connection ∇^{j^1}:

$$\nabla^{j^1} = (\pi^1)^*\nabla^{\mathrm{ad}} .$$

Note that this example shows that a non-linear connection can pull-back to a linear connection.

These remarks immediately yield the theorem. In fact, we have:

$$
\begin{aligned}
u_{2k-1}(A, \nabla^{j^1}) &= i^{k+1} \operatorname{cs}_k(\nabla^{j^1}, (\nabla^{j^1})^g) \\
&= i^{k+1} \operatorname{cs}_k((\pi^1)^*\nabla^{\mathrm{ad}}, ((\pi^1)^*\nabla^{\mathrm{ad}})^g) \\
&= i^{k+1} \operatorname{cs}_k((\pi^1)^*\nabla^{\mathrm{ad}}, (\pi^1)^*(\nabla^{\mathrm{ad}})^g) \\
&= (\pi^1)^*(i^{k+1} \operatorname{cs}_k(\nabla^{\mathrm{ad}}, (\nabla^{\mathrm{ad}})^g) \\
&= (\pi^1)^* u_{2k-1}(A, \partial, \nabla^{\mathrm{ad}}) = (\pi^1)^* u_{2k-1}(\operatorname{Ad} A) ,
\end{aligned}
$$

where the last equality holds by Theorem 1. $\square$

References

1. R. Almeida and A. Kumpera, Structure produit dans la catégorie des algèbroïdes de Lie, *An. Acad. Bra. Ciênc.* **53** (1981), 247–250.

2. J.M. Bismut and J. Lott, Flat vector bundles, direct images and higher real analytic torsion, *J. Amer. Math. Soc.* **8** (1995), 291–363.

3. R. Bott, *Lectures on Characteristic Classes and Foliations*, in Lectures on Algebraic and Differential Topology, Lec. Notes in Mathematics, vol. **279**, Springer-Verlag, Berlin, 1972.

4. A. Cannas da Silva and A. Weinstein, *Geometric Models for Noncommutative Algebras*, Berkeley Mathematics Lectures, vol. **10**, American Math. Soc., Providence, 1999.

5. M. Crainic, Differentiable and algebroid cohomology, Van Est isomorphisms, and characteristic classes, to appear in *Comment. Math. Helv.* (preprint *math.DG/0008064*).

6. M. Crainic, Chern characters via connections up to homotopy, Preprint *math.DG/0008064*.

7. S. Evens, J.-H. Lu and A. Weinstein, Transverse measures, the modular class and a cohomology pairing for Lie algebroids, *Quart. J. Math. Oxford* (2) **50** (1999), 417–436.

8. R.L. Fernandes, Lie algebroids, holonomy and characteristic class, *Adv. in Math.* **170** (2002), 119–179.

9. J. Huebschmann, Duality for Lie-Rinehart algebras and the modular class, *J. reine angew. Math.* **510** (1999), 103–159.

10. F. Kamber and P. Tondeur, *Foliated Bundles and Characteristic Classes*, Springer Lecture Notes in Mathematics **493**, Springer-Verlag, Berlin, 1975.

11. S. Morita, *Geometry of characteristic classes*, Translations of Mathematical Monographs, **199**, American Math. Soc., Providence, 2001.

12. D.J. Saunders, *The Geometry of Jet Bundles*, Cambridge University Press, Cambridge, 1989.

13. D. Quillen, Superconnections and the Chern character, *Topology* **24** (1985), 89–95.

14. A. Weinstein, The modular automorphism group of a Poisson manifold, *J. Geom. Phys.* **23** (1997), 379–394.

15. P. Xu, Gerstenhaber algebras and BV-algebras in Poisson geometry, *Comm. Math. Phys.* **200** (1999), 545–560.

Part III

Applications in Physics

In this part, some physical aspects of noncommutative geometry are discussed. The ideas of the field theory on the noncommutative space together with the results from the effective theory of some solitonic mode in the superstring, such as D-brain, brought remarkable developments concerning the understanding of the role of noncommutative geometry in physics. In this context, the Seiberg-Witten map plays an important role. This is a consequence of the requirement that the field theory, especially the Yang-Mills theory on noncommutative space, is a deformation of the original theory on commutative space. It means that the freedom of the fields appearing in the theory is the same in both cases.

The first lecture describes an approach based on the Seiberg-Witten map, which leads to a systematic construction of the non-commutative gauge theories with general gauge symmetry. It is also shown how this construction leads to gauge theories with an arbitrary gauge group.

The second lecture contains the mathematical aspects of the Seiberg-Witten map and its application is used to clearify the structure of noncommutative line bundles.

The third contribution is a self-contained lecture about 2-dimensional noncommutative gauge theory and exact solutions on it. It contains the quantum gauge theory on the noncommutative torus, its vacuum amplitude, and the classification of instanton contributions. A new solution of gauge theory on a two-dimensional fuzzy torus is also presented.

Gauge Theories on Noncommutative Spacetime Treated by the Seiberg-Witten Method*

J. Wess

Sektion Physik der Ludwig-Maximillians-Universität, Theresienstr. 37, 80333 München, Germany, and Max-Planck-Institut für Physik (Werner-Heisenberg-Institut), Föhringer Ring 6, 80805 München, Germany

The idea of noncommutative coordinates (NCC) is almost as old as quantum field theory (QFT) itself. It was W.Heisenberg who proposed NCC in 1930 in a letter to Peierls [1]. He expressed the hope that uncertainty relations of the coordinates, derived from NCC, might provide a natural cut-off for divergent integrals in QFT. This idea propagated via W. Pauli, R. Oppenheimer and Oppenheimer's student H. S. Snyder [2]. He then published the first analysis of a quantum thoery on NCC. Pauli [3] called this work mathematically ingenious but rejected it for reasons of physics, arguing that an effective cut-off would act like a universal length and thus lead to strange consequences for large momenta of order $\hbar/l_0$.

The mathematical success of quantum groups and quantum spaces as they were introduced by V. G. Drinfeld, L. Faddeev, M. Jimbo and Y. I. Manin [4] brought back to physics the interest in NCC. The question arose if QFT can be deformed to a QFT on NCC that we shall call NCQFT (noncommutative quantum field theory), similar to the deformation of groups to quantum groups.

At the same time NCC and NCQFT appeared in string theory [5] and received an appreciable interest there.

In this lecture I will try to give some systematic approach to the construction of NCQFT and NCGT (noncommutative gauge theories) based on the Seiberg Witten map [6]. This approach was followed by our research group in Munich [7], [8]. It leads to the construction of gauge theories with an arbitrary gauge group. Starting from an ordinary gauge theory, e.g. the Standard Model, in ordinary commuting space the formalism that is going to be developed leads to a deformed gauge theory (NCGT) with the same numbers of degrees of freedom (fields) in the same multiplets of the gauge group as the original theory. The theory can be expanded in a parameter that characterizes the noncommutativity and this parameter enters the dynamics as

---

* Based on the lectures given at the *International Workshop on Quantum Field Theory and Noncommutative Geometry*, November 26–30 2002, Tohoku University, Sendai, Japan.

J. Wess: *Gauge Theories on Noncommutative Spacetime Treated by the Seiberg-Witten Method*, Lect. Notes Phys. **662**, 179–192 (2005)
www.springerlink.com

a new coupling. It has consequences for the renormalizability of the theory and contributes effectively to the phenomenology.

1 The Algebra

The noncommutativity of the space variables is best formulated in terms of commutators

$$[\hat{x}^\mu, \hat{x}^\nu] = i\theta^{\mu\nu}(\hat{x}) , \qquad \mu, \nu = 1, \ldots N .$$

Various conditions like the Jacobi identity restrict the choice of θ. We shall give examples now.

1. The commutative case:

$$[x^\mu, x^\nu] = 0$$

These are the commuting coordinates, we omit the hat in this case.

2. The canonical case:

$$[\hat{x}^\mu, \hat{x}^\nu] = i\theta^{\mu\nu}$$

$\theta^{\mu\nu}$ is $\hat{x}$-independent. The commutation relations are the same as in canonical quantum mechanics. Here they are valid for the coordinate space by itself. This is the structure of noncommutativity that arose in string theory first. The more detailed constructions shall only be discussed for this case.

3. The Lie algebra case:

$$[\hat{x}^\mu, \hat{x}^\nu] = i\theta^{\mu\nu}_\rho \hat{x}^\rho$$

The right-hand side is linear in $\hat{x}$. The first model by H.S.Snyder was of this type [2]. Among these algebras there is the interesting case of the κ-quantum space. This space has a quantum group, the κ-Lorentz group, as a symmetry structure [9], [10]. Its relations are

$$[\hat{x}^\mu, \hat{x}^\nu] = i \left(a^\mu \eta_\rho{}^\nu - a^\nu \eta_\rho{}^\mu \right) \hat{x}^\rho,$$

where a^μ is like a structure constant of a Lie algebra.

4. The quantum spaces:

$$[\hat{x}^\mu, \hat{x}^\nu] = i\theta^{\mu\nu}_{\rho\sigma} \hat{x}^\rho \hat{x}^\sigma$$

The right-hand side is quadratic in $\hat{x}$. This case has been thoroughly investigated in the context of quantum groups. For consistency it has to be of the form

$$\hat{x}^\mu \hat{x}^\nu = \hat{R}^{\mu\nu}_{\rho\sigma} \hat{x}^\rho \hat{x}^\sigma ,$$

where $\hat{R}$ is an $\hat{R}$-matrix satisfying the Yang-Baxter equation [11].

To construct the algebra of NCC we proceed as follows. We first consider the algebra freely generated by $\hat{x}^1, \ldots \hat{x}^N$. In this algebra the relations generate a two-sided ideal $I_{\mathcal{R}}$. We just have to multiply the relations by arbitrary elements of the freely generated algebra from the right and from the left. Finally we factor out this ideal from the freely generated algebra and are left with the coordinate algebra

$$\hat{\mathcal{A}}_{\hat{x}} = [[\hat{x}^1, \ldots \hat{x}^N]]/I_{\mathcal{R}} .$$

We allow formal power series in $\hat{\mathcal{A}}_{\hat{x}}$. We assume that dynamical variables of the field theory to be constructed are elements of this algebra:

$$\hat{\phi}(\hat{x}) \in \hat{\mathcal{A}}_{\hat{x}} .$$

2 The ⋆-Product

Measurements produce real numbers. Theories should do as well. The standard way that we learned from quantum mechanics how to relate an algebra to numbers is to construct Hilbert space representations. This can be done for the algebras we have listed in the previous chapter but it is an unsolved problem how to construct representations where the dyamical variables are represented by (essentially) selfadjoint operators [12].

Fortunately there is another way how to relate the elements of the algebra $\hat{\mathcal{A}}_{\hat{x}}$ to complex valued functions of commuting variables

$$\hat{\phi}(\hat{x}) \sim \phi(x) .$$

For the algebra $\hat{\mathcal{A}}_{\hat{x}}$ we can construct a basis in terms of the homogeneous polynomials. They form a vector space of finite dimension. It turns out that all the examples we studied have the Poincare-Birkhoff-Witt property, that is the dimension of the above mentioned finite-dimensional vector space is exactly the same as for commuting variables. Being finite-dimensional these vector spaces are isomorphic as vector spaces. Through the grading in homogeneous polynomials of fixed degree we obtain a vector space isomorphism

$$\hat{\phi}(\hat{x}) \sim \phi(x)$$

by postulating

$$\hat{\phi}(\hat{x}) = \sum_{r=0}^{\infty} C^r_{j_1 \ldots j_r} : \hat{x}^{j_1} \ldots \hat{x}^{j_r} :$$

$$\rightarrow \phi(x) = \sum_{r=0}^{\infty} C^r_{j_1 \ldots j_r} \, x^{j_1} \ldots x^{j_r} .$$

The dots in the first equation mean that we have chosen a definite ordering in the basis of $\hat{A}_{\hat{x}}$. Complete symmetrization is a possibility, it is frequently used, but any other ordering could be used as well, the explicit isomorphism of course depends on the ordering chosen. We shall work with the symmetric ordering if not stated otherwise.

The vector space isomorphism can be extended to an algebra isomorphism by the following construction: We take two elements $\hat{\phi}$ and $\hat{\psi}$ of the algebra $\hat{A}_{\hat{x}}$, multiply them and expand the product in the chosen (symmetric) basis

$$\hat{\phi}\hat{\psi} = \sum_{r=0}^{\infty} d^{(r)}_{j_1 \dots j_r} : \hat{x}^{j_1} \dots \hat{x}^{j_r} : .$$

The product of the functions ϕ and ψ (they are functions of commuting variables) we denote by a star and define it

$$\phi \star \psi = \sum_{r=0}^{\infty} d^{(r)}_{j_1 \dots j_r} x^{j_1} \dots x^{j_r} .$$

Thus, we have constructed an algebra isomorphism

$$\hat{\phi} \sim \phi ,$$
$$\hat{\phi}\hat{\psi} \sim \phi \star \psi .$$

This $\star$-product will not be commutative, it encodes all the noncommutative properties of the NCC. As an example:

$$x^\mu \star x^\nu - x^\nu \star x^\mu = i\theta^{\mu\nu}$$

where $\theta^{\mu\nu}$ corresponds to the respective algebra.

For the canonical case the $\star$-product with symmetrized basis is just the Moyal-Weyl product

$$\phi \star_c \psi(x) = e^{\frac{i}{2}\frac{\partial}{\partial x^\mu}\theta^{\mu\nu}\frac{\partial}{\partial y^\nu}}\phi(x)\psi(y)|_{y\to x} .$$

For the Lie algebra case we can write down a closed version of the $\star$-product in the symmetrized basis. For this purpose we have to use the Baker-Campbell-Hausdorff formula

$$e^{i\alpha_\mu \hat{x}^\mu}e^{i\beta_\nu \hat{x}^\nu} = e^{i(\alpha_\rho + \beta_\rho + \gamma_\rho(\alpha,\beta))\hat{x}^\rho} .$$

We obtain

$$\phi \star \psi(x) = e^{\frac{1}{2}x^\rho \gamma_\rho(i\frac{\partial}{\partial y},i\frac{\partial}{\partial z})}\phi(y)\psi(z)|_{z\to y\to x} .$$

Here the exponent depends linear on x.

In the quantum plane case we do not yet know a closed form for the $\star$-product. It can be computed in a power series expansion in q.

For the Manin plane

$$x^1 x^2 = q x^2 x^1$$

we can give the product explicitly

$$\phi \star_M \psi(x) = q^{\frac{1}{2}(-x'^1 \frac{\partial}{\partial x'^1} x^2 \frac{\partial}{\partial x^2} + x^1 \frac{\partial}{\partial x^1} x'^2 \frac{\partial}{\partial x'^2})} \phi(x^1, x^2) \psi(x'^1, x'^2)\big|_{\substack{x'^1 \to x^1 \\ x'^2 \to x^2}} .$$

Here the exponent is quadratic in the coordinates x.

The $\star$-product is the NCC approach to deformation quantization as it was developed by Flato and Sternheimer [13]. In the deformation approach it is difficult to prove the associativity of the algebra, in the NCC approach it is difficult to prove the Poincaré-Birkhoff-Witt property.

The physically relevant objects like fields will be represented by the functions $\phi(x)$ that depend on commuting variables. We can, however, multiply them only with the product that in general introduces a coordinate dependent differential operator. It can be expanded in the parameters that characterize the noncommutativity.

3 Gauge Theories

It is possible to realize gauge transformations on NCC. We first define a gauge theory by its Lie group and its action on a differentiable manifold. In addition we introduce a vector potential with transformation properties that allow the construction of a covariant derivative on the differentiable manifold:

Lie algebra:

$$[T^a, T^b] = i f^{ab}_c T^c$$

Gauge transformations:

$$\delta_\alpha \psi(x) = i\alpha(x)\psi(x)$$

where $\alpha(x)$ is Lie algebra valued,

$$\alpha(x) = T^a \alpha_a(x),$$

such that

$$(\delta_\alpha \delta_\beta - \delta_\beta \delta_\alpha)\psi(x) = [\alpha(x), \beta(x)]\,\psi(x) = \delta_{\alpha \times \beta}\,\psi(x)$$

and

$$\alpha \times \beta = i\alpha_a \beta_b f^{ab}_c T^c.$$

The vector potential is Lie algebra valued

$$a_\mu(x) = a_{\mu\,b}(x)T^b \,,$$
$$\delta a_\mu(x) = \partial_\mu \alpha + i[\alpha(x), a_\mu(x)] \,.$$

These are the ingredients for a gauge theory as we use them in physics. Now we are going to show that they are sufficient for the construction of a gauge theory on NCC as well.

We define gauge transformations, using the $\star$-product formalism for the NCC

$$\delta_\alpha \psi(x) = i\Lambda_\alpha(x, a_\mu(x)) \star \psi(x) \,.$$

But now the transformation parameter is enveloping algebra valued and depends on the vector field. This transforms under the gauge transformations. These are new concepts in gauge theories. If we accept them we are able to define the transformation parameters Λ_α such that

$$(\delta_\alpha \delta_\beta - \delta_\beta \delta_\alpha)\psi(x) = \delta_{\alpha \times \beta}\ \psi(x).$$

The gauge parameter Λ_α is enveloping algebra valued

$$\Lambda_\alpha(x) = \alpha_a(x)T^a(x) + \alpha_{ab}(x, a_\mu) : T^a T^b :$$
$$+ \ldots \alpha_{a_1 \ldots a_n}(x, a_\mu) : T^{a_1} \ldots T^{a_n} : + \ldots \,.$$

We assume that Λ depends on higher products of the generators and we have chosen a basis. For the linear term in T we asume that the coefficient is just the transformation parameter α, no additional dependence on a_μ is assumed.

Lie algebra and enveloping algebra are two representation-independent concepts.

Taking into account the dependence of Λ_α on the vector field a_μ we obtain

$$(\delta_\alpha \delta_\beta - \delta_\beta \delta_\alpha)\psi(x) = (\Lambda_\alpha \star \Lambda_\beta - \Lambda_\beta \star \Lambda_\alpha) \star \psi(x) + i(\delta_\alpha \Lambda_\beta - \delta_\beta \star \Lambda_\alpha) \star \psi(x)$$
$$= i\Lambda_{\alpha \times \beta} \star \psi.$$

The variation $\delta_\alpha \Lambda_\beta$ refers to the a_μ-dependence of Λ_β. We are going to show that parameters Λ_α can be constructed such that

$$\Lambda_\alpha \star \Lambda_\beta - \Lambda_\beta \star \Lambda_\alpha + i(\delta_\alpha \Lambda_\beta - \delta_\beta \star \Lambda_\alpha) = i\Lambda_{\alpha \times \beta} \,.$$

We construct Λ_α in a power series expansion in the parameter that characterizes the noncommutativity. To facilitate the notation we introduce a parameter h there:

$$[\hat{x}^\mu, \hat{x}^\nu] = ih\theta^{\mu\nu}(\hat{x})$$

and we expand in h:

$$\Lambda_{\alpha, a_1 \ldots a_n} = \sum_l h^l \Lambda^l_{\alpha, a_1, \ldots a_n} \,.$$

For $\Lambda^0_{\alpha,a}$ we have made the choice already

$$\Lambda^0_{\alpha,a} = \alpha_a$$
$$\Lambda^0_{\alpha,a_1,\ldots a_N} = 0 \qquad n > 1.$$

We have to expand the $\star$-product as well. There we restrict ourselves to the canonical case. In the following we shall work out the formalism for this case only, the generalization to the other cases is possible [10]

$$f \star_C g = f(x)g(x) + \frac{i}{2}h\theta^{\mu\nu}\partial_\mu f(x)\partial_\nu g(x)$$
$$- \frac{h^2}{8}\theta^{\mu\nu}\theta^{\rho\sigma}\partial_\mu\partial_\rho f(x)\partial_\nu\partial_\sigma g(x) + \ldots .$$

The zeroth order in the expansion in h reproduces the usual gauge transformations. In first order we obtain

$$(\delta_\alpha\Lambda^1_\beta - \delta_\beta\Lambda^1_\alpha) - i[\alpha, \Lambda^1_\beta] + i[\beta, \Lambda^1_\alpha] + \frac{i}{2}h\theta^{\mu\nu}\partial_\mu\alpha_a\partial_\nu\beta_b : T^aT^b : = \Lambda^1_{\alpha\times\beta} .$$

This is an inhomogeneous linear equation for Λ^1. The inhomogeneous term is known as a function of α, β

$$\frac{h}{2}\theta^{\mu\nu}\partial_\mu\alpha_a\partial_\nu\beta_b : T^aT^b : .$$

A particular solution can be found

$$\Lambda^1_\alpha = \frac{h}{2}\theta^{\mu\nu}\partial_\mu\alpha_a a_{\nu\,b} : T^aT^b : .$$

Solutions of the homogeneous equation can be found and added to the particular solution.

We can proceed order by order n h, the structure of the equation will always be the same. It will be an inhomogeneous linear equation, the homogeneous part remains the same, the inhomogeneous part will contain known quantities only. Such a construction of the transformation parameter first occurred in the context of the Seiberg-Witten map. A very systematic investigation has been carried out in [7].

The transformation property of a field

$$\delta_\alpha\psi = i\Lambda_\alpha \star \psi$$

can be expanded in h as well. It turns out that ψ can be found as a function of ψ^0 and the ordinary gauge field a_μ

$$\delta_\alpha\psi^0(x) = i\alpha(x)\psi^0(x) ,$$

such that ψ transforms as above by transforming a_μ as well. To first order in h for constant θ

$$\psi = \psi^0 - \frac{1}{2}\theta^{\mu\nu}a_{\mu,b}T^b\partial_\nu\psi^0 + \dots .$$

This allows us to express all the matter fields in a gauge theory in terms of the usual matter fields ψ^0. No new matter fields have to be added. The field ψ^0 transforms in the representation of the T^a used for the construction of Λ_α.

4 Covariant Coordinates

To proceed with a gauge theory a vector potential has to be introduced and a gauge invariant action for this vector potential has to be formulated. This is usually done through the concept of a covariant derivative. Derivatives, however, are not a very natural and general concept for algebras.

It is obvious that coordinates do not commute with gauge transformations for NCC, but covariant coordinates can be introduced in analogy to covariant derivatives

$$X^\mu = x^\mu + A^\mu(x)$$
$$\delta_\alpha\psi = i\Lambda_\alpha \star \psi$$
$$\delta_\alpha X^\mu \star \psi = i\Lambda_\alpha \star X^\mu \star \psi .$$

This leads to the following transformation law of the vector potential

$$\delta_\alpha A^\mu = -i[x^\mu \stackrel{\star}{,} \Lambda_\alpha] + i[\Lambda_\alpha \stackrel{\star}{,} A^\mu].$$

To satisfy such a transformation law we have again to assume that A^μ is enveloping algebra valued. This, at first sight, seems to introduce infinitely many gauge fileds. Analogous to the matter fields, however, it is possible to express the enveloping algebra valued vector potential A^μ in terms of the usual vector potential a^μ and its derivatives. This is the main achievement of the Seiberg-Witten map. Once more, there is an expression for A^μ in terms of a^μ and its derivatives such that A^μ has the above transformation property as a consequence of the well known transformation property of a^μ.

For the canonical case such an expression is

$$A^\mu(x) = \theta^{\mu\nu}a_\nu - \frac{1}{2}\theta^{\mu\nu}\theta^{\rho\sigma}a_{\rho,b}(\partial_\sigma a_{\nu,c} + F_{\sigma\nu,c}) : T^bT^c : + \dots$$
$$F_{\sigma\nu,d} = \partial_\sigma a_{\nu,d} - \partial_\nu a_{\sigma,d} + f_d^{ce}a_{\sigma,c}a_{\nu,e} .$$

A direct calculation shows that the above statement about the transformation law of A^μ is correct. A^μ starts with a linear term in θ. For $\theta \to 0$ A^μ vanishes, the coordinates commute and are covariant.

The construction of tensor fields, the so called field strengths, follows the usual concept. We define

$$\tilde{F}^{\mu\nu} = X^\mu \star X^\nu - X^\nu \star X^\mu - i\theta^{\mu\nu}$$

and obtain for $\tilde{F}^{\mu\nu}$ the transformation properties

$$\delta_\alpha \tilde{F}^{\mu\nu} = [\Lambda_\alpha \overset{\star}{,} \tilde{F}^{\mu\nu}] .$$

The trace on the representation space of the Lie algebra of such a tensor is not an invariant because the product is not commutative. An invariant action can only be defined if we succeed in defining an integral with a trace property

$$\int f \star g = \int g \star f .$$

Integration, however, is not a natural concept for an algebra. For the examples we have listed such integrals can be defined, for constant θ it is the usual integral that has the trace property [14].

5 Derivatives

Derivatives are, as was mentioned above, not a very natural concept for algebras. To define them we first abstract the algebraic properties of the derivatives of commuting variables and then try to deform these to NCC.
    The properties

$$\frac{\partial}{\partial x^l} x^m = \delta_l^m + x^m \frac{\partial}{\partial x^l}$$

and

$$\frac{\partial}{\partial x^l} f(x) g(x) = \left(\frac{\partial}{\partial x^l} f(x) \right) g(x) + f(x) \left(\frac{\partial}{\partial x^l} g(x) \right)$$

can be seen purely algebraically if we define the derivatives as maps of the algebra $\mathcal{A}_{\hat{x}}$. Derivatives are deformed to maps of the algebra $\hat{\mathcal{A}}_{\hat{x}}$ and as such have to be consistent with the relations. To mimic the first condition we make an Ansatz

$$\hat{\partial}_\rho \hat{x}^\mu = \delta_\rho^\mu + \hat{O}_{\rho\sigma}^{\mu\nu} \hat{x}^\sigma \hat{\partial}_\nu ,$$

where $\hat{O}_{\rho\sigma}^{\mu\nu}$ is a system of numbers that are determined by the commutation relations of the space variables and the corresponding consistency relations.
    For the canonical case we find consistency for

$$\hat{\partial}_\mu \hat{x}^\nu = \delta_\mu^\nu + \hat{x}^\nu \hat{\partial}_\mu ,$$

$$\hat{\partial}_\rho(\hat{x}^\mu\hat{x}^\nu - \hat{x}^\nu\hat{x}^\mu - i\theta^{\mu\nu}) = \delta_\rho^\mu\hat{x}^\nu + \hat{x}^\mu\delta_\rho^\nu - \delta_\rho^\nu\hat{x}^\mu$$
$$-\delta_\rho^\mu\hat{x}^\nu + (\hat{x}^\mu\hat{x}^\nu - \hat{x}^\nu\hat{x}^\mu - i\theta^{\mu\nu})\hat{\partial}_\rho$$
$$= (\hat{x}^\mu\hat{x}^\nu - \hat{x}^\nu\hat{x}^\mu - i\theta^{\mu\nu})\hat{\partial}_\rho.$$

This proves consistency. Obviously, the relation

$$\hat{\partial}_\mu\hat{\partial}_\nu - \hat{\partial}_\nu\hat{\partial}_\mu = 0$$

is consistent as well. From the derivatives we can compute the Leibniz rule

$$\hat{\partial}_\mu\hat{f}(\hat{x})\hat{g}(\hat{x}) = (\hat{\partial}_\mu\hat{f})\hat{g} + \hat{f}(\hat{\partial}_\mu\hat{g}).$$

For the κ-Lorentz case the situation is less trivial. To facilitate the calculation we put the vector a^μ in the nth direction

$$[\hat{x}^i, \hat{x}^j] = 0 \quad i, j = 0, \ldots n - 1$$
$$[\hat{x}^n, \hat{x}^i] = ia\hat{x}^i.$$

We make the Ansatz from above and find as a possible solution

$$[\hat{\partial}_n, \hat{x}^\mu] = \delta_n{}^\mu,$$
$$[\hat{\partial}_i, \hat{x}^\mu] = \delta_i{}^\mu - ia\delta_n^\mu\hat{\partial}_i$$

and

$$\hat{\partial}_\mu\hat{\partial}_\nu - \hat{\partial}_\nu\hat{\partial}_\mu = 0 .$$

This is not the only solution but it is the solution that is compatible with the κ-Lorentz group, as we shall see later. Again, we derive the nontrivially deformed Leibniz rule

$$\hat{\partial}_n(\hat{f} \cdot \hat{g}) = (\hat{\partial}_n\hat{f}) \cdot \hat{g} + \hat{f} \cdot \hat{\partial}_n\hat{g} ,$$
$$\hat{\partial}_i(\hat{f} \cdot \hat{g}) = (\hat{\partial}_i\hat{f}) \cdot \hat{g} + (e^{ia\hat{\partial}_n}\hat{f}) \cdot \hat{\partial}_i\hat{g} .$$

For the quantum group case the derivatives have been studied extensively, results can be found in [15].

The Leibniz rule can also be interpreted as a comultiplication rule for the derivative operator:

Commuting case:

$$\Delta\hat{\partial}_\mu = \hat{\partial}_\mu \otimes 1 + 1 \otimes \hat{\partial}_\mu$$

In the product fg the first factor acts on f, the second acts on g.

κ-Poincaré case:

$$\Delta\hat{\partial}_n = \hat{\partial}_n \otimes 1 + 1 \otimes \hat{\partial}_n ,$$
$$\Delta\hat{\partial}_i = \hat{\partial}_i \otimes 1 + e^{ia\hat{\partial}_n} \otimes \hat{\partial}_i .$$

The derivative map can be expressed in the $\star$-product formulation as well. This has been done systematically in [9].

A further development of a κ-deformed field theory in the $\star$-formalism can be found in [10].

6 Deformed Symmetries

Quantum spaces have been introduced as representation spaces of quantum groups. These are deformations of symmetry groups into the category of Hopf algebras. The question arises if NCC can carry such a Hopf algebra structure as well. For the canonical case (constant $\theta^{\mu\nu}$) no such structure is known. For the Lie algebra case there is the example of the κ-deformation. The coordinate space introduced in the previous chapter carries a Hopf algebra that we shall call κ-Euclidean or, in a modification, κ-Lorentz group. We always assume that the deformation vector a^μ points into the space like direction x^n. The relevant formulas should demonstrate the concept.

There are generators $M^{\mu\nu}$, defined by the commutation relations

$$[M^{ij}, \hat{x}^\mu] = \eta^{\mu j}\hat{x}^i - \eta^{\mu i}\hat{x}^j,$$
$$[M^{in}, \hat{x}^\mu] = \eta^{\mu n}\hat{x}^i - \eta^{\mu i}\hat{x}^n + iaM^{i\mu}.$$

The term $iaM^{i\mu}$ makes these commutation relations different from the corresponding Euclidean relations. It is necessary for the consistency with the commutation relations of the coordinates. Making use of the above $M, \hat{x}$ commutators we compute

$$[M^{\mu n}, [\hat{x}^n, \hat{x}^s]] = [M^{\mu n}, ia\hat{x}^s].$$

This guarantees the consistency

$$[M^{\mu n}, [\hat{x}^n, \hat{x}^s] - ia\hat{x}^s] = 0.$$

When we now calculate the commutator of the generators M we find as a specific solution

$$[M^{\mu\nu}, M^{\rho\sigma}] = \eta^{\mu\sigma}M^{\nu\rho} + \eta^{\nu\rho}M^{\mu\sigma} - \eta^{\mu\rho}M^{\nu\sigma} - \eta^{\nu\sigma}M^{\mu\rho}.$$

They are a-independent and they are exactly the defining relations of the Euclidean or Lorentz algebra. The algebraic relations are not deformed, but the comultiplication rules are when acting on functions of $\hat{x}$

$$\Delta M^{ij} = M^{ij} \otimes \mathbf{1} + \mathbf{1} \otimes M^{ij},$$
$$\Delta M^{in} = M^{in} \otimes \mathbf{1} + e^{ia\hat{\partial}_n} \otimes M^{in} + ia\hat{\partial}_k \otimes M^{ik}.$$

As a Hopf algebra, the Euclidean or Lorentz algebra is deformed.

For the derivatives we find a transformation property that is consistent with all the $\hat{x}, \hat{x}$; $\hat{x}, \hat{\partial}$ and $\hat{\partial}, \hat{\partial}$ commutation relations

$$[M^{rs}, \hat{\partial}_\mu] = \eta^s{}_\mu \hat{\partial}^r - \eta^r{}_\mu \hat{\partial}^s,$$
$$[M^{rn}, \hat{\partial}_n] = \hat{\partial}^l,$$
$$[M^{rn}, \hat{\partial}_i] = \eta^r{}_i \frac{e^{2ia\hat{\partial}_n} - 1}{2ia} - \frac{ia}{2}\eta^r{}_i \sum_{j=1}^{n-1} \hat{\partial}^j\hat{\partial}_j + ia\hat{\partial}^r\hat{\partial}_i.$$

The comultiplication rules for the M-generators do not change.

A κ-deformed D'Alembert operator exists as well. It is defined by the property

$$[M^{\mu\nu}, \Box] = 0 \,, \quad \Box \overset{a\to 0}{\to} \Box.$$

The corresponding operator is

$$\Box = e^{-ia\hat{\partial}_n} \hat{\partial}^j \hat{\partial}_j - \frac{2}{a^2}\left(1 - \cos(a\hat{\partial}_n)\right).$$

For the symmetric $\star$-product as it was introduced for the κ-deformed NCC, this operator becomes

$$\Box^\star = \frac{2}{(a\partial_n)^2}(1 - \cos(a\partial_n))\Box.$$

On the right-hand side the derivatives commute.

7 Gauge-Covariant Derivatives

Field theories are formulated in terms of fields and their derivatives. To obtain gauge invariance the derivatives have to be replaced by gauge-covariant derivatives. This leads to a coupling of matter fields to gauge fields, the so called minimal coupling.

We define gauge covariant derivatives

$$\mathcal{D}_\mu \star \psi = (D_\mu - iV_\mu) \star \psi \,,$$
$$\delta_\alpha \mathcal{D}_\mu \star \psi = i\Lambda_\alpha \star \mathcal{D}_\mu \star \psi \,.$$

This leads to the transformation law of the vector potential V_μ

$$\delta_\alpha V_\mu = \partial_\mu \Lambda_\alpha - i[V_\mu \overset{\star}{,} \Lambda_\alpha]$$

The vector potential can again be expressed in terms of the usual vector potential a_μ and its derivatives. For constant θ, the vector potential A^μ for covariant coordinates is simply related to V_μ

$$A^\mu = \theta^{\mu\nu} V_\nu \,.$$

For $\theta \to 0$ we obtain $V_\mu \to a_\mu$ and $A_\mu \to 0$. A^μ vanishes because commuting coordinates are gauge covariant.

Now we proceed as for the usual gauge theory. We will list the fomulas for constant θ only. We define

$$\mathcal{F}_{\mu\nu} \star \psi = (\mathcal{D}_\mu \star \mathcal{D}_\nu - \mathcal{D}_\nu \star \mathcal{D}_\mu) \star \psi$$

and obtain

$$\mathcal{F}_{\mu\rho} = F_{\mu\rho\ a}T^a + \theta^{\lambda\nu}(F_{\mu\lambda\ a}F_{\rho\nu\ b}$$
$$-\frac{1}{2}a_{\lambda\ a}(2\partial_\nu F_{\mu\rho\ b} + a_{\nu\ c}F_{\mu\rho\ d}f_b^{cd})) : T^aT^b : + \dots .$$

The invariant action is given by

$$W = \frac{1}{4}\int d^nx Tr\mathcal{F}_{\mu\nu} \star \mathcal{F}^{\mu\nu} .$$

This leads to additional coupling with the coupling constant $\theta^{\mu\nu}$.
As an example we show a coupling term for a $Z \to \gamma\gamma$ decay

$$\mathcal{L}_{Z\gamma\gamma} = \frac{e}{4}\sin 2\theta_W \, \mathrm{K}_{Z\gamma\gamma}\,\theta^{\rho\tau}\left[2Z^{\mu\nu}\left(2A_{\mu\rho}A_{\nu\tau} - A_{\mu\nu}A_{\rho\tau}\right)\right] .$$

Finally, we show the coupling of a Dirac field to the gauge potential

$$\int \bar{\psi} \star (\gamma^\mu D_\mu \star -m)\psi dx = \int \bar{\psi}^0(\gamma^\mu D_\mu - m)\psi^0 dx$$
$$-\frac{1}{2}\theta^{\nu\lambda}\int \bar{\psi}^0 F_{\nu\lambda}^0(\gamma^\mu D_\mu - m)\psi^0 dx$$
$$-\frac{1}{4}\theta^{\sigma\lambda}\int \bar{\psi}^0\gamma^\mu F_{\mu\sigma}^0 D_\lambda\psi^0 dx.$$

References

1. *Letter of Heisenberg to Peierls* (1930), in: Wolfgang Pauli, Scientific Correspondence, vol. II, 15, Ed. Karl von Meyenn, Springer-Verlag 1985.
2. H.S. Snyder, *Quantized space-time*, Phys. Rev. **71**, 38 (1947).
3. *Letter of Pauli to Bohr* (1947), in: Wolfgang Pauli, Scientific Correspondence, vol. II, 414, Ed. Karl von Meyenn, Springer-Verlag 1985.
4. M. Jimbo, *A q-difference analogue of U(g) and the Yang-Baxter equation*, Lett. Math. Phys. **10**, 63 (1985).
   V.G. Drinfel'd, *Hopf algebras and the quantum Yang-Baxter equation*, Sov. Math. Dokl. **32**, 254 (1985).
   L.D. Faddeev, N.Y. Reshetikhin and L.A. Takhtadzhyan, *Quantisation of Lie groups and Lie algebras*, Leningrad Math. J. **1**, 193 (1990).
   Y. I. Manin, *Multiparametric quantum deformation of the general linear supergroup*, Commun. Math. Phys. **123**, 163 (1989).
5. Chong-Sun Chu, Pei-Ming Ho, *Noncommutative Open String and D-brane*, Nucl.Phys. B550 (1999) 151–168
   V. Schomerus, *D-branes and Deformation Quantization*, JHEP 9906 (1999) 030.
6. N. Seiberg and E. Witten, *String Theory and Noncommutative Geometry*,JHEP 9909 (1999) 032.
7. J. Madore, S. Schraml, P. Schupp and J. Wess, *Gauge theory on noncommutative spaces*, Eur. Phys. J. **C16**, 161 (2000) [hep-th/0001203].

8. B. Jurčo, S. Schraml, P. Schupp and J. Wess, *Enveloping algebra valued gauge transformations for non-Abelian gauge groups on non-commutative spaces*, Eur. Phys. J. **C17**, 521 (2000) [hep-th/0006246].
   B. Jurčo, L. Möller, S. Schraml, P. Schupp and J. Wess, *Construction of non-Abelian gauge theories on noncommutative spaces*, Eur. Phys. J. **C21**, 383 (2001) [hep-th/0104153].
9. M. Dimitrijević, L. Jonke, L. Möller, E. Tsouchnika, J. Wess and M. Wohlgenannt, *Deformed field theory on κ-spacetime*, submitted to Eur. Phys. J. **C31**, 129 (2003) [hep-th/0307149].
10. M. Dimitrijević, F. Meyer, L. Möller and J. Wess, *Gauge theories on the κ-Minkowski spacetime*, submitted to Eur. Phys. J. (2003) [hep-th/0310116].
11. J. Wess, *q-Deformed Heisenberg Algebras*, [math-ph/9910013].
12. W. Weich, *The Hilbert Space Representations for $SO_q(3)$-symmetric quantum mechanics*, [hep-th/9404029].
13. F. Bayen, M. Flato, C. Fronsdal, A. Lichnerowicz and D. Sternheimer, *Deformation theory and quantization*, Ann. Phys. **111**, 61 (1978).
14. M.A. Dietz, *Symmetrische Formen auf Quantenalgebren*, Diploma thesis at the University of Hamburg (2001).
15. J. Wess and B. Zumino, *Covariant differential calculus on the quantum hyperplane*, Nucl. Phys. Proc. Suppl. **18B**, 302–312 (1991).

Noncommutative Line Bundles and Gerbes

B. Jurčo

Theoretische Physik, Universität München, Theresienstr. 37, 80333 München,
Germany
jurco@theorie.physik.uni-muenchen.de

Summary. We introduce noncommutative line bundles and gerbes within the
framework of deformation quantization. The Seiberg-Witten map is used to construct the corresponding noncommutative Čech cocycles. Morita equivalence of star
products and quantization of twisted Poisson structures are discussed from this
point of view.

1 Seiberg-Witten Map and Noncommutative Gauge Transformations

Let $\mathfrak{A} = (C^\infty(M)[[\hbar]], \star)$ be an associative algebra that is the deformation
quantization of a Poisson structure θ over some manifold M [1]. For an arbitrary Poisson bivector θ a star product $\star$ exists and can be expressed in terms
of the Kontsevich formality map [14], so we restrict ourselves to this case. In
this section $M = \mathbb{R}^n$. We need also a 1-form a on $\mathbb{R}^n$. a is a connection on
the trivial line bundle over $\mathbb{R}^n$.

Seberg-Witten (SW) map $\mathcal{D} \equiv \mathcal{D}_{[a]}$ is a formal linear differential operator
on $C^\infty(\mathbb{R}^n)[[\hbar]]$ starting with the identity that satisfies the gauge equivalence
condition

$$\delta_\lambda \mathcal{D}_{[a]}(f) = i[\Lambda_\lambda[a] \overset{\star}{,} \mathcal{D}_{[a]}(f)] , \tag{1}$$

where $\delta_\lambda a = d\lambda$ is an ordinary "commutative" gauge transformation and
$\Lambda_\lambda[a]$ is a solution to the consistency condition (cocyle condition)

$$[\Lambda_\alpha[a] \overset{\star}{,} \Lambda_\beta[a]] + i\delta_\alpha \Lambda_\beta[a] - i\delta_\beta \Lambda_\alpha[a] = 0 . \tag{2}$$

Also we require that $\mathcal{D}_{[a]}$ and $\Lambda_\lambda[a]$ are a local functions of θ and a.

It is easy to see that $\mathcal{D}_{[a]}$ has the form $\mathcal{D}_{[a]}(f) = f + \hbar\theta(a, df) + \mathcal{O}(\hbar^2)$,
the map $\mathcal{D}_{[a]}$ is formally invertible and defines an equivalent star product $\star'$
via

$$\mathcal{D}_{[a]}(f \star' g) = \mathcal{D}_{[a]}f \star \mathcal{D}_{[a]}g . \tag{3}$$

The $\star'$ depends now on the gauge field $F = da$ only.

In [8] we have used this structure of equivalent star products, to find
the general solution for both $\mathcal{D}_{[a]}$ and $\Lambda_\lambda[a]$ and we shall briefly review the
construction here. We start by defining a one-parameter family of Poisson

B. Jurčo: *Noncommutative Line Bundles and Gerbes*, Lect. Notes Phys. **662**, 193–204 (2005)
www.springerlink.com

bivectors θ_t, $t \in [0, 1]$ by the differential equation $\partial_t \theta_t = \hbar F_{\theta_t}$ with $\theta_0 \equiv \theta$, where F_{θ_t} is the contravariant curvature of the contravariant connection D^{θ_t}, with

$$D^\theta_{df} \psi = \{f, \psi\}_\theta + ia_\theta(df)\psi, \quad \text{where} \quad a_\theta(df) \equiv \theta(a, df) . \tag{4}$$

In matrix notation $\partial_t \theta_t = -\hbar \theta_t F_{\theta_t}$ with the following formal solution:

$$\theta_t = \sum_{n=0}^{\infty} (-t)^n \theta(\hbar F\theta)^n . \tag{5}$$

Via the Kontsevich formality maps U_n [14], this defines also a one-parameter family of star products

$$\star_t = \sum_{n=0}^{\infty} \frac{(i\hbar)^n}{n!} U_n(\theta_t, \ldots, \theta_t) . \tag{6}$$

It can be shown that $\star$ and $\star' \equiv \star_1$ are related by (3), with

$$\mathcal{D}_{[a]} = e^{a_{\star_t} + \partial_t} e^{-\partial_t} \Big|_{t=0}, \quad \text{where} \quad a_{\star_t} = \sum_{n=0}^{\infty} \frac{(i\hbar)^{(n+1)}}{n!} U_{n+1}(a_{\theta_t}, \theta_t, \ldots, \theta_t) , \tag{7}$$

which satisfies (1) with

$$\Lambda_\lambda[a] = \sum_{n=0}^{\infty} \frac{(a_{\star_t} + \partial_t)^n(\tilde{\lambda})}{(n+1)!} \Big|_{t=0}, \quad \text{where} \quad \tilde{\lambda} = \sum_{n=0}^{\infty} \frac{(i\hbar)^n}{n!} U_{n+1}(\lambda, \theta_t, \ldots, \theta_t) . \tag{8}$$

The Kontsevich formality maps that appear in these formulas and in the formula of the star product can be computed explicitly on $\mathbb{R}^n$. Formal geometry arguments can be used to construct the global formality [14] on any Poisson manifold M. Here we will consider only global formalities having the following important property. Their restriction to an open subset gives a corresponding local Kontsevich formality. The existence and construction of such formalities is nontrivial. Appendix A 3 in [15] appears to contain all ingredients needed for that.

In the case of a general Poisson manifold M with a line bundle L the SW map $\mathcal{D}_{[a]}$ is only defined on the patch where the local gauge potential a is defined. The new star product $\star'$, however, is defined globally since it only depends on a via the gauge-invariant field strength $F = da$. The star product $\star'$ is in fact the deformation quantization by Kontsevich's formula (6) of a Poisson structure θ' which, for some choice of local coordinates and using matrix notation has the explicit form $\theta' = \sum_{n=0}^{\infty}(-1)^n \theta(\hbar F\theta)^n$. The star products $\star$ and $\star'$ are "patch-wise" equivalent. The corresponding algebras $\mathfrak{A}$, $\mathfrak{A}'$ are in fact Morita equivalent [7].

Let us consider finite classical gauge transformations

$$\psi \mapsto \psi_g = g\psi, \quad a \mapsto a_g = a + igdg^{-1} . \tag{9}$$

The finite version of the gauge equivalence condition (1) is

$$\mathcal{D}_{[a_g]}(f) \star G_g[a] = G_g[a] \star \mathcal{D}_{[a]}(f) \tag{10}$$

and the finite gauge consistency condition (cocycle condition) then takes the form

$$G_{g_1}[a_{g_2}] \star G_{g_2}[a] = G_{g_1 \cdot g_2}[a] . \tag{11}$$

The finite noncommutative gauge transformation corresponding to $g = e^{i\lambda}$ can be contructed explicitly [7] however as in the case of $\star$ and $\mathcal{D}_{[a]}$ the existence of $G_g[a]$ is more important for the following then the explicit formula for it. The $G_g[a]$ play the role of "noncommutative group elements".

2 Noncommutative Line Bundles

Let us recall that a classical (complex) line bundle is uniquely determined by a covering $\{U^k\}$ of a Manifold M and a collection of transition functions $g^{jk} \in C^\infty_{\mathbb{C}}(U^j \cap U^k)$ satisfying relations

$$g^{ij}g^{jk} = g^{ik}, \tag{12}$$

$$g^{jk}g^{kj} = 1 , \tag{13}$$

on all intersections $U^i \cap U^j \cap U^k$ and $U^j \cap U^k$ respectively.

A collection of functions $\psi^k \in C^\infty_{\mathbb{C}}(U^k)$ satisfying

$$\psi^j = g^{jk}\psi^k \tag{14}$$

on the overlaps $U^j \cap U^k$ define a section $\psi = (\psi^k)$. A set of local 1-forms a^k, satisfying

$$a^j = a^k + d\lambda^{jk}, \qquad d\lambda^{jk} \equiv ig^{jk}dg^{kj} \tag{15}$$

defines a connection on the line bundle.

The cohomological class $[da'] = [da]$ in $H^2(M)$ is the Chern class of the line bundle.

We choose a covering $\{U^k\}$ of M such that the patches and all their non-empty intersections are diffeomorphic to $\mathbb{R}^n$. The C^∞-functions on all these open subsets of M become formal power series in a deformation parameter. Working on any intersection of local patches U^i we assume that all noncommutative transition functions and all local equivalence maps on this intersection are constructed using the corresponding restriction of the global Kontsevich formality [14, 15]. The consistency of this follows from the "locality" property of Kontsevich formality mentioned in the introduction.

Choosing $g_1 = g^{ij}$, $g_2 = g^{jk}$, $g_1 \cdot g_2 = g^{ik}$ and $a = a^k$ in the consistency relation (11) gives the following relation in the intersection $U^i \cap U^j \cap U^k$:

$$G_{g^{ij}}[(a^k)_{g^{jk}}] \star G_{g^{jk}}[a^k] = G_{g^{ij}}[a^j] \star G_{g^{jk}}[a^k] = G_{g^{ik}}[a^k] \tag{16}$$

where we have used $(a^k)_{g^{jk}} = a^k + d\lambda^{jk} = a^j$. For the special case $g_1 = g^{kj}$, $g_2 = g^{jk}$ with $g^{kj}g^{jk} = 1$ we find an expression for the inverse of $G_{g^{jk}}[a^k]$:

$$G_{g^{kj}}[(a^k)_{g^{kj}}] \star G_{g^{jk}}[a^k] = G_{g^{kj}}[a^j] \star G_{g^{jk}}[a^k] = 1 \qquad (17)$$

Similarly in the gauge equivalence relation (10) put $g = g^{jk}$ and $a = a^k$, then

$$\mathcal{D}_{[(a^k)_{g^{jk}}]}(f) \star G_{g^{jk}}[a^k] = \mathcal{D}_{[a^j]}(f) \star G_{g^{jk}}[a^k] = G_{g^{jk}}[a^k] \star \mathcal{D}_{[a^k]}(f) \qquad (18)$$

for any function $f \in C^\infty_{\mathbb{C}}(U^j \cap U^k)[[\hbar]]$. So we can use an abbreviated notation

$$G^{jk} \equiv G_{g^{jk}}[a^k], \qquad \mathcal{D}^k \equiv \mathcal{D}_{[a^k]} . \qquad (19)$$

The fundamental relations are then

$$G^{ij} \star G^{jk} = G^{ik}, \qquad G^{kj} \star G^{jk} = 1 \qquad (20)$$

and

$$\mathcal{D}^j(f) \star G^{jk} = G^{jk} \star \mathcal{D}^k(f) . \qquad (21)$$

In view of (20), the G^{jk} play the role of noncommutative transition functions. The collection of $G^{jk} \in C^\infty_{\mathbb{C}}(U^j \cap U^k)[[\hbar]]$ is a good candidate for a noncommutative line bundle in the sense of deformation quantization. As we will see later the local SW maps D^k play the role of a local connection.

The G^{jk} depend explicitly on a classical connection a^k. For a given classical line bundle, i.e. fixed g^{jk}, the choice of different a^k only changes the star product on to one in the same equivalence class. The reason is that the new a^k differ from the old ones by a global one-form b. The equivalence is given by $\mathcal{D}_{[b]}$. For an equivalent classical line bundle with transition functions $\tilde{g}^{jk} = \zeta^j g^{jk} \zeta^{k^{-1}}$ and local connection forms $\tilde{a}^k = a^k + i\zeta^k d\zeta^{k^{-1}}$ we find new transition functions for the noncommutative line bundle of the form

$$\tilde{G}^{jk} = G_{\zeta^j}[a^j] \star G^{jk} \star (G_{\zeta^k}[a^k])^{-1} . \qquad (22)$$

Here we have twice used the consistency relation (11).

For the rest of this contribution a noncommutative line bundle $\mathcal{L}$ is defined by a collection of local transition functions $G^{ij} \in C^\infty(U^i \cap U^j)[[\hbar]]$, and a collection of maps $\mathcal{D}^i : C^\infty(U^i)[[\hbar]] \to C^\infty(U^i)[[\hbar]]$, formal power series in $\hbar$ starting with identity and with coefficients being differential operators such that

$$G^{ij} \star G^{jk} = G^{ik} \qquad (23)$$

on $U^i \cap U^j \cap U^k$, $G^{ii} = 1$ on U^i, and

$$\text{Ad}_\star G^{ij} = \mathcal{D}^i \circ (\mathcal{D}^j)^{-1} \qquad (24)$$

on $U^i \cap U^j$ or, equivalently, $\mathcal{D}^i(f) \star G^{ij} = G^{ij} \star \mathcal{D}^j(f)$ for all $f \in C^\infty(U^i \cap U^j)[[\hbar]]$. Obviously the local maps $\mathcal{D}^i$ define globally a new star product $\star'$ (because the inner automorphisms $\text{Ad}_\star G^{ij}$ do not affect $\star'$)

$$\mathcal{D}^i(f \star' g) = \mathcal{D}^i f \star \mathcal{D}^i g \,. \tag{25}$$

We say that two line bundles $\mathcal{L}_1 = \{G_1^{ij}, \mathcal{D}_1^i, \star\}$ and $\mathcal{L}_2 = \{G_2^{ij}, \mathcal{D}_2^i, \star\}$ are equivalent if there exist a collection of invertible local functions $H^i \in C^\infty(U^i)[[\hbar]]$ such that

$$G_1^{ij} = H^i \star G_2^{ij} \star (H^j)^{-1} \tag{26}$$

and

$$\mathcal{D}_1^i = \mathrm{Ad}_\star H^i \circ \mathcal{D}_2^i \,. \tag{27}$$

The tensor product of two line bundles $\mathcal{L}_1 = \{G_1^{ij}, \mathcal{D}_1^i, \star_1\}$ and $\mathcal{L}_2 = \{G_2^{ij}, \mathcal{D}_2^i, \star_2\}$ is well defined if $\star_2 = \star_1'$ (or $\star_1 = \star_2'$.) Then the corresponding tensor product is a line bundle $\mathcal{L}_2 \otimes \mathcal{L}_1 = \mathcal{L}_{21} = \{G_{12}^{ij}, \mathcal{D}_{12}^{ij}, \star_1\}$ defined as

$$G_{12}^{ij} = \mathcal{D}_1^i(G_2^{ij}) \star_1 G_1^{ij} = G_1^{ij} \star_1 \mathcal{D}_1^j(G_2^{ij}) \tag{28}$$

and

$$\mathcal{D}_{12}^i = \mathcal{D}_1^i \circ \mathcal{D}_2^i \,. \tag{29}$$

The order of indices of $\mathcal{L}_{21}$ indicates the bimodule structure of the corresponding space of sections to be defined later, whereas the first index on the G_{12}'s and $\mathcal{D}_{12}$'s indicates the star product (here: $\star_1$) by which the objects multiply.

A section $\Psi = (\Psi^i)$ is a collection of functions $\Psi^i \in C_\mathbb{C}^\infty(U^i)[[\hbar]]$ satisfying consistency relations

$$\Psi^i = G^{ij} \star \Psi^i \tag{30}$$

on all intersections $U^i \cap U^j$. With this definition the space of sections $\mathcal{E}$ is a right $\mathfrak{A} = (C^\infty(M)[[\hbar]], \star)$ module. We shall use the notation $\mathcal{E}_\mathfrak{A}$ for it. The right action of the function $f \in \mathfrak{A}$ is the regular one

$$\Psi.f = (\Psi^k \star f) \,. \tag{31}$$

Using the maps $\mathcal{D}^i$ it is easy to turn $\mathcal{E}$ also into a left $\mathfrak{A}' = (C^\infty(M)[[\hbar]], \star')$ module $_{\mathfrak{A}'}\mathcal{E}$. The left action of $\mathfrak{A}'$ is given by

$$f.\Psi = (\mathcal{D}^i(f) \star \Psi^i) \,. \tag{32}$$

It is easy to check, using (24), that the left action (32) is compatible with (30). From the property (25) of the maps $\mathcal{D}^i$ we find

$$f.(g.\Psi) = (f \star' g).\Psi \,. \tag{33}$$

Together we have a bimodule structure $_{\mathfrak{A}'}\mathcal{E}_\mathfrak{A}$ on the space of sections. There is an obvious way of tensoring sections. The section

$$\Psi_{12}^i = \mathcal{D}_1^i(\Psi_2^i) \star_1 \Psi_1^i \tag{34}$$

is a section of the tensor product line bundle (28), (29). Tensoring of line bundles naturally corresponds to tensoring of bimodules.

Using for example the Hochschild complex we can introduce a natural differential calculus on the algebra $\mathfrak{A}$. The p-cochains, elements of $C^p = \mathrm{Hom}_{\mathbb{C}}(\mathfrak{A}^{\otimes p}, \mathfrak{A})$, play the role of p-forms and the derivation $\mathrm{d} : C^p \to C^{p+1}$ is given on $C \in C^p$ as

$$
\begin{aligned}
\mathrm{d}C(f_1, f_2, \ldots, f_{p+1}) &= f_1 \star C(f_2, \ldots, f_{p+1}) - C(f_1 \star f_2, \ldots, f_{p+1}) \\
&+ C(f_1, f_2 \star f_3, \ldots, f_{p+1}) - \cdots + (-1)^p C(f_1, f_2, \ldots, f_p \star f_{p+1}) \\
&+ (-1)^{p+1} C(f_1, f_2, \ldots, f_p) \star f_{p+1} .
\end{aligned}
\tag{35}
$$

A (contravariant) connection $\nabla : \mathcal{E} \otimes_{\mathfrak{A}} C^p \to \mathcal{E} \otimes_{\mathfrak{A}} C^{p+1}$ can now be defined by a formula similar to (35) using the natural extension of the left and right module structure of $\mathcal{E}$ to $\mathcal{E} \otimes_{\mathfrak{A}} C^p$. Namely, for a $\Phi \in \mathcal{E} \otimes_{\mathfrak{A}} C^p$ we have

$$
\begin{aligned}
\nabla \Phi(f_1, f_2, \ldots, f_{p+1}) &= f_1.\Phi(f_2, \ldots, f_{p+1}) - \Phi(f_1 \star f_2, \ldots, f_{p+1}) \\
&+ \Phi(f_1, f_2 \star f_3, \ldots, f_{p+1}) - \cdots + (-1)^p \Phi(f_1, f_2, \ldots, f_p \star f_{p+1}) \\
&+ (-1)^{p+1} \Phi(f_1, f_2, \ldots, f_p).f_{p+1} .
\end{aligned}
\tag{36}
$$

The cup product $C_1 \cup C_2$ of two cochains $C_1 \in C^p$ and $C_2 \in C^q$;

$$
(C_1 \cup C_2)(f_1, \ldots, f_{p+q}) = C_1(f_1, \ldots, f_p) \star C_2(f_{p+1}, \ldots, f_q)
\tag{37}
$$

extends to a map from $(\mathcal{E} \otimes_{\mathfrak{A}} C^p) \otimes_{\mathfrak{A}} C^q$ to $\mathcal{E} \otimes_{\mathfrak{A}} C^{p+q}$. The connection ∇ satisfies the graded Leibniz rule with respect to the cup product and thus defines a bona fide connection on the module $\mathcal{E}_{\mathfrak{A}}$. On the sections the connection ∇ introduced here is simply the difference between the two actions of $C^\infty(M)[[\hbar]]$ on $\mathcal{E}$:

$$
\nabla \Psi(f) = f.\Psi - \Psi.f = \left(\nabla^i \Psi^i(f)\right) = \left(\mathcal{D}^i(f) \star \Psi^i - \Psi^i \star f\right) .
\tag{38}
$$

As in [8] we define the gauge potential $\mathcal{A} = (\mathcal{A}^i)$, where the $\mathcal{A}^i : C^\infty(U^i)[[\hbar]] \to C^\infty(U^i)[[\hbar]]$ are local 1-cochains, by

$$
\mathcal{A}^i \equiv \mathcal{D}^i - \mathrm{id} .
\tag{39}
$$

Then we have for a section $\Psi = (\Psi^i)$, where the $\Psi^i \in C^\infty_{\mathbb{C}}(U^i)[[\hbar]]$ are local 0-cochains,

$$
\nabla^i \Psi^i(f) = \mathrm{d}\Psi^i(f) + \mathcal{A}^i(f) \star \Psi^i ,
\tag{40}
$$

and more generally $\nabla^i \Phi^i = \mathrm{d}\Phi^i + \mathcal{A}^i \cup \Phi^i$ with $\Phi = (\Phi^i) \in \mathcal{E} \otimes_{\mathfrak{A}} C^p$. In the intersections $U^i \cap U^j$ we have the gauge transformation (cf. (24))

$$
\mathcal{A}^i = \mathrm{Ad}_\star G^{ij} \circ \mathcal{A}^j + G^{ij} \star \mathrm{d}(G^{ij})^{-1} .
\tag{41}
$$

The curvature $K_\nabla \equiv \nabla^2 : \mathcal{E} \otimes_{\mathfrak{A}} C^p \to \mathcal{E} \otimes_{\mathfrak{A}} C^{p+2}$ corresponding to the connection ∇, measures the difference between the two star products $\star'$ and $\star$. On a section Ψ, it is given by

$$(K_\nabla \Psi)(f,g) = \left(\mathcal{D}^i(f \star' g - f \star g) \star \Psi^i\right) . \tag{42}$$

The connection for the tensor product line bundle (28) is given on sections as

$$\nabla_{12}\Psi_{12}^i = \mathcal{D}_1^i(\nabla_2\Psi_2^i) \star_1 \Psi_1^i + \mathcal{D}_1^i(\Psi_2) \star_1 \nabla_1\Psi_1^i . \tag{43}$$

Symbolically,

$$\nabla_{12} = \nabla_1 + \mathcal{D}_1(\nabla_2) . \tag{44}$$

Let us note that the space of sections $\mathcal{E}$ as a right $\mathfrak{A}$-module is projective of finite type. Of course, the same holds if $\mathcal{E}$ is considered as a left $\mathfrak{A}'$ module. Also let us note that the two algebras $\mathfrak{A}$ and $\mathfrak{A}'$ are Morita equivalent. The Morita inverse bimodule $_{\mathfrak{A}}\overline{\mathcal{E}}_{\mathfrak{A}'}$ is defined by changing the order of multiplication in (30). Up to a global isomorphism $\star$ and $\star'$ are related by an action of the Picard group $\mathrm{Pic}(M) \cong H^2(M,\mathbb{Z})$ as follows. Let $L \in \mathrm{Pic}(M)$ be a (complex) line bundle on M and F its Chern class. Consider the formal Poisson structure θ' given by the geometric series

$$\theta' = \theta(1 + \hbar F\theta)^{-1} . \tag{45}$$

In this formula θ and F are understood as maps $\theta : T^*M \to TM$, $F : TM \to T^*M$ and θ' is the result of the indicated map compositions. Then $\star'$ must (up to a global isomorphism) be the deformation quantization of θ' corresponding to some $F \in H^2(M,\mathbb{Z})$. If $F = da$ then the corresponding quantum line bundle is trivial, i.e.,

$$G^{ij} = (H^i)^{-1} \star H^j \tag{46}$$

and the linear map

$$\mathcal{D} = \mathrm{Ad}_\star H^i \circ \mathcal{D}^i \tag{47}$$

defines a global equivalence of $\star$ and $\star'$.

We also recommend papers [12] for an intersting discussion of deformation quantization of vector bundles and Morita equivalence of star products.

3 Noncommutative Gerbes

This and the next section are based on [9]. Let us consider any covering $\{U_\alpha\}$ (not necessarily a good one) of a manifold M. We switch from upper Latin to lower Greek indices to label the local patches. Consider each local patch equipped with its own star product $\star_\alpha$ the deformation quantization of a local Poisson structure θ_α. We assume that on each double intersection $U_{\alpha\beta} = U_\alpha \cap U_\beta$ the local Poisson structures θ_α and θ_β are related similarly as in the previous section via some integral closed two form $F_{\beta\alpha}$, which is the curvature of a line bundle $L_{\beta\alpha} \in \mathrm{Pic}(U_{\alpha\beta})$

$$\theta_\alpha = \theta_\beta(1 + \hbar F_{\beta\alpha}\theta_\beta)^{-1}. \tag{48}$$

Next consider a good covering $U_{\alpha\beta}^i$ of each double intersection $U_\alpha \cap U_\beta$ with a noncommutative line bundle $\mathcal{L}_{\beta\alpha} = \{G_{\alpha\beta}^{ij}, \mathcal{D}_{\alpha\beta}^i, \star_\alpha\}$,

$$G_{\alpha\beta}^{ij} \star_\alpha G_{\alpha\beta}^{jk} = G_{\alpha\beta}^{ik}, \qquad G_{\alpha\beta}^{ii} = 1, \tag{49}$$

$$\mathcal{D}_{\alpha\beta}^i(f) \star_\alpha G_{\alpha\beta}^{ij} = G_{\alpha\beta}^{ij} \star_\alpha \mathcal{D}_{\alpha\beta}^j(f) \tag{50}$$

and

$$\mathcal{D}_{\alpha\beta}^i(f \star_\beta g) = \mathcal{D}_{\alpha\beta}^i(f) \star_\alpha \mathcal{D}_{\alpha\beta}^i(g). \tag{51}$$

As in the previous section the order of indices of $\mathcal{L}_{\alpha\beta}$ indicates the bimodule structure of the corresponding space of sections.

A noncommutative gerbe is characterised by the following axioms:

Axiom 1 $\mathcal{L}_{\alpha\beta} = \{G_{\beta\alpha}^{ij}, \mathcal{D}_{\beta\alpha}^i, \star_\beta\}$ and $\mathcal{L}_{\beta\alpha} = \{G_{\alpha\beta}^{ij}, \mathcal{D}_{\alpha\beta}^i, \star_\alpha\}$ are related as follows

$$\{G_{\beta\alpha}^{ij}, \mathcal{D}_{\beta\alpha}^i, \star_\beta\} = \{(\mathcal{D}_{\alpha\beta}^j)^{-1}(G_{\alpha\beta}^{ji}), (\mathcal{D}_{\alpha\beta}^i)^{-1}, \star_\beta\} \tag{52}$$

i.e. $\mathcal{L}_{\alpha\beta} = \mathcal{L}_{\beta\alpha}^{-1}$. (Notice also that $(\mathcal{D}_{\alpha\beta}^j)^{-1}(G_{\alpha\beta}^{ji}) = (\mathcal{D}_{\alpha\beta}^i)^{-1}(G_{\alpha\beta}^{ji})$.)

Axiom 2 On the triple intersection $U_\alpha \cap U_\beta \cap U_\gamma$ the tensor product $\mathcal{L}_{\gamma\beta} \otimes \mathcal{L}_{\beta\alpha}$ is equivalent to the line bundle $\mathcal{L}_{\gamma\alpha}$. Explicitly

$$G_{\alpha\beta}^{ij} \star_\alpha \mathcal{D}_{\alpha\beta}^j(G_{\beta\gamma}^{ij}) = \Lambda_{\alpha\beta\gamma}^i \star_\alpha G_{\alpha\gamma}^{ij} \star_\alpha (\Lambda^j)_{\alpha\beta\gamma}^{-1}, \tag{53}$$

$$\mathcal{D}_{\alpha\beta}^i \circ \mathcal{D}_{\beta\gamma}^i = \mathrm{Ad}_{\star_\alpha} \Lambda_{\alpha\beta\gamma}^i \circ \mathcal{D}_{\alpha\gamma}^i. \tag{54}$$

Axiom 3 On the quadruple intersection $U_\alpha \cap U_\beta \cap U_\gamma \cap U_\delta$

$$\Lambda_{\alpha\beta\gamma}^i \star_\alpha \Lambda_{\alpha\gamma\delta}^i = \mathcal{D}_{\alpha\beta}^i(\Lambda_{\beta\gamma\delta}^i) \star_\alpha \Lambda_{\alpha\beta\delta}^i, \tag{55}$$

$$\Lambda_{\alpha\beta\gamma}^i = (\Lambda_{\alpha\gamma\beta}^i)^{-1} \quad \text{and} \quad \mathcal{D}_{\alpha\beta}^i(\Lambda_{\beta\gamma\alpha}^i) = \Lambda_{\alpha\beta\gamma}^i. \tag{56}$$

With slight abuse of notation we have used Latin indices $\{i, j, ..\}$ to label both the good coverings of the intersection of the local patches U_α and the corresponding transition functions of the consistent restrictions of line bundles $\mathcal{L}_{\alpha\beta}$ to these intersections. A short comment on the consistency of Axiom 3 is in order. Let us define

$$\mathcal{D}_{\alpha\beta\gamma}^i = \mathcal{D}_{\alpha\beta}^i \circ \mathcal{D}_{\beta\gamma}^i \circ \mathcal{D}_{\gamma\alpha}^i. \tag{57}$$

Then it is easy to see that

$$\mathcal{D}_{\alpha\beta\gamma}^i \circ \mathcal{D}_{\alpha\gamma\delta}^i \circ \mathcal{D}_{\alpha\delta\beta}^i = \mathcal{D}_{\alpha\beta}^i \circ \mathcal{D}_{\beta\gamma\delta}^i \circ \mathcal{D}_{\beta\alpha}^i. \tag{58}$$

In view of (54) this implies that

$$\Lambda_{\alpha\beta\gamma\delta}^i \equiv \mathcal{D}_{\alpha\beta}^i(\Lambda_{\beta\gamma\delta}^i) \star_\alpha \Lambda_{\alpha\beta\delta}^i \star_\alpha \Lambda_{\alpha\delta\gamma}^i \star_\alpha \Lambda_{\alpha\gamma\beta}^i$$

is central. Using this and the associativity of $\star_\alpha$ together with (53) applied to the triple tensor product $\mathcal{L}_{\delta\gamma} \otimes \mathcal{L}_{\gamma\beta} \otimes \mathcal{L}_{\beta\alpha}$ transition functions

$$G^{ij}_{\alpha\beta\gamma} \equiv G^{ij}_{\alpha\beta} \star_\alpha \mathcal{D}^j_{\alpha\beta}(G^{ij}_{\beta\gamma}) \star_\alpha \mathcal{D}^j_{\alpha\beta}(\mathcal{D}^j_{\beta\gamma}(G^{ij}_{\gamma\delta})) \tag{59}$$

reveals that $\Lambda^i_{\alpha\beta\gamma\delta}$ is independent of i. It is therefore consistent to set $\Lambda^i_{\alpha\beta\gamma\delta}$ equal to 1. A similar consistency check works also for (56). If we replace all noncommutative line bundles $\mathcal{L}_{\alpha\beta}$ in Axioms 1–3 by equivalent ones, we get by definition an equivalent noncommutative gerbe.

There is a natural (contravariant) connection on a quantum gerbe. It is defined using the (contravariant) connections $\nabla_{\alpha\beta} = (\nabla^i_{\alpha\beta})$ (cf. (36), (38)) on quantum line bundles $\mathcal{L}_{\beta\alpha}$. Let us denote by $\nabla_{\alpha\beta\gamma}$ the contravariant connection formed on the triple tensor product $\mathcal{L}_{\alpha\gamma\beta} \equiv \mathcal{L}_{\alpha\gamma} \otimes \mathcal{L}_{\gamma\beta} \otimes \mathcal{L}_{\beta\alpha}$ with maps $\mathcal{D}^i_{\alpha\beta\gamma}$ and transition functions (59) according to the rule (44). Axiom 2 states that $\Lambda^i_{\alpha\beta\gamma}$ is a trivialization of $\mathcal{L}_{\alpha\gamma\beta}$ and that

$$\nabla^i_{\alpha\beta\gamma}\Lambda^i_{\alpha\beta\gamma} = 0 . \tag{60}$$

Using Axiom 2 one can show that the product bundle

$$\mathcal{L}_{\alpha\beta\gamma\delta} = \mathcal{L}_{\alpha\beta\gamma} \otimes \mathcal{L}_{\alpha\gamma\delta} \otimes \mathcal{L}_{\alpha\delta\beta} \otimes \mathcal{L}_{\alpha\beta} \otimes \mathcal{L}_{\beta\delta\gamma} \otimes \mathcal{L}_{\beta\alpha} \tag{61}$$

is trivial: it has transition functions $G^{ij}_{\alpha\beta\gamma\delta} = 1$ and maps $\mathcal{D}^i_{\alpha\beta\gamma\delta} = \mathrm{id}$. The constant unit section is thus well defined on this bundle. On $\mathcal{L}_{\alpha\beta\gamma\delta}$ we also have the section $(\Lambda^i_{\alpha\beta\gamma\delta})$. Axiom 3 implies $(\Lambda^i_{\alpha\beta\gamma\delta})$ to be the unit section. If two of the indices α, β, γ, δ are equal, triviality of the bundle $\mathcal{L}_{\alpha\beta\gamma\delta}$ implies (56). Using for example the first relation in (56) one can show that (55) written in the form $\mathcal{D}^i_{\alpha\beta}(\Lambda^i_{\beta\gamma\delta}) \star_\alpha \Lambda^i_{\alpha\beta\delta} \star_\alpha \Lambda^i_{\alpha\delta\gamma} \star_\alpha \Lambda^i_{\alpha\gamma\beta} = 1$ is invariant under cyclic permutations of any three of the four factors appearing on the l.h.s..

If we now assume that $F_{\alpha\beta} = da_{\alpha\beta}$ for each $U_\alpha \cap U_\beta$ then

$$G^{ij}_{\alpha\beta} = (H^i_{\alpha\beta})^{-1} \star_\alpha H^j_{\alpha\beta}$$

and

$$\mathcal{D}_{\alpha\beta} \equiv \mathrm{Ad}_{\star_\alpha} H^i_{\alpha\beta} \circ \mathcal{D}^i_{\alpha\beta} = \mathrm{Ad}_{\star_\alpha} H^j_{\alpha\beta} \circ \mathcal{D}^j_{\alpha\beta} .$$

It then easily follows that

$$\Lambda_{\alpha\beta\gamma} \equiv H^i_{\alpha\beta} \star_\alpha \mathcal{D}^i_{\alpha\beta}(H^i_{\beta\gamma}) \star_\alpha \mathcal{D}^i_{\alpha\beta}\mathcal{D}^i_{\beta\gamma}(H^i_{\gamma\alpha}) \star_\alpha \Lambda^i_{\alpha\beta\gamma} \tag{62}$$

defines a global function on the triple intersection $U_\alpha \cap U_\beta \cap U_\gamma$. $\Lambda_{\alpha\beta\gamma}$ is just the quotient of the two sections $\left(H^i_{\alpha\beta} \star_\alpha \mathcal{D}^i_{\alpha\beta}(H^i_{\beta\gamma}) \star_\alpha \mathcal{D}^i_{\alpha\beta}\mathcal{D}^i_{\beta\gamma}(H^i_{\gamma\alpha})\right)^{-1}$ and $\Lambda^i_{\alpha\beta\gamma}$ of the triple tensor product $\mathcal{L}_{\alpha\gamma} \otimes \mathcal{L}_{\gamma\beta} \otimes \mathcal{L}_{\beta\alpha}$. On the quadruple overlap $U_\alpha \cap U_\beta \cap U_\gamma \cap U_\delta$ it satisfies conditions analogous to (55) and (56)

$$\Lambda_{\alpha\beta\gamma} \star_\alpha \Lambda_{\alpha\gamma\delta} = \mathcal{D}_{\alpha\beta}(\Lambda_{\beta\gamma\delta}) \star_\alpha \Lambda_{\alpha\beta\delta} , \tag{63}$$

$$\Lambda_{\alpha\beta\gamma} = (\Lambda_{\alpha\gamma\beta})^{-1} \quad \text{and} \quad \mathcal{D}_{\alpha\beta}(\Lambda_{\beta\gamma\alpha}) = \Lambda_{\alpha\beta\gamma} . \tag{64}$$

Also

$$\mathcal{D}_{\alpha\beta} \circ \mathcal{D}_{\beta\gamma} \circ \mathcal{D}_{\gamma\alpha} = \mathrm{Ad}_{\star_\alpha} \Lambda_{\alpha\beta\gamma} . \tag{65}$$

So we can take formulas (63)–(65) as a definition of a gerbe in the case of a good covering $\{U_\alpha\}$. The collection of local equivalences $\mathcal{D}_{\alpha\beta}$ satisfying (65) with $\Lambda_{\alpha\beta\gamma}$ fulfilling (63), (64) defines on M a stack of algebras [13].

From now on we shall consider only good coverings. A noncommutative gerbe defined by $\Lambda_{\alpha\beta\gamma}$ and $\mathcal{D}_{\alpha\beta}$ is said to be trivial if there exist a global star product $\star$ on M and a collection of "twisted" transition functions $G_{\alpha\beta}$ defined on each overlap $U_\alpha \cap U_\beta$ and a collection $\mathcal{D}_\alpha$ of local equivalences between the global product $\star$ and the local products $\star_\alpha$

$$\mathcal{D}_\alpha(f) \star \mathcal{D}_\alpha(g) = \mathcal{D}_\alpha(f \star_\alpha g)$$

satisfying the following two conditions:

$$G_{\alpha\beta} \star G_{\beta\gamma} = \mathcal{D}_\alpha(\Lambda_{\alpha\beta\gamma}) \star G_{\alpha\gamma} \tag{66}$$

and

$$\mathrm{Ad}_\star G_{\alpha\beta} \circ \mathcal{D}_\beta = \mathcal{D}_\alpha \circ \mathcal{D}_{\alpha\beta} . \tag{67}$$

Locally, every noncommutative gerbe is trivial as is easily seen from (63), (64) and (65) by fixing the index α. Defining as in (39), $\mathcal{A}_\alpha = \mathcal{D}_\alpha - \mathrm{id}$, $\mathcal{A}_{\alpha\beta} = \mathcal{D}_{\alpha\beta} - \mathrm{id}$ we obtain the "twisted" gauge transformations

$$\mathcal{A}_\alpha = \mathrm{Ad}_\star G_{\alpha\beta} \circ \mathcal{A}_\beta + G_{\alpha\beta} \star \mathrm{d}(G_{\alpha\beta})^{-1} - \mathcal{D}_\alpha \circ \mathcal{A}_{\alpha\beta} . \tag{68}$$

4 Quantization of Twisted Poisson Structures

Let $H \in H^3(M, \mathbb{Z})$ be a closed integral three form on M. We can find a good covering $\{U_\alpha\}$ and local potentials B_α with $H = dB_\alpha$ for H. On $U_\alpha \cap U_\beta$ the difference of the two local potentials $B_\alpha - B_\beta$ is closed and hence exact: $B_\alpha - B_\beta = da_{\alpha\beta}$. On a triple intersection $U_\alpha \cap U_\beta \cap U_\gamma$ we have

$$a_{\alpha\beta} + a_{\beta\gamma} + a_{\gamma\alpha} = -i\lambda_{\alpha\beta\gamma}d\lambda_{\alpha\beta\gamma}^{-1} . \tag{69}$$

The collection of local functions $\lambda_{\alpha\beta\gamma}$ defines the gerbe.

Consider a formal antisymmetric bivector field $\theta = \theta^{(0)} + \hbar\theta^{(1)} + \dots$ on M such that

$$[\theta, \theta] = \hbar \, \theta^* H , \tag{70}$$

where $[\,,\,]$ is the Schouten-Nijenhuis bracket and θ^* denotes the natural map sending n-forms to n-vector fields. In local coordinates, $\theta^* H^{ijk} = \theta^{im}\theta^{jn}\theta^{ko}H_{mno}$. We call θ a Poisson structure twisted by H [5, 6, 10]. On each U_α we have a local formal Poisson structure $\theta_\alpha = \theta(1 - \hbar B_\alpha \theta)^{-1}$, $[\theta_\alpha, \theta_\alpha] = 0$. The Poisson structures θ_α and θ_β are related on the intersection $U_\alpha \cap U_\beta$ as in (48)

$$\theta_\alpha = \theta_\beta (1 + \hbar F_{\beta\alpha}\theta_\beta)^{-1} , \tag{71}$$

with an exact $F_{\beta\alpha} = da_{\beta\alpha}$. Now we can use Formality [14] to obtain local star products $\star_\alpha$ and to construct for each intersection $U_\alpha \cap U_\beta$ the corresponding equivalence maps $\mathcal{D}_{\alpha\beta}$. These $\mathcal{D}_{\alpha\beta}$, supplemented by trivial transition functions, define a collection of noncommutative line bundles $\mathcal{L}_{\beta\alpha}$. On each triple intersection we then have

$$\mathcal{D}_{\alpha\beta} \circ \mathcal{D}_{\beta\gamma} \circ \mathcal{D}_{\gamma\alpha} = \mathrm{Ad}_{\star_\alpha} \Lambda_{\alpha\beta\gamma} . \tag{72}$$

It follows from the discussion after formula (56) that $\Lambda_{\alpha\beta\gamma}$ defines a quantum gerbe (a deformation quantization of the classical gerbe $\lambda_{\alpha\beta\gamma}$) if each of the central functions $\Lambda_{\alpha\beta\gamma\delta}$ introduced there can be chosen to be equal to 1. See [11], Sect. 5 and [15] that this is really the case.

Acknowledgement

I would like to thank the organizers for hospitality and the participants for interesting discussions.

References

1. F. Bayen, M. Flato, C. Frønsdal, A. Lichnerowicz and D. Sternheimer, "Deformation Theory And Quantization. 1. Deformations of Symplectic Structures," Annals Phys. **111**, 61 (1978).
2. J. Giraud, "Cohomologie non-abélienne," Grundlehren der mathematischen Wissenschaften, Band 179, Springer Verlag, Berlin (1971).
3. J.L. Brylinski, "Loop Spaces, Characteristic Classes And Geometric Quantization," *Boston, USA: Birkhaeuser (1993) (Progress in mathematics, 107)*.
4. N. Hitchin, "Lectures on special Lagrangian submanifolds," arXiv:math.dg/9907034.
5. C. Klimčík and T. Strobl, "WZW-Poisson manifolds," J. Geom. Phys. 43 (2002) 341–344.
6. J.S. Park, "Topological open p-branes," arXiv:hep-th/0012141.
7. B. Jurčo, P. Schupp and J. Wess, "Noncommutative line bundle and Morita equivalence," Lett. Math. Phys. **61**, 171–186 (2002).
8. B. Jurčo, P. Schupp and J. Wess, "Noncommutative gauge theory for Poisson manifolds," Nucl. Phys. B **584**, 784 (2000).
   "Nonabelian noncommutative gauge theory via noncommutative extra dimensions," Nucl. Phys. B **604**, 148 (2001).
9. P. Aschieri, I. Baković, B. Jurčo, P. Schupp, "Noncommutative gerbes and deformation quantization", hep-th/0206101
10. P. Ševera and A. Weinstein, "Poisson geometry with a 3-form background," Prog. Theor. Phys. Suppl. 144 (2001) 145–154.
11. P. Ševera, "Quantization of Poisson Families and of twisted Poisson structures," arXiv:math.qa/0205294.

12. H. Bursztyn, S. Waldmann, "Deformation Quantization of Hermitian Vector Bundles," Lett. Math. Phys. **53**, 349 (2000); H. Bursztyn, "Semiclassical Geometry of Quantum Line Bundles and Morita Equivalence of Star Products," Int. Math. Res. Not. **16**, 821 (2002); H. Bursztyn, S. Waldmann, "The characteristic classes of Morita equivalent star products on symplectic manifolds," Comm. Math. Phys. **228** 103 (2002).

13. M. Kashiwara, "Quantization of contact manifolds" Publ. Res. Inst. Math. Sci. **32**, 1 (1996).

14. M. Kontsevich, "Deformation quantization of Poisson manifolds, I," arXiv:q-alg/9709040.

15. M. Kontsevich, "Deformation quantization of algebraic varietes," Lett. Math. Phys. **56**, 271 (2001).

Lectures on Two-Dimensional Noncommutative Gauge Theory Quantization

L.D. Paniak[1] and R.J. Szabo[2]

[1] Michigan Center for Theoretical Physics, University of Michigan, Ann Arbor, Michigan 48109-1120, U.S.A.
paniak@umich.edu

[2] Department of Mathematics, Heriot-Watt University, Scott Russell Building, Riccarton, Edinburgh EH14 4AS, U.K.
R.J.Szabo@ma.hw.ac.uk

Summary. These notes are devoted to the construction of exact solutions of noncommutative gauge theory in two spacetime dimensions. Here we shall deal with the quantum field theory. Topics covered include an investigation of the symmetries of quantum gauge theory on the noncommutative torus within the path integral formalism, the derivation of the exact expression for the vacuum amplitude, and the classification of instanton contributions. A section dealing with a new, exact combinatorial solution of gauge theory on a two-dimensional fuzzy torus is also included.

1 Introduction

These lecture notes continue the study of noncommutative gauge theory in two dimensions which was begun in [1] at the classical level. In this second part we shall deal with matters concerning the quantization of these gauge theories, and in particular demonstrate how to explicitly obtain nonperturbative solutions. Some background and motivation for dealing with this particular class of models may be found in [1] and won't be repeated here. Various aspects of two-dimensional noncommutative gauge theory have been studied over the past few years in [2]–[14]. In the present article we shall only analyze the vacuum amplitudes of these theories. More general gauge invariant correlation functions are studied in [6, 8, 9],[11]–[14]. Reviews on noncommutative field theory pertinent to the present material may be found in [15]–[17]. A detailed review of ordinary Yang-Mills theory in two dimensions is given in [18]. All relevant mathematical details and properties of the classical noncommutative gauge theory may be found in [1] and are briefly reviewed in Sect. 1.2 below.

1.1 How to Solve Yang-Mills Theory in Two Dimensions

When one comes to the issue of quantizing noncommutative gauge theory in two dimensions, one is naively faced with a plethora of possibilities. The commutative version of this theory has a long history as an exactly solvable

L.D. Paniak and R.J. Szabo: *Lectures on Two-Dimensional Noncommutative Gauge Theory Quantization*, Lect. Notes Phys. **662**, 205–237 (2005)
www.springerlink.com

quantum field theory, and as such is explicitly solvable by many different techniques. We will therefore begin with a brief run through of the various methods that may be used to solve ordinary Yang-Mills theory, and elucidate on the possibilities of extending them to the noncommutative setting.

Heat Kernel/Group Theory Methods

One of the most profound features of two-dimensional Yang-Mills theory is the interplay between the two-dimensional geometry on which it is defined and the representation theory of its structure group [18]. As will be reviewed in Sect. 3.1, the propagator between two states may be easily written down in terms of the standard heat kernel on the group manifold of the structure group, and from this the vacuum amplitude and Wilson loops on arbitrary geometries may be extracted. However, these techniques are not readily available in the noncommutative case for several reasons. First and foremost is the lack of a notion of structure group in the noncommutative setting. While there is a well-defined gauge group, it mixes spacetime and internal colour symmetries through noncommutative gauge transformations and there is no clear separation of spacetime and internal degrees of freedom. Secondly, a Hamiltonian formalism is not available because making time a noncommutative coordinate causes problems with unitarity and the overall interpretation of time-evolution in these systems. While this approach in the commutative case will play a crucial role in the foregoing line of development, it is not the one that will be *a priori* used analyse the quantum field theory. The group theory approach in the noncommutative setting has been analysed recently in [14].

Integrability

The fact that Yang-Mills theory is exactly solvable in two-dimensions is intimately connected with the fact that it is related to an integrable system [19]. It is possible to relate dynamics in this theory to that of certain one-dimensional gauged matrix models which are related to Calogero-Moser systems [20, 21]. While the integrability of the noncommutative counterpart may be established to a certain extent [1], it is not clear what integrable structure underlies this system. This line of attack therefore does not immediately lead to an appropriate generalization.

Semi-Classical Methods

One way to understand the exact solvability of the two-dimensional gauge theory is through the observation that its partition function and observables are given exactly by their semi-classical approximation [22]. This is related to the fact that ordinary Yang-Mills theory can be recast as a cohomological field theory in two dimensions. These properties *do* generalize to the noncommutative setting with some care, as we discuss in Sect. 2. In fact, these techniques will be the focal point of much of this article. They have been recently applied in [14] to explicitly compute the correlation functions of open Wilson line operators.

Lattice Regularization

Discretizing spacetime also provides a fruitful way of tackling the problem and is at the very heart of the group theory methods mentioned above [23, 24]. While a lattice formulation of noncommutative gauge theory is available [25], it is much more complicated than its commutative version because the nice self-similarity property possessed by the latter is ruined by the inherent nonlocality of the former. Nonetheless, we have succeeded in explicitly solving the lattice model in two dimensions at finite cutoff, and this is new material which will be presented in detail in Sect. 5. We shall therefore postpone further discussion of this approach until then.

Relations to Other Field Theories

Besides its relationship with a cohomological gauge theory, two-dimensional quantum Yang-Mills theory may also be related to various other field theories in certain limits, such as three-dimensional Chern-Simons theory and two dimensional conformal field theory [26]. These connections can be used to give explicit formulas for the volumes of the moduli spaces of representations of fundamental groups of two-dimensional surfaces. As we discuss in Sect. 4, some of these volumes are also effectively computable in the noncommutative setting. These ideas can also all be cast into the formalism of abelianization [21], a technique that relies heavily upon the presence of a well-defined structure group. However, it is not clear what sort of mathematical structures one should find in general and these further connections remain an interesting, as yet unexplored area of this subject.

1.2 Background

As we have mentioned above, the solution of the quantum gauge theory will be determined in large part by the very structure of the classical solutions of the field theory. This is described at length in [1]. To keep the presentation of the present article reasonably self-contained and to set some notation, we shall briefly summarize the classical solutions of gauge theory on a noncommutative torus in two-dimensions that were obtained in [1]. The classical action is defined on a fixed Heisenberg module $\mathcal{E}_{p,q}$ over the algebra $\mathcal{A}_\theta$ of functions on the noncommutative torus of fixed topological numbers $(p, q) \in \mathbb{Z}^2$, with q the Chern number, $\dim \mathcal{E}_{p,q} = p - q\theta > 0$, and $N = \gcd(p, q)$ the rank of the gauge theory. It is given explicitly by

$$S[A] = \frac{1}{2g^2} \, \text{Tr} \, [\nabla_1, \nabla_2]^2 = \frac{1}{2g^2} \int d^2x \, \text{tr}_N \left(F_A(x) - \frac{2\pi q}{p - q\theta} \right)^2 , \quad (1)$$

where $\nabla = \partial + A$ is a connection on $\mathcal{E}_{p,q}$ and F_A is the corresponding field strength. In the first equality of (1), Tr is the canonical trace on the endomorphism algebra $\text{End}(\mathcal{E}_{p,q})$. In the second equality the integration extends over the two-dimensional, unit area square torus $\mathbf{T}^2$, tr_N is the usual $N \times N$ matrix trace, and the constant subtraction corresponds to the constant curvature of the module $\mathcal{E}_{p,q}$.

Classical solutions of this gauge theory are in a one-to-one correspondence with the direct sum decompositions

$$\mathcal{E}_{p,q} = \bigoplus_k \mathcal{E}_{p_k, q_k} \quad (2)$$

of the given Heisenberg module into projective submodules. These are characterized by *partitions* $(\boldsymbol{p}, \boldsymbol{q}) = \{(p_k, q_k)\}$ of the topological numbers (p, q) satisfying the constraints

$$p_k - q_k\theta > 0 ,$$
$$\sum_k (p_k - q_k\theta) = p - q\theta ,$$
$$\sum_k q_k = q . \quad (3)$$

In addition, to avoid overcounting, it is sometimes useful to impose a further ordering constraint $p_k - q_k\theta \leq p_{k+1} - q_{k+1}\theta \, \forall k$, and regard any two partitions as the same if they coincide after rearranging their components according to this ordering. We may then characterize the components of a partition by integers $\nu_a > 0$ which are defined as the number of partition components that have the ath least dimension $p_a - q_a\theta$. The integer

$$|\boldsymbol{\nu}| = \sum_a \nu_a \quad (4)$$

is then the total number of components in the given partition. The noncommutative Yang-Mills action (1) evaluated on a classical solution, with corresponding partition $(\boldsymbol{p}, \boldsymbol{q})$, is given by

$$S(\boldsymbol{p}, \boldsymbol{q}) = \frac{2\pi^2}{g^2} \sum_k (p_k - q_k\theta) \left(\frac{q_k}{p_k - q_k\theta} - \frac{q}{p - q\theta} \right)^2 . \quad (5)$$

1.3 Outline

The outline of material in the remainder of this paper is as follows. In Sect. 2 we will carefully define the quantum theory, examine a deep "hidden" supersymmetry of it, and prove that the partition function and observables are all given exactly by their semi-classical approximation. In Sect. 3 we will derive an exact, analytical expression for the partition function of gauge theory on the noncommutative torus in two dimensions, and use it to analyse precisely how noncommutativity alters the properties of Yang-Mills theory on $\mathbf{T}^2$. In Sect. 4 we will describe how to organize the non-perturbative expression for the vacuum amplitude into a sum over contributions from (unstable) instantons of the two-dimensional gauge theory, and compare with analogous expressions obtained on the noncommutative plane. This paves the way for our analysis in Sect. 5 which deals with the matrix model/lattice formulation of noncommutative gauge theory in two dimensions. We will present here a new, exact expression for the partition function on the fuzzy torus, and describe the scaling limits which map this model onto the continuum gauge theory.

2 Quantum Gauge Theory on the Noncommutative Torus

In this section we will carefully define the quantum gauge theory within the path integral formalism. We will show that it admits a natural interpretation as a phase space path integral of an infinite-dimensional Hamiltonian system, the noncommutative Yang-Mills system. In this formulation a particular cohomological symmetry of the quantum theory is manifest, which leads immediately to the property that the partition function is given exactly by its semi-classical approximation. While naively the vacuum amplitude may seem to merely produce uninteresting determinants, the non-trivial topology of the torus provides a rich analytic structure (through large gauge transformations). For some time we will neglect the constant curvature subtraction in the action (1), and simply reinstate it when we come to the derivation of the exact formula for the partition function. This is possible to do because of the Morita invariance of the gauge theory [1].

2.1 Definition

The quantum field theory is defined formally through the functional integral

$$Z = \int \mathrm{D}A \ \mathrm{e}^{-S[A]} , \tag{6}$$

where $S[A]$ is the noncommutative Yang-Mills action (1). The integration in (6) is over the space $\mathcal{C} = \mathcal{C}(\mathcal{E})$ of compatible connections on a given fixed Heisenberg module $\mathcal{E} = \mathcal{E}_{p,q}$. Since the action is gauge invariant, the integration measure $\mathrm{D}A$ must be carefully defined so as to select only gauge orbits of the field configurations. There is a very natural way to define this measure in the present situation. As discussed in [1], the noncommutative

Yang-Mills system naturally defines a Hamiltonian system with moment map $\mu[A] = F_A$, so that the Yang-Mills action is the square of the moment map, $S[A] = \operatorname{Tr} \mu[A]^2/2g^2$. The gauge-invariant symplectic form

$$\omega[\alpha, \beta] = \operatorname{Tr} \alpha \wedge \beta , \quad \alpha, \beta \in \Omega^1(\mathcal{E}) , \tag{7}$$

is defined on the tangent space to $\mathcal{C}$, which is identified as the space $\Omega^1(\mathcal{E}) = \operatorname{End}(\mathcal{E}) \otimes \bigwedge^1 \mathcal{L}^*$ with $\mathcal{L}$ the (centrally extended) Lie algebra of the translation group acting on $\mathbf{T}^2$. We have defined $\alpha \wedge \beta \equiv \alpha_1 \beta_2 - \alpha_2 \beta_1$ with respect to an orthonormal basis of $\mathcal{L}$.

We now let

$$dA = \prod_{a,b=1}^{N} \prod_{x \in \mathbf{T}^2} dA_1^{ab}(x) \, dA_2^{ab}(x) \tag{8}$$

be the usual, formal (gauge non-invariant) Feynman path integral measure on $\mathcal{C}$, and let ψ be the odd generators of the infinite-dimensional superspace $\mathcal{C} \oplus \Pi \Omega^1(\mathcal{E})$ with corresponding functional Berezin measure $dA \, d\psi$, where Π is the parity reversion operator. We may then define

$$DA = dA \int d\psi \, e^{-i\omega[\psi,\psi]} , \tag{9}$$

where here and in the following we will absorb the infinite volume of the group $\mathcal{G} = \mathcal{G}(\mathcal{E})$ of gauge transformations on $\mathcal{C}$ (determined by the Haar measure $d\nu$ on $\mathcal{G}$ induced by the inner product $(\lambda, \lambda') = \operatorname{Tr} \lambda \lambda'$, $\lambda, \lambda' \in \operatorname{End}(\mathcal{E})$), by which (9) should be divided. By construction, this measure is gauge-invariant and coincides with the functional Liouville measure associated to the infinite-dimensional dynamical system. An infinitesimal gauge transformation $A \mapsto A + [\nabla, \lambda]$, $\lambda \in \operatorname{End}(\mathcal{E})$ on $\mathcal{C}$ naturally induces the transformation $\psi \mapsto \psi + [\lambda, \psi]$ on its tangent space, under which (7) is invariant. In this setting the partition function (6) is naturally defined as a phase space path integral. Note that, since the fermion fields ψ appear only quadratically in (9), this measure coincides with that of its commutative counterpart at $\theta = 0$.

While this definition is very natural in the present context, we should demonstrate explicitly that it coincides with the more conventional gauge field measure obtained from the standard Faddeev-Popov gauge fixing procedure. The basic point is that the measure (9) has the following requisite property. Let $\pi : \mathcal{C} \to \mathcal{C}/\mathcal{G}$ be the projection onto the quotient space of $\mathcal{C}$ by the gauge group $\mathcal{G}$. Then the quotient measure DA on $\mathcal{C}/\mathcal{G}$ is the measure which satisfies $dA = \pi^*(DA) \, d\nu$. The Faddeev-Popov procedure constructs DA by introducing the standard fermionic ghost field $c \in \Pi \Omega^0(\mathcal{E}) = \Pi \operatorname{End}(\mathcal{E})$, and the anti-ghost multiplet consisting of a fermionic field $\bar{c} \in \Pi \Omega^0(\mathcal{E})$ and a bosonic field $w \in \Omega^0(\mathcal{E})$, along with the BRST transformation laws

$$\delta A = -[\nabla, c] \ ,$$
$$\delta c = \tfrac{1}{2} [c, c] \ ,$$
$$\delta \bar{c} = \mathrm{i}\, w \ ,$$
$$\delta w = 0 \tag{10}$$

obeying $\delta^2 = 0$.

The gauge-fixing term is given by $I = -\delta V$ for a suitable functional V of the BRST field multiplet. For this, we write a generic connection ∇ in the neighbourhood of a representative $\nabla^0 = \partial + A^0 \in \mathcal{C}$ of its gauge orbit as $\nabla = \nabla^0 + B$, and make the local choice $V = -\operatorname{Tr} \bar{c}\, \nabla^0 \cdot B$, where we have defined $\alpha \cdot \beta \equiv \alpha_1 \beta_1 + \alpha_2 \beta_2$ for cotangent vectors $\alpha, \beta \in \Omega^1(\mathcal{E})$. This produces

$$I = \operatorname{Tr} \left(\mathrm{i}\, w\, \nabla^0 \cdot B - \bar{c}\, \nabla^0 \cdot \nabla c \right) \ , \tag{11}$$

and the gauge fixed path integral measure is then defined by

$$DA = dA^0 \int dB \ dc \ d\bar{c} \ dw \ \mathrm{e}^{-I} \ . \tag{12}$$

Formally integrating over the bosonic field w and the Grassmann fields $c, \bar{c}$ gives

$$DA = dA^0 \int dB \ \delta \left(\nabla^0 \cdot B \right) \det \nabla^0 \cdot \nabla \ . \tag{13}$$

The integration over B enforces the gauge condition $\nabla^0 \cdot B = 0$ on the quantum field theory with the choice (11) of gauge-fixing term. Since $\delta(\nabla^0 \cdot B) = \delta(B)/|\det \nabla^0 \cdot \nabla^0|$, the resulting ratio of determinants after integrating out B in (13) coincides exactly with the determinant induced by integrating out ψ in (9). Thus the elementary measure defined by the symplectic structure of $\mathcal{C}$ coincides with that of the usual Faddeev-Popov gauge-fixing procedure.

2.2 The Cohomological Gauge Theory

We will now describe a remarkable cohomological symmetry of the partition function (6), with path integration measure (9), which will be the crux of much of our ensuing analysis of the quantum gauge theory. For this, we linearize the Yang-Mills action in the field strength F_A via a functional Gaussian integral transformation defined by an auxilliary field $\phi \in \Omega^0(\mathcal{E})$ as

$$Z = \int d\phi \ \mathrm{e}^{-\frac{g^2}{2} \operatorname{Tr} \phi^2} \int dA \ d\psi \ \mathrm{e}^{-\mathrm{i} \operatorname{Tr}(\psi \wedge \psi - \phi\, F_A)} \ . \tag{14}$$

Note that because of the quadratic form of the action in (14), the only place where noncommutativity is buried is in F_A. This is one of the features that makes this quantum field theory effectively solvable.

The basic field multiplet (A, ψ, ϕ) possesses a "hidden supersymmetry" that resides in the cohomology of the operator

$$\mathbf{Q}_\phi = \mathrm{Tr}\left(\psi \cdot \tfrac{\delta}{\delta A} + [\nabla, \phi] \cdot \tfrac{\delta}{\delta \psi}\right) \tag{15}$$

which generates the transformations

$$
\begin{aligned}
[\mathbf{Q}_\phi, A] &= \psi \ , \\
\{\mathbf{Q}_\phi, \psi\} &= [\nabla, \phi] \ , \\
[\mathbf{Q}_\phi, \phi] &= 0 \ .
\end{aligned}
\tag{16}
$$

The crucial property of the operator $\mathbf{Q}_\phi$ is that it is nilpotent precisely on gauge invariant field configurations,

$$(\mathbf{Q}_\phi)^2 = \delta_\phi \ , \tag{17}$$

where $\delta_\lambda A = [\nabla, \lambda]$ is an infinitesimal gauge transformation with gauge parameter $\lambda \in \mathrm{End}(\mathcal{E})$. Furthermore, the linearized action on $\mathcal{C} \oplus \Pi \Omega^1(\mathcal{E})$, for fixed ϕ, is closed under $\mathbf{Q}_\phi$,

$$\mathbf{Q}_\phi \, \mathrm{Tr}\left(\psi \wedge \psi - \phi \, F_A\right) = 0 \ , \tag{18}$$

which follows from the Hamiltonian flow equations for the moment map $\mu[A] = F_A$ and symplectic structure (7) [1]. Thus the partition function (14) defines a cohomological gauge theory with supersymmetry (16) and (A, ψ, ϕ) the basic field multiplet of topological Yang-Mills theory in two dimensions [22].

The nilpotency property (17) implies that the operator $\mathbf{Q}_\phi$ is simply the BRST supercharge, acting in the quantum field theory (14), which generates the transformations (10). Gauge fixing in this setting amounts to introducing additional anti-ghost multiplets analogous to those that were used in the previous subsection for BRST quantization. We shall return to this point in Sect. 4.1. From a more formal perspective, $\mathbf{Q}_\phi$ is the Cartan model differential for the $\mathcal{G}$-equivariant cohomology of $\mathcal{C}$ [27]. The second term in the action of (14) is the $\mathcal{G}$-equivariant extension of the moment map on $\mathcal{C}$, the integration over A, ψ defines an equivariant differential form in $\Omega_{\mathcal{G}}(\mathcal{C})$, and the integral over ϕ defines equivariant integration of such forms. In this way, as we explain in the next subsection, the cohomological symmetry of the quantum field theory will lead to a localization theorem for the partition function. Fundamentally, the localization points correspond to the BRST fixed points of the anti-ghost multiplets.

2.3 Localization of the Partition Function

We now come to the fundamental consequence of the hidden supersymmetry of the previous subsection. Let α be any gauge invariant functional of the fields of (14), i.e. $(\mathbf{Q}_\phi)^2 \alpha = 0$, and consider the one-parameter family of partition functions defined by

$$Z_t = \int d\phi \ e^{-\frac{g^2}{2} \operatorname{Tr} \phi^2} \int dA \ d\psi \ e^{-i \operatorname{Tr}(\psi \wedge \psi - \phi F_A) - t \, \mathbf{Q}_\phi \alpha} \qquad (19)$$

with $t \in \mathbb{R}$. The $t \to 0$ limit of (19) is just the partition function (14) of interest, $Z = Z_0$. The remarkable feature of (19) is that it is independent of the parameter $t \in \mathbb{R}$. This follows from the Leibnitz rule for the functional derivative operator $\mathbf{Q}_\phi$, the supersymmetry (18) of the (equivariantly extended) action, and the gauge invariance of α which, along with a formal functional integration by parts over the superspace $\mathcal{C} \oplus \Pi\Omega^1(\mathcal{E})$, can be easily used to show that $\partial Z_t / \partial t = 0$. This is a basic cohomological property of the noncommutative quantum field theory. Adding a supersymmetric $\mathbf{Q}_\phi$-exact term to the action deforms it without changing the value of the functional integral. It follows that the path integral (14) can be alternatively evaluated as the $t \to \infty$ limit of the expression (19). It thereby receives contributions from only those field configurations which obey the equations $\mathbf{Q}_\phi \alpha = 0$.

At this stage we need to specify an explicit form for α. Different choices will localize the partition function onto different components in field space, but the final results are (at least superficially) all formally identical. A convenient choice is $\alpha = \operatorname{Tr} \psi \cdot [\nabla, F_A]$, for which the $t \to \infty$ limit of (19) yields

$$Z = \int dA \ d\psi \ e^{-\operatorname{Tr}\left(i \psi \wedge \psi + \frac{1}{2g^2}(F_A)^2\right)} \lim_{t \to \infty} e^{-\frac{t^2}{2g^2} \operatorname{Tr}\left[\nabla \, ; \, [\nabla, F_A]\right]^2}$$
$$\times (\text{fermions}) \qquad (20)$$

after performing the functional Gaussian integration over ϕ. These arguments of course assume formally that the original action has no flat directions, but in the present case this is not a problem since it has a nondegenerate kinetic energy. The additional terms involving the Grassmann fields ψ in (20) formally yield a polynomial function in the parameter t after integration, and their precise form is not important. What is important here is the quadratic term in t, which in the limit implies that the functional integral vanishes everywhere except near those points in $\mathcal{C}$ which are solutions of the equations

$$[\nabla \, ; \, [\nabla, F_A]] = 0 \,, \qquad (21)$$

where we have used positivity of the trace Tr on $\operatorname{End}(\mathcal{E})$. By using the Leibnitz rule and the integration by parts property $\operatorname{Tr}[\nabla, \lambda] = 0$ this equation implies

$$0 = \operatorname{Tr} F_A[\nabla \, ; \, [\nabla, F_A]]^2 = -\operatorname{Tr}[\nabla, F_A] \cdot [\nabla, F_A] \,. \qquad (22)$$

By using non-degeneracy of the trace on $\operatorname{End}(\mathcal{E})$ we arrive finally at

$$[\nabla, F_A] = 0 \,, \qquad (23)$$

which are just the classical equations of motion of the original noncommutative gauge theory.

We have thereby formally shown that the partition function of noncommutative gauge theory in two dimensions receives contributions *only* from the space of solutions of the noncommutative Yang-Mills equations. As we reviewed in Sect. 1.2, each such solution corresponds to a partition (p, q), obeying the constraints (3), of the topological numbers (p, q) of the given Heisenberg module $\mathcal{E} = \mathcal{E}_{p,q}$ on which the gauge theory is defined. Symbolically, the partition function is therefore given by

$$Z = Z_{p,q} = \sum_{\substack{\text{partitions} \\ (p,q)}} W(p,q) \; e^{-S(p,q)} . \tag{24}$$

This result expresses the fact that quantum noncommutative gauge theory in two dimensions is given exactly by a sum over contributions from neighbourhoods of the stationary points of the Yang-Mills action (1). The Boltzmann weight $e^{-S(p,q)}$ involving the action (5) gives the contribution to the path integral (6) from a classical solution, while $W(p,q)$ encode the quantum fluctuations about each stationary point. These latter terms may in principle be determined from (20) by carefully integrating out the fermion fields and evaluating the functional fluctuation determinants that arise. However, these determinants are not effectively computable and are rather cumbersome to deal with. In the next section our main goal will be to devise an alternative method to extract these quantum fluctuation terms and hence the exact solution of the noncommutative quantum field theory.

3 Exact Solution

In this section we will present the exact solution of gauge theory on the noncommutative torus in two dimensions. We will start by recalling some well-known facts about ordinary Yang-Mills gauge theory in two dimensions, and show how it can be cast precisely into the form (24). From this we will then extract the exact expression in the general noncommutative case. Our techniques will rely heavily on the full machinery of the geometry of the noncommutative torus.

3.1 The Torus Amplitude

The vacuum amplitude for ordinary Yang-Mills theory on the torus $\mathbf{T}^2$ with structure group $U(p)$ and generators T^a, $a = 1, \ldots, p^2$ may be obtained as follows [18]. Let us consider the physical Hilbert space, in canonical quantization, for gauge theory on a cylinder $\mathbb{R} \times \mathbf{S}^1$ (Fig. 1). In two dimensions, Gauss' law implies that the physical state wavefunctionals $\Psi_{\text{phys}}[A] = \Psi[U]$ depend only on the holonomy $U = \mathrm{P} \exp \mathrm{i} \int_0^L \mathrm{d}x \; A_1(x)$ of the gauge connection around the cycle of the cylinder. By gauge invariance, Ψ furthermore

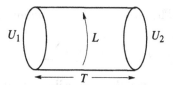

Fig. 1. Quantization of Yang-Mills theory on a spatial circle of circumference L yields the propagation amplitude between two states characterized by holonomies U_1 and U_2 in time T

depends only on the conjugacy class of U. It follows that the Hilbert space of physical states is the space of L^2-class functions, invariant under conjugation, with respect to the invariant Haar measure $[dU]$ on the unitary group $U(p)$,

$$\mathcal{H}_{\text{phys}} = L^2\big(U(p)\big)^{\text{Ad}\big(U(p)\big)} . \tag{25}$$

By the Peter-Weyl theorem, it may be decomposed into the unitary irreducible representations R of $U(p)$ as $\mathcal{H}_{\text{phys}} \cong \bigoplus_R R \otimes \overline{R}$. The representation basis of this Hilbert space is thereby provided by characters in the unitary representations, such that the states $|R\rangle$ have wavefunctions

$$\langle U|R\rangle = \chi_R(U) = \text{tr}_R U . \tag{26}$$

The Hamiltonian acting on the physical state wavefunctions $\Psi[U]$ is given by the Laplacian on the group manifold of $U(p)$,

$$H = \tfrac{g^2}{2} L \ \text{tr} \left(U \tfrac{\partial}{\partial U} \right)^2 , \tag{27}$$

and it is thereby diagonalized in the representation basis as

$$H\chi_R(U) = \tfrac{g^2}{2} L C_2(R) \, \chi_R(U) , \tag{28}$$

where $C_2(R)$ is the eigenvalue of the quadratic Casimir operator $C_2 = \sum_a T^a T^a$ in the representation R. From these facts it is straightforward to write down the cylinder amplitude corresponding to propagation of the system between two states with holonomies U_1 and U_2 in the form (Fig. 1)

$$Z_p(T; U_1, U_2) = \langle U_1|e^{-TH}|U_2\rangle = \sum_R \chi_R(U_1) \chi_R(U_2^\dagger) \ e^{-\frac{g^2}{2} LT C_2(R)} . \tag{29}$$

This is just the standard heat kernel on the $U(p)$ group. In keeping with our previous normalizations, we shall set the area of the cylinder to unity, $LT = 1$. To extract from (29) the partition function of $U(p)$ Yang-Mills theory on the torus, we glue the two ends of the cylinder together by setting $U_1 = U_2 = U$ and integrate over all U by using the fusion rule for the $U(p)$ characters,

$$\int [dU] \ \chi_{R_1}(VU) \chi_{R_2}(U^\dagger W) = \delta_{R_1, R_2} \frac{\chi_{R_1}(VW)}{\dim R_1} , \tag{30}$$

where $\dim R = \chi_R(\mathbb{1})$. This yields the torus vacuum amplitude

$$Z_p = \int [dU] \, Z_p(T; U, U) = \sum_R e^{-\frac{g^2}{2} C_2(R)} . \tag{31}$$

We can make the sum over the irreducible unitary representations R of $U(p)$ in (31) explicit by using the fact that each R is labelled by a decreasing set $\boldsymbol{n} = (n_1, \ldots, n_p)$ of p integers

$$+\infty > n_1 > n_2 > \cdots > n_p > -\infty \tag{32}$$

which are shifted highest weights parametrizing the lengths of the rows of the corresponding Young tableaux. Up to an irrelevant constant, the quadratic Casimir can be written in terms of these integers as

$$C_2(R) = C_2(\boldsymbol{n}) = \sum_{a=1}^{p} \left(n_a - \frac{p-1}{2} \right)^2 . \tag{33}$$

Since (33) is symmetric under permutations of the n_a's, it follows that the ordering restriction (32) can be removed in the partition function (31) to write it as a sum over non-coincident integers as (always up to inconsequential constants)

$$Z_p = \sum_{n_1 \neq \cdots \neq n_p} e^{-\frac{g^2}{2} C_2(\boldsymbol{n})} . \tag{34}$$

We may extend the sums in (34) over *all* $\boldsymbol{n} \in \mathbb{Z}^p$ by inserting the products of delta-functions

$$\det_{1 \leq a,b \leq p} (\delta_{n_a, n_b}) = \sum_{\sigma \in S_p} (-1)^{|\sigma|} \prod_{a=1}^{p} \delta_{n_a, n_{\sigma(a)}} . \tag{35}$$

The vanishing of the determinant for coincident rows prevents any two n_a's from coinciding when inserted into the sum.

Because of the permutation symmetry of (33), when inserted into the partition function (34) the sum in (35) truncates to a sum over conjugacy classes $[1^{\nu_1} 2^{\nu_2} \cdots p^{\nu_p}]$ of the symmetric group S_p. They are labelled by *partitions* of the rank p of the gauge theory,

$$\nu_1 + 2\nu_2 + \cdots + p\nu_p = p , \tag{36}$$

where ν_a is the number of elementary cycles of length a in $[1^{\nu_1} 2^{\nu_2} \cdots p^{\nu_p}]$. The sign of such a conjugacy class is $(-1)^{p+|\nu|}$ and its order is $p! / \prod_a a^{\nu_a} \nu_a!$, where $|\nu| = \nu_1 + \nu_2 + \cdots + \nu_p$ is the total number of cycles in the class. By using this, along with the Poisson resummation formula

$$\sum_{n=-\infty}^{\infty} e^{-\pi g n^2 - 2\pi i b n} = \frac{1}{\sqrt{g}} \sum_{q=-\infty}^{\infty} e^{-\pi(q-b)^2/g} , \tag{37}$$

we may bring the vacuum amplitude (34) after some work into the form [10]

$$
Z_p = \sum_{\boldsymbol{\nu}: \sum_a a\nu_a = p} \sum_{q_1, \ldots, q_{|\nu|} = -\infty}^{\infty} e^{i\pi\left(|\boldsymbol{\nu}| + (p-1)q\right)}
$$
$$
\times \prod_{a=1}^{p} \frac{\left(g^2 a^3 / 2\pi^2\right)^{-\nu_a/2}}{\nu_a!} \, e^{-S(\boldsymbol{\nu}, q)} , \tag{38}
$$

where $q = q_1 + q_2 + \cdots + q_{|\nu|}$ and

$$
S(\boldsymbol{\nu}, q) = \frac{2\pi^2}{g^2} \left(\sum_{k_1=1}^{\nu_1} \frac{q_{k_1}^2}{1} + \sum_{k_2=\nu_1+1}^{\nu_1+\nu_2} \frac{q_{k_2}^2}{2} + \sum_{k_3=\nu_1+\nu_2+1}^{\nu_1+\nu_2+\nu_3} \frac{q_{k_3}^2}{3} \right.
$$
$$
\left. + \cdots + \sum_{k_p=\nu_1+\cdots+\nu_{p-1}+1}^{|\nu|} \frac{q_{k_p}^2}{p} \right) . \tag{39}
$$

The important feature of the final expression (38) is that it agrees with the expected sum (24) over classical solutions of the commutative gauge theory on $\mathbf{T}^2$. For this, we note that the K-theory group of the ordinary torus is $K_0(C(\mathbf{T}^2)) = \mathbb{Z} \oplus \mathbb{Z}$, so that any projective module $\mathcal{E} = \mathcal{E}_{p,q}$ over the algebra $C(\mathbf{T}^2)$ of functions on the torus is determined by a pair of integers (p, q), with $\dim \mathcal{E}_{p,q} = p > 0$ and constant curvature q/p. Geometrically, any such module is the space of sections $\mathcal{E}_{p,q} = \Gamma(\mathbf{T}^2, E_{p,q})$ of a complex vector bundle $E_{p,q} \to \mathbf{T}^2$ of rank p, Chern number q, and structure group $U(p)$. The direct sum decompositions (2) correspond to the usual Atiyah-Bott bundle splittings [28]

$$
E_{p,q} = \oplus_k E_{p_k, q_k} \tag{40}
$$

into sub-bundles $E_{p_k, q_k} \subset E_{p,q}$ about each Yang-Mills critical point on $\mathbf{T}^2$. The first two partition constraints in (3) for $\theta = 0$ correspond to those on the rank of (40), $p = \sum_k p_k$ with $p_k > 0$. This condition coincides with (36), where ν_a is the number of submodules $\mathcal{E}_{p_k, q_k}$ of dimension a (equivalently the number of sub-bundles E_{p_k, q_k} of rank a). The action (39) is precisely of the form (5) at $\theta = 0$ and without the background flux subtraction, while the exponential prefactors in (38) correspond to the fluctuation determinants $W(\boldsymbol{p}, \boldsymbol{q})$ in (24).

The third constraint in (3) on the magnetic charges q_k, which are dual to the lengths of the rows of the Young tableaux of $U(p)$, restricts the gauge theory to a particular isomorphism class of bundles over the torus. It is straightforward to rewrite the partition function (38) of *physical* Yang-Mills theory, defined as a weighted sum over contributions from topologically distinct vector bundles over $\mathbf{T}^2$, in terms of that of Yang-Mills theory defined on a particular isomorphism class $\mathcal{E}_{p,q}$ of projective modules over $\mathcal{A}_\theta$ up to irrelevant constants as

$$Z_p = \sum_{q=-\infty}^{\infty} (-1)^{(p-1)q} \, Z_{p,q} \, , \qquad (41)$$

where

$$Z_{p,q} = \sum_{\substack{\text{partitions} \\ (p,q)}} (-1)^{|\nu|} \prod_{a=1}^{p} \frac{\left(g^2 a^3/2\pi^2\right)^{-\nu_a/2}}{\nu_a!} \; \mathrm{e}^{-S(p,q)} \, . \qquad (42)$$

The partition sum here arises from the sum over cycle decompositions that appears in the group theoretic setting above, with the number of partition components $|\nu|$ (or cycles) given by (4).

3.2 The Exact Vacuum Amplitude

From the commutative partition function (42) we may now extract the exact expression for the noncommutative field theory defined for *any* θ in the following manner. We use the fact, reviewed in Sect. 5 of [1], that gauge Morita equivalence provides a one-to-one correspondence between projective modules over different noncommutative tori (i.e. for different θ's) associated with different topological numbers, augmented with transformations of connections between the modules. It is an exact symmetry of the noncommutative Yang-Mills action (1) which is firmly believed to extend to the full quantum level. There are many good pieces of evidence in support of this assumption [25, 29, 30].

In the present case, we will use the fact that Morita duality can be used to map the quantum partition function of *ordinary* $\theta' = 0$ Yang-Mills theory on $\mathbf{T}^2$ onto noncommutative gauge theory with deformation parameter $\theta = n/s$. The dimensionless coupling constant and module dimensions in (42) transform in this case as $g^2 = |s|^3 g'^2$ and $\dim \mathcal{E} = \dim \mathcal{E}'/|s|$. The equivalence provides a one-to-one correspondence between classical solutions in the two field theories, i.e. their partitions. The symmetry factors $\nu_a!$ in (42) corresponding to permutation of partition components of identical dimension are preserved, as is the total number $|\nu|$ of submodules in any given partition (p, q). From these facts it follows that the fluctuation factors in (42) are invariant under this Morita duality *only* if the indices a transform as

$$a = a'/|s| \, , \qquad (43)$$

which is equivalent to the expected requirement that the cycle lengths a be interpreted as the dimensions of submodules in the commutative gauge theory.

With these identifications we can now straightforwardly map (42) onto the exact partition function of the $\theta = n/s$ Morita equivalent noncommutative gauge theory. The key point is that the localization arguments which led to (24) do not distinguish between the commutative, rational or irrational cases there. All of the analysis and formulas of the previous section hold universally for any value of θ, and hence so should the exact expression for the vacuum amplitude. Thus, given the generic structure of partitions as outlined in Sect. 1.3, including the general definition of ν_a, the final analytic expression for the partition function of gauge theory on a fixed projective module over the noncommutative torus, for any value of the noncommutativity parameter θ, is given by

$$Z_{p,q} = \sum_{\substack{\text{partitions} \\ (p,q)}} \prod_a \frac{(-1)^{\nu_a}}{\nu_a!} \left(\frac{g^2}{2\pi^2} (p_a - q_a \theta)^3 \right)^{-\nu_a/2}$$

$$\times \exp\left[-\frac{2\pi^2}{g^2} \sum_k (p_k - q_k \theta) \left(\frac{q_k}{p_k - q_k \theta} - \frac{q}{p - q\theta} \right)^2 \right]. \quad (44)$$

We have reinstated the constant curvature of $\mathcal{E}_{p,q}$, as it is required to ensure that the Yang-Mills action transforms homogeneously under Morita duality. This technique thereby explicitly determines the fluctuation determinants $W(p,q)$ of the semi-classical expansion (24).

We close this section with a brief description of how the expansion (44) elucidates the relations with and modifications of ordinary Yang-Mills theory on the torus:

- It can be shown [10] that the partition function (44) is a smooth function of θ, even about $\theta = 0$. At least at the level of two-dimensional noncommutative gauge theory, violations of θ-smoothness in the quantum theory disappear at the non-perturbative level.
- The Morita equivalence between rational noncommutative Yang-Mills theory on a projective module $\mathcal{E}_{p,q}$ with deformation parameter $\theta = n/s$, $n, s > 0$ relatively prime, and ordinary non-abelian gauge theory is particularly transparent in this formalism. As mentioned above, for $\theta' = 0$ the module dimensions transform as $\dim \mathcal{E} = \dim \mathcal{E}'/s$, and since in the commutative theory the bundle ranks are always positive integers, any module $\mathcal{E}$ in the rational theory has dimension bounded as $\dim \mathcal{E} \geq 1/s$. Since $\dim \mathcal{E}_{p,q} = p - nq/s$, it follows that any partition (p,q) of the rational theory consisting of submodules of dimension $\geq 1/s$ has at most $\frac{p-nq/s}{1/s} = ps - qn$ components. Thus any gauge theory dual to this one admits partitions with $ps - qn$ components. In particular, as we have seen in the previous subsection, for $U(N)$ commutative Yang-Mills theory the maximum number of components is precisely the rank N, corresponding to the cycle decomposition with $\nu_1 = N$ and $\nu_a = 0 \ \forall a > 1$. Putting these facts together we arrive at the well-known result that noncommutative Yang-Mills theory with $\theta = n/s$ on a module $\mathcal{E}_{p,q}$ is Morita equivalent to $U(N)$ commutative gauge theory on $\mathbf{T}^2$ with rank $N = ps - qn$.
- The expansion (44) clearly shows the differences between the commutative and noncommutative gauge theories. In the rational case $\theta = n/s$,

all partitions contain at most $ps - qn$ submodules of $\mathcal{E}_{p,q}$ of dimension $\geq 1/s$. But for θ irrational, there is no *a priori* bound on the number of submodules in a partition (although it is always finite) and submodules of arbitrarily small dimension can contribute to the partition function (44). In particular, in this case we can approximate θ by a sequence of rational numbers, $\theta = \lim_m n_m/s_m$ with both $n_m, s_m \to \infty$ as $m \to \infty$. The rigorous way to take the limit of the noncommutative field theory is described in [31]. In the rational gauge theory with noncommutativity parameter $\theta_m = n_m/s_m$, the dimension of any submodule is bounded from below by $1/s_m$. It follows that any rational approximation to the vacuum amplitude $Z_{p,q}$ contains contributions from partitions of arbitrarily small dimension. Thus although formally similar, the exact expansion (44) of the partition function has drastically different analytic properties in the commutative and noncommutative cases.

4 Instanton Contributions

The fact that gauge theory on the noncommutative torus has an exact semi-classical expansion in powers of e^{-1/g^2} suggests that it should admit an interpretation in terms of non-perturbative contributions from instantons of the two-dimensional gauge theory. By an instanton we mean a finite action solution of the Euclidean Yang-Mills equations (23) which is not a gauge transformation of the trivial gauge field configuration $A = 0$. Interpreting (44) in terms of such configurations is not as straightforward as it may seem, because the contributions to the sum as they stand are not arranged into gauge equivalence classes. In this section we will briefly describe how to rearrange the semi-classical expansion (44) into a sum over (unstable) instantons. This will entail a deep analysis of the moduli spaces of the noncommutative gauge theory and will also naturally motivate, via a comparison with corresponding structures on the noncommutative plane, a matrix model analysis of the field theory which will be carried out in the next section.

4.1 Topological Yang-Mills Theory

We will begin by studying the weak-coupling limit of the noncommutative gauge theory as it is the simplest case to describe. In the limit $g^2 \to 0$, the only non-vanishing contribution to (44) comes from those partitions for which the Yang-Mills action attains its global minimum of 0. The only partition for which this happens is the trivial one $(p, q) = (p, q)$ associated to the original Heisenberg module $\mathcal{E}_{p,q}$ itself. The corresponding moduli space of classical solutions is the space of constant curvature connections on $\mathcal{E}_{p,q}$ modulo gauge transformations. Such classical configurations preserve $\frac{1}{2}$ of the supersymmetries in an appropriate supersymmetric extension of the gauge theory [32, 33]. In this context, the classical solutions live in a Higgs branch of the $\frac{1}{2}$-BPS moduli space, with the whole moduli space determined by a fibration over the Higgs branch.

As described in detail in [1], as a vector space the Heisenberg module is given by $\mathcal{E}_{p,q} = L^2(\mathbb{R}) \otimes \mathbb{C}^q$, where $L^2(\mathbb{R})$ is the irreducible Schrödinger representation of the constant curvature condition, and $\mathbb{C}^q$ is the $q \times q$ representation of the Weyl-'t Hooft algebra in two dimensions. The latter algebra is known to possess a unique irreducible unitary representation of dimension q/N, $N = \gcd(p,q)$, so that module decomposes into irreducible components as

$$\mathcal{E}_{p,q} = L^2(\mathbb{R}) \otimes \left(\mathcal{W}_{\zeta_1} \oplus \cdots \oplus \mathcal{W}_{\zeta_N} \right) , \tag{45}$$

where $\mathcal{W}_\zeta \subset \mathbb{C}^q$ are the irreducible representations of the Weyl-'t Hooft algebra and $\zeta \in \tilde{\mathbf{T}}^2$ generate its center, with values in a dual torus to the original one $\mathbf{T}^2$. The only gauge transformations which act trivially on (45) are those which live in the Weyl subgroup of $U(N)$, and dividing by this we find that the moduli space of constant curvature connections on $\mathcal{E}_{p,q}$ is the Nth symmetric product

$$\mathcal{M}_{p,q} = \text{Sym}^N \tilde{\mathbf{T}}^2 \equiv \left(\tilde{\mathbf{T}}^2 \right)^N / S_N . \tag{46}$$

Remarkably, this space coincides with $\text{Hom}(\pi_1(\mathbf{T}^2), U(N))/U(N)$, the moduli space of *flat* $U(N)$ bundles over the torus $\mathbf{T}^2$ in commutative gauge theory [28].

Now let us examine more closely the partition function (44) in the limit $g^2 \to 0$. After using Morita duality to remove the background flux contribution, the series receives contributions only from partitions with vanishing magnetic charges $q_k = 0 \; \forall k$, and we find

$$Z_{p,q}\big|_{g^2=0} = \sum_{\nu : \sum_a a\nu_a = N} \prod_{a=1}^{N} \frac{(-1)^{\nu_a}}{\nu_a!} \left(\frac{g^2 a^3}{2\pi^2} \right)^{-\nu_a/2} + O\left(\mathrm{e}^{-1/g^2} \right) . \tag{47}$$

We thereby find that the weak coupling limit is independent of the noncommutativity parameter θ, and in particular it coincides with the commutative version of the theory with structure group $U(N)$. This is easiest to see from the form (14), whose $g^2 = 0$ limit gives explicitly

$$Z_{p,q}\big|_{g^2=0} = \int \mathrm{d}\phi \int \mathrm{d}A \; \mathrm{d}\psi \; \mathrm{e}^{-\,\mathrm{i}\,\mathrm{Tr}\,(\psi \wedge \psi - \phi\, F_A)} . \tag{48}$$

The integration over ϕ, after reinstating the proper constant curvature subtraction in (48), localizes this functional integral onto gauge field configurations of constant curvature, and the partition function thereby computes the symplectic volume of the moduli space (46) with respect to the symplectic structure on $\mathcal{M}_{p,q}$ inherited from the one (7) on $\mathcal{C}(\mathcal{E}_{p,q})$. It is formally the same as that of topological Yang-Mills theory on $\mathbf{T}^2$, except that now the noncommutativity (through Morita equivalence) identifies (46) as the space of *all* constant curvature connections, in contrast to the usual case where it only corresponds to flat gauge connections.

In this case, the gauge theory is BRST equivalent (in the sense described in Sect. 2.3 for $t \to \infty$) to that with gauge fixing functional $V = \text{Tr} \{ \frac{1}{2} (H - 4F_A) + \nabla\lambda \cdot \psi \}$, where we have introduced pairs (λ, η) and (χ, H) of anti-ghost multiplets of ghost numbers $(-2, -1)$ and $(-1, 0)$, respectively, with λ, H bosonic and η, χ Grassmann-valued fields. Their BRST transformation rules are

$$[\mathbf{Q}_\phi, \lambda] = \mathrm{i}\,\eta \,,$$
$$\{\mathbf{Q}_\phi, \eta\} = [\phi, \lambda] \,,$$
$$\{\mathbf{Q}_\phi, \chi\} = H \,,$$
$$[\mathbf{Q}_\phi, H] = \mathrm{i}\,[\phi, \chi] \,, \tag{49}$$

and the $\mathbf{Q}_\phi$-invariant action $S_\text{D} \equiv -\mathrm{i}\,\{\mathbf{Q}_\phi, V\}$ is given by

$$S_\text{D} = \text{Tr} \left\{ \tfrac{1}{2} (H - F_A)^2 - \tfrac{1}{2} (F_A)^2 - \mathrm{i}\chi \nabla \wedge \psi + \mathrm{i}\nabla\eta \cdot \psi \right.$$
$$\left. + \tfrac{1}{2} \chi [\chi, \phi] + \nabla\lambda \cdot \nabla\phi + \mathrm{i}[\psi, \lambda] \cdot \psi \right\} \,. \tag{50}$$

The functional V conserves ghost number and the action (50) has non-degenerate kinetic energy, as in the case of the original Yang-Mills system of Sect. 2.3. It gives the action of two-dimensional Donaldson theory, and in this way the full noncommutative gauge theory can be used to extract information about the intersection pairings on the moduli space $\mathcal{M}_{p,q}$ [22].

Going back to the formula (47), we see that it involves a sum over cycles ν of terms which are singular at $g^2 = 0$. These terms represent contributions to the symplectic volume from the conical orbifold singularities of the moduli space (46), which arise due to the existence of reducible connections. For this, we note that the fixed point locus of a conjugacy class element $\sigma \in [1^{\nu_1} 2^{\nu_2} \cdots p^{\nu_p}]$ acting on $(\zeta_1, \ldots, \zeta_N) \in (\tilde{\mathbf{T}}^2)^N$ is $\prod_a (\tilde{\mathbf{T}}^2)^{\nu_a}$. The action of the corresponding stabilizer subgroup of S_N is $\prod_a S_{\nu_a} \ltimes (\mathbb{Z}_a)^{\nu_a}$, where the symmetric group S_{ν_a} permutes coordinates in the factor $(\tilde{\mathbf{T}}^2)^{\nu_a}$ while the cyclic group $\mathbb{Z}_a$ acts in each cycle of length a. Only the S_{ν_a} factors act non-trivially, and so the singular locus of $\mathcal{M}_{p,q}$ is a disjoint union over the conjugacy classes $[1^{\nu_1} 2^{\nu_2} \cdots p^{\nu_p}] \subset S_N$ of the strata $\prod_a \text{Sym}^{\nu_a} \tilde{\mathbf{T}}^2$, as reflected by the expansion (47).

4.2 Instanton Partitions

Let us now consider the general case. The basic problem is that there is an iso-morphism $\mathcal{E}_{mp,mq} \cong \oplus^m \mathcal{E}_{p,q}$ of Heisenberg modules, owing to the reducibility of the Weyl-'t Hooft algebra, with $\mathcal{E}_{mp,mq}$ and $\mathcal{E}_{p,q}$ both possessing the *same* constant curvature. We circumvent this problem by writing each component of a given partition as $(p_k, q_k) = N_k(p'_k, q'_k)$, with $N_k = \gcd(p_k, q_k)$ and p'_k, q'_k relatively prime, and restrict the sum over partitions $(\boldsymbol{p}, \boldsymbol{q})$ to those with *distinct* K-theory charges (p'_k, q'_k). We call such partitions "instanton partitions" [10], as they each represent distinct, gauge equivalence classes of classical solutions to the noncommutative Yang-Mills equations. Then the direct sum decomposition (2) is modified to

$$\mathcal{E}_{p,q} = \oplus_a \, \mathcal{E}_{N_a p'_a, N_a q'_a} \,, \tag{51}$$

and the corresponding moduli space of classical solutions is [10]

$$\mathcal{M}'_{p,q} = \prod_a \mathcal{M}_{N_a p'_a, N_a q'_a} = \prod_a \mathrm{Sym}^{N_a} \, \tilde{\mathbf{T}}^2 \,. \tag{52}$$

The orbifold singularities present in (52) can now be used to systematically construct the gauge inequivalent contributions to noncommutative Yang-Mills theory. In this way one may rewrite the expansion (44) as a sum over instantons along with a finite number of quantum fluctuations about each instanton, representing a finite, non-trivial perturbative expansion in $1/g$. For more details, see [10].

4.3 Fluxon Contributions

The instanton solutions that we have found for gauge theory on the noncommutative torus bear a surprising relationship to soliton solutions of gauge theory on the noncommutative *plane* [2]–[4]. The classical solutions of the noncommutative Yang-Mills equations in this latter case are labelled by two integers, the rank of the gauge group and the magnetic charge, similarly to the case of the torus. These noncommutative solitons are termed "fluxons" and they are finite energy instanton solutions which carry quantized magnetic flux. The classical action evaluated on a fluxon of charge q is given by [4]

$$S(q) = \frac{2\pi^2 q}{g^2 \theta} \,. \tag{53}$$

This action is very similar to (5) in the limit $g^2 \theta \to \infty$, and in [8] it was described how to map the instanton expansion on the noncommutative torus to one on the noncommutative plane by using Morita equivalence and taking a suitable large area limit. In terms of the partition sum (44), a fluxon of charge q is composed of ν_a elementary vortices of charges $a = 1, 2, \ldots$. The symmetry factors $\nu_a!$ appear in (44) to account for the fact that vortices of equal charge inside the fluxon are identical, while the moduli dependence (through the vortex positions) is accompanied by the anticipated exponent $|\boldsymbol{\nu}|$, the total number of elementary vortex constituents of the fluxon. The remaining terms correspond to quantum fluctuations about each fluxon in the following manner.

The basic fluxon solution corresponds to the elementary vortex configuration $\nu_1 = q$, $\nu_a = 0$ $\forall a > 1$. In the large area limit, the semi-classical expansion (44) can be interpreted in terms of the contributions from basic fluxons of charge q and classical action (53), along with fluctuations around the soliton solution, leading to the partition function [8]

$$
\mathcal{Z}_q = \frac{e^{-2\pi^2 q/g^2\theta}}{N\sqrt{g^2\theta}} \sum_{\nu : \sum_a a\nu_a = q} \prod_{a=1}^{q} \left(-\frac{1}{\nu_a!} \sqrt{\frac{2\pi^2}{a^3 g^2\theta^3}} \right)^{\nu_a} . \tag{54}
$$

The (unweighted) sum over topological charges can be performed exactly and the result is

$$
\mathcal{Z} \equiv \sum_{q=0}^{\infty} \mathcal{Z}_q = \exp\left[-\frac{2\pi\, e^{-2\pi^2/g^2\theta}}{\sqrt{g^2\theta^3}}\, \Phi\left(e^{-2\pi^2/g^2\theta} ; \tfrac{3}{2} ; 1 \right) \right] , \tag{55}
$$

where the function

$$
\Phi(z; s; \mu) = \sum_{k=0}^{\infty} \frac{z^k}{(k+\mu)^s} \tag{56}
$$

is analytic in $z \in \mathbb{C}$ with a branch cut from $z = 1$ to $z = \infty$. The instanton series has been resummed in (55) into the non-perturbative exponential, which is typical of a dilute instanton gas. This is not surprising, given that fluxons are non-interacting objects and thereby lead to an extensive partition function. It would be interesting to examine the dynamics of all the instantons described in this picture on the moduli spaces (46) and (52), using the Kähler structure inherited from the symplectic structure (7) and metric $\mathrm{Tr}\,\alpha \cdot \beta$ on the space $\mathcal{C}$ of compatible gauge connections.

The non-trivial results obtained for the noncommutative plane suggest another way of tackling two-dimensional noncommutative gauge theory in general [4]. Since the planar algebra of functions is generated by the coordinate operators x^1, x^2 obeying the Heisenberg algebra $[x^1, x^2] = \mathrm{i}\theta$, gauge connections act by inner automorphisms and may be written as

$$
D_i = \frac{\mathrm{i}}{\theta}\, \epsilon_{ij}\, x^j + A_i \tag{57}
$$

for $i = 1, 2$. The curvature is given by

$$
F_A = [D_1, D_2] - \frac{\mathrm{i}}{\theta} , \tag{58}
$$

and after a rescaling of fields the partition function is defined by the infinite dimensional matrix model

$$
\mathcal{Z} = \lim_{\varepsilon \to 0^+} \int \mathrm{d}D_1\, \mathrm{d}D_2\, \exp\left[-\frac{\pi\theta}{2g^2}\, \mathrm{Tr}\left([D_1, D_2] - 1 \right)^2 - \varepsilon\, \mathrm{Tr}\, D \cdot D \right] . \tag{59}
$$

The second term in the action of (59) regulates the partition function and is a gauge-invariant analog of the infrared regularization provided by the area of the torus. It is required to ensure that the semi-classical approximation to the functional integral exists. The classical fluxon solutions are unstable critical points whose moduli are the positions of the vortices [4]. The Yang-Mills energy density of the vortices is independent of these positions and integrating along these moduli would lead to a divergent path integral in (59). While this may seem like a fruitful line of attack, it presents many difficulties. Foremost among these is the fact that finite action configurations would require the field strength F_A to be a compact operator. Since there are no bounded operators D_i for which (58) is compact, the effective gauge configuration space consists only of *unbounded* operators and the partition function (59) is not naturally realized as the large N limit of a finite dimensional matrix model. This makes an exact solution intractable. In the next section we shall present a matrix model formulation of noncommutative gauge theory in two dimensions which circumvents these difficulties, and enables an explicit and exact analysis of the configurations described here.

5 Combinatorial Quantization

In this final section we will show how a combinatorial approach can be used to explicitly compute the partition function of noncommutative gauge theory in two dimensions. Part of the motivation for doing this was explained at the end of the last section. Another reason is to make sense of the Feynman path integral over the space $\mathcal{C}$ of compatible connections. We will approximate $\mathcal{C}$ by a finite-dimensional $N \times N$ matrix group and then analyse the partition function in the limit $N \to \infty$. The hope is then that this procedure yields a concrete, non-perturbative definition of the noncommutative field theory. This matrix model is intimately connected with a lattice regularization of the noncommutative gauge theory obtained by triangulating $\mathbf{T}^2$, and restricting to modules over the finite-dimensional matrix algebras. In this setting the non-trivial K-theory of the torus algebra $\mathcal{A}_\theta$ is lost, and as in Sect. 3.1 the computation will give the Yang-Mills partition function summed over all topological types of projective modules over $\mathcal{A}_\theta$. We will begin by recalling some salient features of commutative lattice gauge theory, and contrast it with what happens in the noncommutative setting. Then we will proceed to define and completely solve the discrete version of noncommutative gauge theory in two dimensions, and describe how it can be used to extract information about the continuum field theory. The material contained in this section is new and presents a novel explicit solution of noncommutative Yang-Mills theory.

5.1 The Local Lattice Regularization

In ordinary two-dimensional Yang-Mills theory, the lattice form of the quantum field theory [34] possesses some very special properties and provides an indispensible tool for obtaining its complete analytic solution [23, 24]. Let us consider the partition function on a disk of area A (Fig. 2). It can be obtained from the cylinder amplitude (29) by pinching the right boundary of the cylinder to a point, so that U_2 becomes the holonomy surrounding a disk of vanishing area. The corresponding physical state wavefunction is the delta-function supported at the identity element $U_2 = 1$ of the unitary group with respect to its Haar measure, $\Psi[U_2] = \delta(U_2, 1)$. Then from (29) with $U_1 = U$ and $U_2 = 1$ we obtain the disk amplitude

$$Z(A, U) = \sum_R \dim R\, \chi_R(U)\, e^{-\frac{g^2 A}{2} C_2(R)} . \tag{60}$$

By using the area-preserving diffeomorphism invariance of the theory, we may interpret (60) as an amplitude for a *plaquette*, i.e. the interior of a simplex in a local triangulation of the spacetime (Fig. 3).

One of the main advantages of the discrete formalism is its self-similarity property [23]. Consider the gluing together of two disk amplitudes along a plaquette link as depicted in Fig. 3. The gluing property follows from (60) and the fusion rule (30) for the characters, which together imply

$$\int [dU]\, Z(A_1, VU)\, Z(A_2, U^\dagger W) = Z(A_1 + A_2, VW) . \tag{61}$$

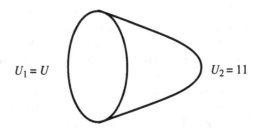

Fig. 2. The disk amplitude

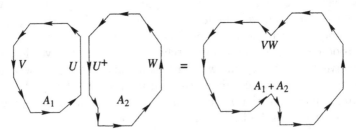

Fig. 3. The partition function of two-dimensional Yang-Mills theory is invariant under subdivision of the plaquettes of the lattice

This result expresses the renormalization group invariance of the basic plaquette Boltzmann weight. Subdivision of the lattice into a very fine lattice
yields a result which converges to that in the continuum theory. But (61) implies that the computation may be carried out on an arbitrarily coarse lattice.
Hence the lattice field theory produces the exact answer and the continuum
limit is trivial. From this treatment it is in fact possible to directly obtain
the torus amplitude (31).

As we will soon see, this self-similarity property under gluing of plaquettes is not shared by the noncommutative version of the lattice gauge theory,
reflecting its inherent non-locality. Noncommutativity introduces long-ranged
interactions between plaquettes of the lattice. A clear way to understand this
breakdown is to recall the Gross-Witten reduction of $U(N)$ Yang-Mills theory
on $\mathbb{R}^2$ [35]. The calculation of the lattice partition function in this case can be
easily reduced to a single unitary matrix integration by exploiting the gauge
invariance of the theory. This is possible to do by fixing an axial gauge and
an appropriate change of variables. If $U_i(x)$ denotes the operator of parallel
transport from a lattice site x to its neighbouring point along a link in direc-
tion $\hat{\imath}$, $i = 1, 2$, then one may fix the gauge $U_1(x) = \mathbb{1}$ $\forall x$. This renders the
theory trivial in the $\hat{1}$ direction. There is a residual gauge symmetry which
may be used to define $U_2(x + \hat{1}) = W(x) U_2(x)$, and the partition function
thereby factorizes into a product of decoupled integrals over the unitary matrices $W(x)$ [35]. This is *not* possible to do in the noncommutative gauge
theory, because in its lattice incarnation it is required to be formulated on
a periodic lattice as a result of UV/IR mixing [25], and large gauge transformations thereby forbid axial gauge choices. As expected, UV/IR mixing
drastically alters the Wilsonian renormalization features of the noncommutative field theory, and it admits non-trivial scaling limits. Later on we will
see how noncommutativity explicitly modifies the Gross-Witten result.

5.2 Noncommutative Lattice Gauge Theory

We will now proceed to formulate and explicitly solve the noncommutative
version of lattice gauge theory, which gives yet another proof of the exact
solvability of the continuum theory. We discretize the torus of the previous
sections as an $L \times L$ periodic square lattice. For convenience, we assume that
L is an odd integer. Let ε be the dimensionful lattice spacing, so that the
area of the discrete torus is

$$A = \varepsilon^2 L^2 .\tag{62}$$

Any function $f(x)$ on the periodic lattice admits a Fourier series expansion
over a Brillouin zone $\mathbb{Z}_L \times \mathbb{Z}_L$,

$$f(x) = \frac{1}{L^2} \sum_{m \in (\mathbb{Z}_L)^2} f_m \, \mathrm{e}^{2\pi \, \mathrm{i} \, m_i \, x^i / \varepsilon L} .\tag{63}$$

A natural lattice star-product may be defined as the proper discretized version of the integral kernel representation of the continuum star product as

$$(f \star g)(x) = \frac{1}{L^2} \sum_{y,z} f(x+y)\, g(x+z)\; e^{2\pi i\, y \wedge z/\varepsilon^2 L} , \tag{64}$$

where the sums run over lattice points. This identifies $\theta = 2/L$ and hence the dimensionful noncommutativity parameter of the commutant algebra as

$$\Theta = \frac{\theta A}{2\pi} = \frac{\varepsilon^2 L}{\pi} . \tag{65}$$

As mentioned in the previous subsection, because of a kinematical version of UV/IR mixing, the lattice regularization of noncommutative field theory *requires* the space to be a torus [25].

We will now write down, in analogy with the commutative case, the natural, nonperturbative lattice regularization of the continuum noncommutative gauge theory. This is provided by the noncommutative version of the standard Wilson plaquette model [34]. The partition function is [25]

$$\mathcal{Z}_r = \int \prod_x \left[dU_1(x) \right]\, \left[dU_2(x) \right]\, \exp\left[\frac{1}{4\lambda^2 L} \sum_\square \operatorname{tr}_N \left(U_\square + U_\square^\dagger \right) \right] , \tag{66}$$

where

$$\lambda = \sqrt{g^2 \varepsilon^2 L} \tag{67}$$

is the 't Hooft coupling constant. Here $\prod_x [dU_i(x)]$ is the normalized, invariant Haar measure on the ordinary $r \times r$ unitary group $U(r)$ with

$$r = L \cdot N , \tag{68}$$

and N is the rank of the given module. The fields $U_i(x)$ are $U(N)$ gauge fields which live at the links (x, i) of the lattice and which are "star-unitary",

$$(U_i \star U_i^\dagger)(x) = (U_i^\dagger \star U_i)(x) = \mathbb{1}_r . \tag{69}$$

In the continuum limit $\varepsilon \to 0$, they are identified with the gauge fields of the previous sections through $U_i = e^{\star \varepsilon A_i}$. The sum in (66) runs through the plaquettes $\square$ of the lattice with $U_\square$ the ordered star-product of gauge fields around the plaquette,

$$U_\square = U_1(x) \star U_2(x + \varepsilon\,\hat{1}) \star U_1(x + \varepsilon\,\hat{2})^\dagger \star U_2(x)^\dagger , \tag{70}$$

where x is the basepoint of the plaquette and $\hat{i}$ denotes the unit vector along the ith direction of the lattice. The lattice gauge theory (66) is invariant under the gauge transformation

$$U_i(x) \longmapsto g(x) \star U_i(x) \star g(x + \varepsilon\,\hat{i})^\dagger , \tag{71}$$

where the gauge function $g(x)$ is star-unitary,

$$(g \star g^\dagger)(x) = (g^\dagger \star g)(x) = \mathbb{1}_r . \tag{72}$$

5.3 Gauge Theory on the Fuzzy Torus

The feature which makes the noncommutative lattice gauge theory (66) exactly solvable is that the entire lattice formalism presented above can be cast into a finite dimensional version of the abstract algebraic description of gauge theory on a projective module over the noncommutative torus [36]. For this, we note that since the noncommutativity parameter of the commutant algebra is the rational number $\theta = 2/L$, the generators Z_i of $\mathcal{A}_\theta$ obey the commutation relations

$$Z_1 Z_2 = e^{4\pi i/L} Z_2 Z_1 \,. \tag{73}$$

This algebra admits a finite dimensional representation which gives the noncommutative space the geometry of a *fuzzy* torus. Namely, $\mathcal{A}_\theta$ can be represented on the finite dimensional Hilbert module $\mathbb{C}^L$, regarded as the space of functions on the finite cyclic group $\mathbb{Z}_L$, as

$$Z_1 = V_L \,, \quad Z_2 = (W_L)^2 \,, \tag{74}$$

where V_L and W_L are the $SU(L)$ shift and clock matrices which obey $V_L W_L = e^{2\pi i/L} W_L V_L$.

Since $(Z_i)^L = \mathbb{1}_L$, the matrices (74) generate the finite-dimensional algebra $\mathbb{M}_L$ of $L \times L$ complex matrices. In fact, they provide a one-to-one correspondence between lattice fields (63) with the star-product (64) and $L \times L$ matrices through

$$\hat{f} = \frac{1}{L^2} \sum_{\boldsymbol{m} \in (\mathbb{Z}_L)^2} f_{\boldsymbol{m}} \, e^{-2\pi i m_1 m_2/L} \, Z_1^{m_1} Z_2^{m_2} \,. \tag{75}$$

It is easy to check that this correspondence possesses the same formal properties as in the continuum, namely

$$\mathrm{tr}_L \, \hat{f} = f_{\boldsymbol{0}} = \frac{1}{L^2} \sum_x f(x) \,, \tag{76}$$

$$\hat{f} \, \hat{g} = \widehat{f \star g} \,. \tag{77}$$

In particular, the star-unitarity condition (69) translates into the requirement

$$\hat{U}_i \, \hat{U}_i^\dagger = \hat{U}_i^\dagger \, \hat{U}_i = \mathbb{1}_r \,. \tag{78}$$

Therefore, there is a one-to-one correspondence between $N \times N$ star-unitary matrix fields $U_i(x)$ and $r \times r$ unitary matrices $\hat{U}_i$. In the parlance of the geometry of the noncommutative torus, we have $\mathcal{A}_\theta \cong \mathbb{M}_L$ and the endomorphism algebra is $\mathrm{End}(\mathcal{E}) \cong \mathcal{A}_\theta \otimes \mathbb{M}_N \cong \mathbb{M}_r$. The gauge fields in the present setting live in the unitary group of this algebra, which is just $U(r)$ as above.

To cast the gauge theory (66) into a form which is the natural nonperturbative version of (6) [36], we introduce connections $V_i = e^{\varepsilon \nabla_i}$ on this discrete geometry which are $r \times r$ unitary matrices that may be decomposed in terms of the gauge fields $\hat{U}_i$ as

$$V_i = \hat{U}_i \, \Gamma_i \,, \tag{79}$$

where the matrices $\Gamma_i = e^{\varepsilon \, \partial_i}$ correspond to lattice shift operators. They thereby satisfy the commutation relations

$$\Gamma_1 \, \Gamma_2 = \zeta \, \Gamma_2 \, \Gamma_1 \,, \tag{80}$$

$$\Gamma_i \, Z_j \, \Gamma_i^\dagger = e^{2\pi \, i \, \delta_{ij}/L} \, Z_j \,, \tag{81}$$

where

$$\zeta = e^{2\pi \, i \, q/L} \tag{82}$$

is a $\mathbb{Z}_L$-valued phase factor whose continuum limit gives the background flux in (1). The integer q is chosen, along with some other integer c, to satisfy the Diophantine equation

$$cL - 2q = 1 \tag{83}$$

for the relatively prime pair of integers $(L, 2)$. The equations (80) and (81) can then be solved by

$$\Gamma_1 = \left(W_L^\dagger \right)^{2q} \,, \quad \Gamma_2 = \left(V_L \right)^q \,. \tag{84}$$

Note that while the Heisenberg commutation relations for constant curvature connections admit no finite dimensional representations, the Weyl-'t Hooft commutation relation (80), which is its exponentiated version, does. In other words, the matrices (84) generate the irreducible action of the Heisenberg-Weyl group on the finite-dimensional algebra $\mathcal{A}_\theta \cong \mathbb{M}_L$. This construction can be generalized to provide discrete versions of the standard Heisenberg modules over the noncommutative torus [36].

We now substitute the matrix-field correspondences (63) and (75)–(77) for the gauge fields into the partition function (66), use the fact that Γ_i generates a lattice shift along direction $\hat{i}$, and use the decomposition (79) to rewrite the action in terms of the finite dimensional connections V_i. By using in addition the invariance of the Haar measure, the Weyl-'t Hooft algebra (80), and the representation of the trace $\mathrm{tr}\,_r = \mathrm{tr}\,_L \otimes \mathrm{tr}\,_N$ on $\mathrm{End}(\mathcal{E}) \cong \mathbb{M}_r$, after some algebra we find that the partition function (66) can be written finally as the unitary two-matrix model [36]

$$\mathcal{Z}_r = \int [dV_1] \, [dV_2] \, e^{\frac{1}{2\lambda^2} \, \mathrm{Re} \, \mathrm{tr}\,_r \zeta \, V_1 \, V_2 \, V_1^\dagger \, V_2^\dagger} \,. \tag{85}$$

This is the partition function of the twisted Eguchi-Kawai model in two dimensions [37, 38], with twist given by the $\mathbb{Z}_L$ phase factor (82), and it coincides with the dimensional reduction of ordinary Wilson lattice gauge theory to a single plaquette [25]. The star-gauge invariance (71) of the plaquette model (66) corresponds to the $U(r)$ invariance

$$V_i \longmapsto \hat{g} \, V_i \, \hat{g}^\dagger \, , \quad \hat{g} \in U(r) \tag{86}$$

of the twisted Eguchi-Kawai model (85). Note that the $U(r)$ gauge symmetry of the matrix model (85) is a mixture of the original $L \times L$ spacetime degrees of freedom of the noncommutative lattice gauge theory (66) and its $U(N)$ colour symmetry. The partition function (85) is a well-defined, finite-dimensional operator version of the noncommutative Wilson lattice gauge theory in two-dimensions, which we will now proceed to compute explicitly.

5.4 Exact Solution

To evaluate the unitary group integrals (85), we insert an extra integration involving the gauge invariant delta-function acting on class functions on $U(r)$ to get

$$\mathcal{Z}_r = \int [dV_1] \, [dV_2] \int [dW] \, \delta \left(W \, , V_1 \, V_2 \, V_1^\dagger \, V_2^\dagger \right) \, e^{\frac{1}{4\lambda^2} \, \mathrm{tr}\,_r(\zeta \, W + \bar{\zeta} \, W^\dagger)} \, . \tag{87}$$

The delta-function in the Haar measure may be expanded in terms of the orthornormal $U(r)$ characters as

$$\delta(W, U) = \sum_R \chi_R(W) \, \chi_R(U^\dagger) \, . \tag{88}$$

As in Sect. 3.1, the unitary irreducible representations R of the Lie group $U(r)$ may be parametrized by partitions $\boldsymbol{n} = (n_1, \ldots, n_r)$ into r parts of decreasing integers as in (32). The character of the unitary matrix W in this representation can then be written explicitly by means of the Weyl formula

$$\chi_R(W) = \chi_{\boldsymbol{n}}(W) = \frac{\det_{a,b} \left[e^{i \, (n_a - b + r)\phi_b} \right]}{\det_{a,b} \left[e^{i \, (a-1)\phi_b} \right]} \, , \tag{89}$$

where $e^{i \, \phi_1}, \ldots, e^{i \, \phi_r}$ are the eigenvalues of W.

On substituting (88) into (87), the integration over V_1 and V_2 can be carried out explicitly by using the fusion rule (30) for the $U(r)$ characters along with the fission relation

$$\int [dU] \, \chi_{\boldsymbol{n}} \left(U \, V \, U^\dagger \, W \right) = \frac{\chi_{\boldsymbol{n}}(V) \, \chi_{\boldsymbol{n}}(W)}{d_{\boldsymbol{n}}} \, , \tag{90}$$

where

$$d_{\boldsymbol{n}} = \chi_{\boldsymbol{n}}(\mathbb{1}_r) = \prod_{a<b} \left(1 + \frac{n_a - n_b}{b - a} \right) \tag{91}$$

is the dimension $\dim R$ of the representation R with highest weight vector $\boldsymbol{n} = (n_1, \ldots, n_r)$. In this way the partition function takes the form

$$\mathcal{Z}_r = \sum_{n_1 > \cdots > n_r} \frac{1}{d_n} \int [\mathrm{d}W] \, \chi_n(W) \, \mathrm{e}^{\frac{1}{4\lambda^2} \, \mathrm{tr}_r(\zeta W + \bar{\zeta} W^\dagger)} \, . \tag{92}$$

The twist factors (82) can be decoupled from the integration in (92) by the rescaling $W \to \bar{\zeta} W$ and by using $U(r)$ invariance of the Haar measure along with the character identity

$$\chi_n(\bar{\zeta} W) = \mathrm{e}^{-2\pi i q \, C_1(n)/L} \, \chi_n(W) \, , \tag{93}$$

where

$$C_1(R) = C_1(n) = \sum_{a=1}^{r} n_a \tag{94}$$

is the linear Casimir invariant of the representation R which counts the total number of boxes in the corresponding $U(r)$ Young tableau.

We now expand the invariant function in (92) which after rescaling is the Boltzmann factor for the one-plaquette $U(r)$ Wilson action. Its character expansion can be given explicitly in terms of modified Bessel functions $I_n(z)$ of the first kind of integer order n which are defined by their generating function as

$$\exp\left[\frac{z}{2}\left(t + \frac{1}{t}\right)\right] = \sum_{n=-\infty}^{\infty} I_n(z) \, t^n \, . \tag{95}$$

By using (89) one finds [39]

$$\mathrm{e}^{\beta \, \mathrm{tr}_r(W + W^\dagger)} = \sum_{n_1 > \cdots > n_r} \det_{a,b} \left[I_{n_a - a + b}(2\beta) \right] \chi_n\left(W^\dagger\right) \, , \tag{96}$$

and, by using the fusion rule (30), substitution of (96) into (92) gives a representation of the lattice partition function as a sum over a single set of partitions alone,

$$\mathcal{Z}_r = \sum_{n_1 > \cdots > n_r} \frac{\mathrm{e}^{-2\pi i q \, C_1(n)/L}}{d_n} \det_{a,b} \left[I_{n_a - a + b}(1/2\lambda^2) \right] \, . \tag{97}$$

To express (97) as a perturbation series in the effective coupling constant $1/\lambda^2$, we substitute into this expression the power series expansion of the modified Bessel functions,

$$I_\nu(z) = \sum_{m=0}^{\infty} \frac{1}{m! \, \Gamma(\nu + m + 1)} \left(\frac{z}{2}\right)^{\nu + 2m} \, , \tag{98}$$

where $\Gamma(z)$ is the Euler function. The infinite sum may then be extracted out line by line from the determinant in (97) by using the multilinearity of the determinant as a function of its r rows, and we find

$$\mathcal{Z}_r = \sum_{n_1 > \cdots > n_r} \frac{e^{-2\pi i q C_1(n)/L}}{d_n} \sum_{m_1=0}^{\infty} \cdots \sum_{m_r=0}^{\infty} \prod_{s=1}^{r} \frac{(1/2\lambda)^{2m_s}}{m_s!}$$

$$\times \det_{a,b} \left[\frac{(1/2\lambda)^{2(m_a+n_a-a+b)}}{\Gamma(m_a + n_a - a + b + 1)} \right]. \tag{99}$$

Note that the total contribution to (99) vanishes from any set of integers for which $m_a + n_a < a - r$ for any *single* index $a = 1, \ldots, r$.

The determinant in (99) can be evaluated explicitly as follows. For any sequence of integers $s_1, \ldots, s_r$, we have

$$\det_{a,b} \left[\frac{z^{s_a-a+b}}{\Gamma(s_a - a + b + 1)} \right] = z^{s_1+\cdots+s_r} \begin{vmatrix} \frac{1}{\Gamma(s_1+1)} & \frac{1}{\Gamma(s_2)} & \cdots & \frac{1}{\Gamma(s_r-r+2)} \\ \frac{1}{\Gamma(s_1+2)} & \frac{1}{\Gamma(s_2+1)} & \cdots & \frac{1}{\Gamma(s_r-r+3)} \\ \vdots & \vdots & \ddots & \vdots \\ \frac{1}{\Gamma(s_1+r)} & \frac{1}{\Gamma(s_2+r-1)} & \cdots & \frac{1}{\Gamma(s_r+1)} \end{vmatrix}. \tag{100}$$

Factorizing $1/\Gamma(s_b - b + r + 1)$ from each column b of the remaining determinant in (100) yields

$$\det_{a,b} \left[\frac{z^{s_a-a+b}}{\Gamma(s_a - a + b + 1)} \right] = z^{s_1+\cdots+s_r} \prod_{b'=1}^{r} \frac{1}{\Gamma(s_{b'} - b' + r + 1)}$$

$$\times \det_{a,b} \left[(s_b - b + a + 1)(s_b - b + a + 2) \cdots (s_b - b + r) \right]. \tag{101}$$

The argument of the determinant in the right-hand side of (101) is a monic polynomial in the variable $\alpha_b = s_b - b$ with highest degree term α_b^{r-a}. By using multilinearity of the determinant, it becomes $\det_{a,b}[\alpha_a^{r-b}] = \prod_{a<b}(\alpha_a - \alpha_b)$, and we arrive finally at

$$\det_{a,b} \left[\frac{z^{s_a-a+b}}{\Gamma(s_a - a + b + 1)} \right] = z^{s_1+\cdots+s_r} \prod_{b=1}^{r} \frac{(r-b)!}{\Gamma(s_b - b + r + 1)}$$

$$\times \prod_{a<b} \left(1 + \frac{s_a - s_b}{b - a} \right). \tag{102}$$

Note that if $s = (s_1, \ldots, s_r)$ is a partition, then the last product in (102) is just the dimension d_s of the corresponding $U(r)$ representation.

The partition function (99) is thereby given as

$$\mathcal{Z}_r = \sum_{n_1 > \cdots > n_r} \frac{e^{-2\pi i q C_1(n)/L}}{d_n (2\lambda)^{2C_1(n)}}$$

$$\times \sum_{m_1=0}^{\infty} \cdots \sum_{m_r=0}^{\infty} \prod_{b=1}^{r} \frac{(r-b)! (1/2\lambda)^{4m_b}}{m_b! \, \Gamma(m_b + n_b - b + r + 1)}$$

$$\times \prod_{a<b} \left(1 + \frac{m_a - m_b + n_a - n_b}{b - a} \right). \tag{103}$$

Finally, we can simplify this expansion for $\mathcal{Z}_r$ even further by decoupling the sum over partitions $\boldsymbol{n} = (n_1, \ldots, n_r)$. For this, we define a new set of integers by

$$p_a = n_a - n_{a+1} + 1 , \quad a = 1, \ldots, r-1 ,$$
$$p_r = n_r . \tag{104}$$

Then the p_a's are all independent variables, constrained only by their ranges which are given by $1 \leq p_a < \infty$ for $a = 1, \ldots, r-1$ and $-\infty < p_r < \infty$.

The decoupled expansion of the partition function is thereby obtained by substituting

$$n_a = p_a + p_{a+1} + \cdots + p_r + a - r , \tag{105}$$

along with the explicit group theoretical formulas (91) and (94), into (103) to get the final result (up to irrelevant numerical factors)

$$\mathcal{Z}_r = \sum_{p_1=1}^{\infty} \cdots \sum_{p_{r-1}=1}^{\infty} \sum_{p_r=-\infty}^{\infty} \cos\left(\frac{2\pi q}{L} \sum_{b=1}^{r} b\, p_b\right)$$
$$\times \sum_{m_1=0}^{\infty} \cdots \sum_{m_r=0}^{\infty} \prod_{b=1}^{r} \frac{(b-1)!\,(2\lambda)^{-4m_b - 2b\, p_b}}{m_b!\; \Gamma(m_b + p_b + p_{b+1} + \cdots + p_r + 1)}$$
$$\times \prod_{a<b} \frac{m_a - m_b + p_a + p_{a+1} + \cdots + p_b}{p_a + p_{a+1} + \cdots + p_b} , \tag{106}$$

where we have used the reality of the left-hand side of (96) to make the expression for the partition function manifestly real by adding its complex conjugate to itself. The partition function (106) is a straightforward expansion in powers of $1/\lambda^2$ over $2r$ independent integers p_a, m_a, $a = 1, \ldots, r$. Note the reduction in the number of dynamical degrees of freedom of the model. The original $2r^2$ degrees of freedom of the two-dimensional lattice gauge theory (66) (or equivalently of the unitary two-matrix model (85)) is reduced to $2r$. This proves that the lattice model is exactly solvable, and thereby gives yet another indication that noncommutative gauge theory in two dimensions is a topological field theory. The sum (106) is formally analogous to the partition expansion of continuum noncommutative Yang-Mills theory.

5.5 Scaling Limits

The final step of this calculation should be to take the continuum limit $\varepsilon \to 0$ of the lattice theory. In order to prevent the spacetime from degenerating to zero area, from (62) we see that we must also take $L \to \infty$, or equivalently $r \to \infty$ in (106). There are different ways of performing these two limits, each of which leads to a different continuum gauge theory. If the limit is taken such that the dimensionful noncommutativity parameter (65) vanishes, then the continuum limit is ordinary Yang-Mills theory in two dimensions. The area (62) may be either finite or infinite in this limit. If $A \to \infty$, then the expansion (97) truncates to the trivial representation for which $n_a = 0 \ \forall a$ and one obtains

$$\left. \mathcal{Z}_r \right|_{\substack{\Theta \to 0 \\ A \to \infty}} = \det_{a,b} \left[I_{b-a}(2/\lambda^2) \right] . \tag{107}$$

This expression is recovered in the naive large r limit due to the suppression of higher representations which is induced by the dimension factors d_n in the denominators of (97). It is just the standard expression for Yang-Mills theory on the plane which arises from the one-plaquette Wilson model in the limit of a large number of colours [40]. Going back to (92), we see that the truncation to $n = 0$ is indeed nothing but the Gross-Witten reduction of commutative lattice gauge theory in two dimensions [35].

The other scaling limit that one can take is $\varepsilon \to 0$, $L \to \infty$ with $\varepsilon^2 L$ finite. Then the noncommutativity parameter (65) is finite, but the area (62) diverges. The resulting continuum limit is gauge theory on the noncommutative plane, and from (92) we see that its partition function generalizes that of ordinary Yang-Mills theory by including a sum over non-trivial representations of the unitary group. This quantitative difference is similar in spirit to that which occurs in the group theory presentation of noncommutative gauge theory [14], which can be thought of as a modification of ordinary gauge theory by the addition of infinitely many higher Casimir operators to the action (equivalently higher powers of the field strength F_A). The inclusion of higher representations in the statistical sum means that this series cannot be expressed in terms of a unitary one-matrix model. Determinants such as (107) whose matrix elements depend only on the difference between row and column labels are called Toeplitz determinants and are known to be equivalent to the evaluation of a related unitary one-matrix integral [41]. In the present noncommutative case, the partition function is not given by a Toeplitz determinant, although it is represented by the unitary two-matrix model (85) and depends only on the eigenvalues of the matrix $W = V_1 V_2 V_1^\dagger V_2^\dagger$.

Unravelling the precise continuum limit of the expansion (106) is one of the important unsolved analytical problems in the combinatorial approach to two-dimensional noncommutative Yang-Mills theory. The noncommutativity parameter Θ enters in the 't Hooft coupling constant as $\lambda^2 = \pi g^2 \Theta$ and implicitly in the factors of $L = r/N$ appearing in (106). It is necessary to identify whether the double-scaling limit required, over and above the naive continuum limit, exists within this framework. Both the naive and non-trivial double-scaling limits have been observed numerically in the Eguchi-Kawai model [42], and more recent numerical investigations indicate that they exist also within the full noncommutative field theory [9],[43]–[45]. The rigorous derivation of this limit is described at the classical level in [31]. Amongst other things, the solution to this system may help in unravelling the mysterious properties of the gauge group of noncommutative gauge theory, which in the present context is formally an $r \to \infty$ limit of $U(r)$, confirming other independent expectations [1, 4, 10],[46]–[48]. It would also be interesting to understand the complete solution of the discrete theory whose continuum spacetime is a torus, which is given by a more general construction [36] to which the present analysis does not apply.

Acknowledgement

R.J.S. would like to thank the organisors and participants of the meetings for the many questions and comments which have helped to improve the material presented here, and also for the very pleasant scientific and social atmospheres. He would also like to thank the School of Theoretical Physics of the Dublin Institute for Advanced Study for its hospitality during the completion of the manuscript. The work of R.J.S. was supported in part by an Advanced Fellowship from the Particle Physics and Astronomy Research Council (U.K.).

References

1. L.D. Paniak and R.J. Szabo, hep-th/0302195.
2. A.P. Polychronakos, Phys. Lett. B **495**, 407 (2000) [hep-th/0007043].
3. D. Bak, Phys. Lett. B **495**, 251 (2000) [hep-th/0008204].
4. D.J. Gross and N.A. Nekrasov, JHEP **0103**, 044 (2001) [hep-th/0010090].
5. D. Bak, K. Lee and J.-H. Park, Phys. Rev. D **63**, 125010 (2001) [hep-th/0011099].
6. A. Bassetto, G. Nardelli and A. Torrielli, Nucl. Phys. B **617**, 308 (2001) [hep-th/0107147].
7. Z. Guralnik, JHEP **0206**, 010 (2002) [hep-th/0109079].
8. L. Griguolo, D. Seminara and P. Valtancoli, JHEP **0112**, 024 (2001) [hep-th/0110293].
9. W. Bietenholz, F. Hofheinz and J. Nishimura, JHEP **0209**, 009 (2002) [hep-th/0203151].
10. L.D. Paniak and R.J. Szabo, hep-th/0203166.
11. A. Bassetto, G. Nardelli and A. Torielli, Phys. Rev. D **66**, 085012 (2002) [hep-th/0205210].
12. A. Bassetto and F. Vian, JHEP **0210**, 004 (2002) [hep-th/0207222].
13. A. Torrielli, hep-th/0301091.
14. L.D. Paniak and R.J. Szabo, JHEP **0305**, 029 (2003) [hep-th/0302162].
15. A. Konechny and A. Schwarz, Phys. Rept. **360**, 353 (2002) [hep-th/0012145 , hep-th/0107251].
16. M.R. Douglas and N.A. Nekrasov, Rev. Mod. Phys. **73**, 977 (2002) [hep-th/0106048].
17. R.J. Szabo, Phys. Rept. **378**, 207 (2003) [hep-th/0109162].
18. S. Cordes, G. Moore and S. Ramgoolam, Nucl. Phys. Proc. Suppl. **41**, 184 (1995) [hep-th/9411210].
19. A. Gorsky and N. A. Nekrasov, Nucl. Phys. B **414**, 213 (1994) [hep-th/9304047].

20. J.A. Minahan and A.P. Polychronakos, Phys. Lett. B **312**, 155 (1993) [hep-th/9303153].
21. G. Grignani, L.D. Paniak, G.W. Semenoff and P. Sodano, Ann. Phys. **260**, 275 (1997) [hep-th/9705102].
22. E. Witten, J. Geom. Phys. **9**, 303 (1992) [hep-th/9204083].
23. A.A. Migdal, Sov. Phys. JETP **42**, 743 (1975).
24. B.E. Rusakov, Mod. Phys. Lett. A **5**, 693 (1990).
25. J. Ambjørn, Y.M. Makeenko, J. Nishimura and R.J. Szabo, JHEP **0005**, 023 (2000) [hep-th/0004147].
26. E. Witten, Commun. Math. Phys. **141**, 153 (1991).
27. R.J. Szabo, *Equivariant Cohomology and Localization of Path Integrals*, Lect. Notes Phys. **m63** (Springer-Verlag, 2000).
28. M.F. Atiyah and R. Bott, Phil. Trans. Roy. Soc. London A **308**, 523 (1982).
29. N. Seiberg and E. Witten, JHEP **9909**, 032 (1999) [hep-th/9908142].
30. L. Alvarez-Gaumé and J.L.F. Barbòn, Nucl. Phys. B **623**, 165 (2002) [hep-th/0109176].
31. G. Landi, F. Lizzi and R.J. Szabo, Commun. Math. Phys. **217**, 181 (2001) [hep-th/9912130].
32. A. Connes, M.R. Douglas and A. Schwarz, JHEP **9802**, 003 (1998) [hep-th/9711162].
33. A. Schwarz, Nucl. Phys. B 534, 720 (1998) [hep-th/9805034].
34. K.G. Wilson, Phys. Rev. D **10**, 2445 (1974).
35. D.J. Gross and E. Witten, Phys. Rev. D **21**, 446 (1980).
36. J. Ambjørn, Y. M. Makeenko, J. Nishimura and R. J. Szabo, JHEP **9911**, 029 (1999) [hep-th/9911041].
37. T. Eguchi and H. Kawai, Phys. Rev. Lett. **48**, 1063 (1982).
38. A. Gonzalez-Arroyo and C.P. Korthals Altes, Phys. Lett. B **131**, 396 (1983).
39. J.-M. Drouffe and J.-B. Zuber, Phys. Rept. **102**, 1 (1983).
40. I. Bars and F. Green, Phys. Rev. D **20**, 3311 (1979).
41. C. Itzykson, in: *Analytic Methods in Mathematical Physics* (Gordon and Breach, 1970), p. 469.
42. T. Nakajima and J. Nishimura, Nucl. Phys. B **528**, 355 (1998) [hep-th/9802082].
43. W. Bietenholz, F. Hofheinz and J. Nishimura, hep-lat/0209021.
44. W. Bietenholz, F. Hofheinz and J. Nishimura, hep-th/0212258.
45. J. Ambjørn and S. Catterall, Phys. Lett. B **549**, 253 (2002) [hep-lat/0209106].
46. V.P. Nair and A.P. Polychronakos, Phys. Rev. Lett. **87**, 030403 (2001) [hep-th/0102181].
47. J.A. Harvey, hep-th/0105242.
48. F. Lizzi, R.J. Szabo and A. Zampini, JHEP **0108**, 032 (2001) [hep-th/0107115].

Part IV

Topological Quantum Field Theory

Topological aspects in quantum field theory are attracting a great deal of attention recently. Especially, the construction of topological invariants is now one of the main subjects in this field, as these quantities carry rich information about geometrical data.

The first contribution of this part gives a detailed introduction to the theory of topological invariants arising from operator algebras and a review of the current status of the theory of quantum topological invariants of 3-manifolds.

The second contribution gives a survey of operads and their relation to topological quantum field theories. It contains a review of the notion of topological quantum field theory as described by Atiyah. The definition of operads and algebras over them is presented and several concrete examples are described. The concept of cohomological field theory is introduced.

The third contribution discusses some topological aspects of the classification of subfactors of von Neumann algebra.

Topological Quantum Field Theories and Operator Algebras

Y. Kawahigashi

Department of Mathematical Sciences, University of Tokyo, Komaba, Tokyo, 153-8914, JAPAN
yasuyuki@ms.u-tokyo.ac.jp

1 Introduction

We have seen much fruitful interactions between 3-dimensional topology and operator algebras since the stunning discovery of the Jones polynomial for links [19] arising from his theory of subfactors [18] in theory of operator algebras. In this paper, we review the current status of theory of "quantum" topological invariants of 3-manifolds arising from operator algebras. The original discovery of topological invariants arising from operator algebras was for knots and links, as above, rather than 3-manifolds, but here we concentrate on invariants for 3-manifolds. On the way of studying such topological invariants, we naturally go through topological invariants of knots and links. From operator algebraic data, we construct not only topological invariants of 3-manifolds, but also topological quantum field theories of dimension 3, in the sense of Atiyah [2], as the title of this paper shows, but for simplicity of expositions, we consider mainly complex number-valued topological invariants of oriented compact manifolds of dimension 3 without boundary.

All the constructions of such topological invariants we discuss here are given in the following steps.

1. Obtain combinatorial data arising from representation theory of an operator algebraic system.
2. Realize a manifold concretely using basic building blocks.
3. Multiply or add the complex numbers appearing in the data in Step 1, in a way specified by how the basic building blocks are composed in Step 2, and compute the resulting complex number.
4. Prove that the complex number in Step 3 is independent of how the basic building blocks are composed, as long as the homeomorphism class of the resulting manifold is fixed.

In Step 1, the prototype of the representation theory for operator algebras is the one for finite groups. That is, for a finite group G, we consider representatives of unitary equivalences classes of irreducible unitary representations. This finite set has an algebraic structure arising from the tensor product operation of representations, and it produces combinatorial data such as fusion rules and $6j$-symbols. In our setting, we work on some form of representation

Y. Kawahigashi: *Topological Quantum Field Theories and Operator Algebras*, Lect. Notes Phys.
662, 241–253 (2005)
www.springerlink.com

theory of operator algebraic systems analogous to this classical representation theory of finite groups.

Steps 2 and 3 already appear in the original definition of the Jones polynomial [19], where each link is represented as a closure of a braid, the Jones polynomial is defined from such a braid through certain representation theory, and then it is proved that this polynomial is independent of a choice of a braid for a fixed link.

This strategy should work, in principle, in any dimension, but so far, most of the interesting constructions arising from operator algebras are for dimension 3, so we concentrate in this case in this survey.

There have been many constructions of such topological invariants for 3-dimensional manifolds and two of them are particularly related to operator algebras. One is a construction of Turaev-Viro [36] in a generalized form due to Ocneanu, and the other is the one by Reshetikhin-Turaev [33]. For these two, the triple of operator algebraic systems, representation theoretic data, and the topological invariants in each case is listed as in Table 1.

Table 1. Topological invariants arising from operator algebras

Operator Algebras	Representation Theory	Combinatorial Construction
Subfactors	Quantum $6j$-symbols	TVO invariants
Nets of factors on S^1	Braided tensor categories	RT invariants

Since both operator algebras and (topological) quantum field theory are of infinite dimensional nature, one expects a direct and purely infinite dimensional construction of the latter from the former, but such a construction has not been known yet. All the constructions below go through representation theoretic combinatorial data who "live in" finite dimensional spaces, so one could eliminate the initial infinite dimensionality entirely, if one is interested in only new constructions and computations of topological invariants of 3-dimensional manifolds. Still, the infinite dimensional framework of operator algebras is useful, as we see below, even in such a case, because it gives a conceptually convenient working place for various constructions and computations.

We also mention one reason we operator algebraists are interested in this type of theory, even purely from a viewpoint of operator algebras. Classification theory is a central topic in theory of operator algebras, and representation theory gives a very important invariant for classification. Since a series of great works of A. Connes in 1970's, it is believed that under some nice analytic condition, generally called "amenability", a certain representation theory should give a complete invariant of operator algebraic systems, such as operator algebras themselves, group actions on them, or certain families of operator algebras. For this reason, studies of representation theories in operator algebraic theory are quite important since old days of theory of operator algebras. What is new after the emergence of the Jones theory is that the representation theory now has a "quantum" nature, whatever it means.

The author thanks R. Longo, N. Sato, and H. Wenzl for comments on this manuscript.

2 Turaev-Viro-Ocneanu Invariants

Here we review the Turaev-Viro-Ocneanu invariants of 3-dimensional manifolds. The book [11] is a basic reference.

Our operator algebra here is a so-called *von Neumann algebra*, which is an algebra of bounded linear operators on a certain Hilbert space that is closed under the ∗-operation and the strong operator topology. (Here we consider only infinite dimensional separable Hilbert spaces, though a general theory exists for other Hilbert spaces.) Requiring closedness under the weak operator topology, we obtain the same class of operator algebras. If we use a norm topology, we have a wider class of operator algebras called C^*-*algebras*. Although von Neumann algebras give a subclass of C^*-algebras, it is not very useful, except for some elementary aspects of the theory, to regard a von Neumann algebra as a C^*-algebra, because a von Neumann algebra is far from being a "typical" C^*-algebra. For example, most of natural C^*-algebras are separable, as Banach spaces, but von Neumann algebras are never separable, unless they are finite dimensional. We assume, as usual, that a von Neumann algebra contains the identity operator, which is the unit of the algebra. A commutative C^*-algebra having a unit is the algebra of all the continuous functions on a compact Hausdorff space, and a commutative von Neumann algebra is the algebra of L^∞-functions on a measure space. This gives a reason for a basic idea that a general C^*-algebra is a "noncommutative topological space" and a general von Neumann algebra is a "noncommutative measure space". A finite dimensional C^*- or von Neumann algebra is isomorphic to a finite direct sum of full matrix algebras $M_n(\mathbf{C})$. In this paper, we deal with only *simple* von Neumann algebras in the sense that they have only trivial two-sided closed ideals in the strong or weak operator topology. This simplicity is equivalent to triviality of the center of the algebra, and we call such a von Neumann algebra a *factor*, rather than a simple von Neumann algebra.

In the Murray-von Neumann classification, factors are classified into type I, type II_1, type II_∞, and type III. Factors of type I are simply all the bounded linear operators on some Hilbert space, and they are not interesting for the purpose of this survey. We are interested in factors of type II_1 in the following two sections and those of type III in the last section. Although technical details on these factors are not necessary for conceptual understanding of the theory, we give brief explanations on how to construct such factors.

We start with a countable group G. The (left) regular representation gives a unitary representation of G on the Hilbert space $\ell^2(G)$. We consider the von Neumann algebra generated by its image. If the group G is commutative, the resulting von Neumann algebra is isomorphic to $L^\infty(\hat{G})$. If the group G is "reasonably noncommutative" in an appropriate sense, the resulting von Neumann algebra is a factor of type II_1. One example of such a group is that of all permutations of a countable set that fix all but finite elements.

Another construction of a factor arises from an infinite tensor product of the $n \times n$-matrix algebra $M_n(\mathbf{C})$. We can define such an infinite tensor product in an appropriate sense, and then this infinite dimensional algebra has a natural representation on a separable Hilbert space. The von Neumann algebra generated by its image is a type II_1 factor and these are all isomorphic, regardless n. This infinite tensor product also has many other representations on Hilbert spaces and "most" of them generate factors of type III.

The most natural starting point of a representation theory for factors is certainly a study of all representations of a fixed factor on Hilbert spaces. (A factor is an algebra of operators on a certain Hilber space by definition, but we consider representations on other Hilbert spaces. In our setting, it is enough to consider only representations on infinite dimensional separable Hilbert spaces.) We certainly have a natural notion of unitary equivalence for representations of factors of type II_1 or III, but this notion is not particularly interesting, as follows. Such representations are never irreducible, and for a fixed type II_1 factor, we can classify representations completely, up to unitary equivalence, with a single invariant, called a coupling constant, due to Murray and von Neumann, having values in $(0, \infty]$. (This invariant produces the Jones index as below, and produces something deep in this sense, but the classification of representations themselves is rather simple and classical.) For factors of type III, the situations are even simpler; they are all unitarily equivalent for a fixed type III factor.

A representation of a factor can be regarded as a (left) module over a factor, trivially. It was Connes who realized first that the right setting for studying representation theory of factors is to study *bimodules*, rather than modules. That is, we consider two factors M and N, which could be equal, and study a Hilbert space H which is a left M-module and a right N-module with the two actions commuting. We call such H an M-N bimodule and write ${}_M H_N$. The situation where both M and N are of type II_1 is technically simpler. We have natural notions of irreducible decomposition, dimensions having values in $(0, \infty]$ which are defined in terms of the coupling constants, contragredient bimodules, and relative tensor products. For example, for two bimodules ${}_M H_N$ and ${}_N K_P$, we can define an M-P bimodule ${}_M H \otimes_N K_P$ and the dimension is multiplicative. For a factor M, the algebra M itself trivially has the left and right actions of M, so it has a bimodule structure, but this M is not a Hilbert space. We have a natural method to put an inner product on M and complete it, and in this way, we obtain an M-M bimodule. By an abuse of notation, we often write $_M M_M$ for this bimodule, by ignoring the completion. This bimodule has dimension one, and plays a role of a trivial representation. In this way, our representation theory is quite analogous to that of a compact group. Connes used a terminology *correspondences* rather than bimodules. See [30] for a general theory on bimodules.

Jones initiated studies of inclusions of factors $N \subset M$ in [18]. Such N is called a *subfactor* of M. By an abuse of terminology, the inclusion $N \subset M$ is often called a subfactor. Technically simpler situations are that both M and N are of type II_1. Then we have an M-M bimodule $_M M_M$ as above, and we restrict the left action to the subalgebra N to obtain $_N M_M$. The dimension of this bimodule is called the *Jones index* of the subfactor $N \subset M$ and written as $[M : N]$. (This terminology and notation come from an analogy to a notion of an index of a subgroup.) Jones proved in [18] an astonishing statement that this index takes values in $\{4\cos^2(\pi/n) \mid n = 3, 4, 5 \ldots\} \cup [4, \infty]$ and all the values in this set are realized. This is in a sharp contrast to the fact that the coupling constant of a type II_1 factor M can take all values in $(0, \infty]$. Jones introduced the *basic construction* whose successive uses produce an increasing sequence $N \subset M \subset M_1 \subset M_2 \subset \cdots$ and using this, he introduced the *higher relative commutants* and the *principal graph* for subfactors. Although we do not give their definitions here, we only mention that if the subfactor has index less than 4, then the principal graph is one of the *A-D-E* Dynkin diagrams, as noted by Jones. (See [11, Chap. 9] for precise definitions.)

It was Ocneanu [27] who realized that these invariants and further finer structures related to them can be captured by theory of bimodules and that they can be characterized by a set of combinatorial axioms. We explain his theory here. See [11, Chap. 9] for more details. We start with a type II_1 subfactor $N \subset M$ with finite Jones index. (If we have a finite index and one of N and M is of type II_1, then the other is also of type II_1 automatically.) Ocneanu's idea was to develop a *representation theory for a pair* $N \subset M$. We start with $_N M_M$ and this plays a role of the fundamental representation. We also have $_M M_N$ and make relative tensor products such as $_N M \otimes_M M \otimes_N M_M$. They are not irreducible in general, so we make irreducible decompositions. We look at all unitary equivalence classes of N-N bimodules arising in this way. In general, we expect to have infinitely many equivalence classes, but it sometimes happens that we have only finitely many equivalence classes. This is the situation we are interested in, and in such a case, we say that the subfactor $N \subset M$ has a *finite depth*. (The terminology "depth" comes from the way of Jones to write higher relative commutants.) This finite depth condition is similar to rationality condition in conformal field theory and quantum group theory. If we have a finite depth, we also have only finitely many equivalence classes of irreducible M-M bimodules arising in the above way. Note that a compact group has only finitely many equivalence classes of irreducible unitary representations if and only if the group is finite. We assume the finite depth condition and fix a finite set of representatives of equivalence classes of irreducible N-N bimodules arising as above from $N \subset M$. Note that it contains a trivial bimodule, that for each bimodule in the set, its contragredient bimodule is equivalent to one in the set, and that a relative tensor product of two in the set decomposes into a finite direct sum of irreducible bimodules each of which is equivalent to one in the set. We say such a finite set of bimodules is a finite system of bimodules. Choose three, not necessarily distinct, irreducible N-N bimodules A, B, C in the system. Then we can decompose $A \otimes_N B \otimes_N C$ in two ways. That is, we first decompose $A \otimes_N B$ in one, and we first decompose $B \otimes_N C$ in the other. In this way, we obtain the "quantum" version of the classical $6j$-symbols which produce a complex number from six bimodules and four intertwiners. Such quantum $6j$-symbols were known in the quantum group theory, and Ocneanu found that a general system of bimodules produce similar $6j$-symbols and that classical properties such as the Frobenius reciprocity also holds in this setting. Associativity of the relative tensor product gives a so-called pentagonal relation as in the classical setting. This finite system of bimodules and quantum $6j$-symbols are the combinatorial data arising from a representation theory of a subfactor $N \subset M$.

Turaev and Viro [36] constructed topological invariants of 3-dimensional manifolds using the quantum $6j$-symbols for the quantum group $U_q(sl_2)$ at roots of unity, and Ocneanu realized that a generalized version of this construction works for general quantum $6j$-symbols arising from a subfactor of finite Jones index and finite depth as above. The construction goes as follows for a fixed finite system of bimodules. (See [11, Chap. 12] for more details.)

We first make a triangulation of a manifold. That is, we regard a manifold made of gluing faces of finitely many tetrahedra so that we have an empty boundary and compatible orientation. Then we label each of the six edges with bimodules in the system and each of the four faces, triangles, with (co-)isometric intertwiners. When all the tetrahedra are labeled in this way, the quantum $6j$-symbol produce a complex number for each labeled tetrahedron. This number is simply a composition of the four intertwiners, up to normalization arising from dimensions of the four bimodules. (The composed intertwiners give a complex number because of irreducibility of the bimodules.) The well-definedness of this number comes from the so-called tetrahedral symmetry of quantum $6j$-symbols. Then we multiply all these numbers over all the tetrahedra in the triangulation, and add these products over all isometric intertwiners in an orthonormal basis for each face and over all labeling of edges with bimodules. With an appropriate normalization arising from dimensions of the bimodules, the resulting number is a topological invariant of the original 3-dimensional manifold. In order to prove this topological invariance, one has to prove that the complex number is independent of triangulations of a manifold. The relations of two triangulation of a manifold have been known by Alexander. That is, one triangulation is obtained from the other by successive applications of finitely many local changes of triangulations, called Alexander moves. (This result of Alexander holds in any dimension.) Pachner has proved that a different set of local moves also gives a similar theorem, and this set is more convenient for our purpose. That is, it is enough for us to prove that the above complex number is invariant under each of the Pachner moves. This invariance follows from properties of the quantum $6j$-symbols, such as the pentagon relation. So we conclude that the above complex number gives a well-defined topological invariant of 3-dimensional closed oriented manifolds. If we reverse the orientation, the topological invariant becomes the complex conjugate of the original value. In the original setting of Turaev-Viro [36] based on the quantum $6j$-symbols of $U_q(sl_2)$, the resulting invariants are real, so they do not detect orientations, but there is an example of a subfactor which produces a non-real invariant for some manifold and thus can detect orientations. (Actually, the original construction of Turaev-Viro [36] works without orientability.) Also note that in our setting, each intertwiner space has a Hilbert space structure and each dimension of a bimodule, which is sometimes called a *quantum dimension*, is positive. Such a feature is called *unitarity* of quantum $6j$-symbols, and this unitarity does not necessarily hold in a purely algebraic setting of quantum $6j$-symbols for quantum groups. We can apply the same construction by using the system of the M-M bimodules instead of that of the N-N bimodules, but the resulting invariant is the same.

A large class of subfactors are constructed with methods related to classical theory of groups and Hopf algebras, and their "quantum" counterparts, that is, quantum group theory and conformal field theory such as the Wess-Zumino-Witten models. For such subfactors, we have various interesting studies from an operator algebraic viewpoint, but if we are interested only in resulting topological invariants through the above machinery, they do not produce really new invariants. It is, however, expected that we have much wider varieties of subfactors in general. One "evidence" for such expectation is study of Haagerup [15]. By purely combinatorial arguments, he found a list of candidates of subfactors of finite depth in the index range $(4, 3 + \sqrt{2})$, and it seems that most of these are indeed realized. None of them seem to be related to conformal field theory or today's theory of quantum groups so far. Haagerup himself proved that the first one in the list is indeed realized, and Asaeda and he further proved that another in the list is also realized in [1]. The nature of topological invariants arising from these two subfactors is not understood yet, but we expect that they contain some interesting information. Since the list of Haagerup is only for a small range of the index values, we expect that we would have by far more examples of "exotic" subfactors as mentioned above, but an explicit construction of even a single example is highly difficult. We know almost nothing about topological meaning of invariants arising from such subfactors. Izumi [17] has some more examples of such interesting subfactors.

3 Reshetikhin-Turaev Invariants

Another construction of topological invariants due to Reshetikhin-Turaev [36] requires a "higher symmetry" for combinatorial data arising from a representation theory. This higher symmetry is called a *modularity* of a tensor category. It is also called a *nondegenerate braiding*.

Wenzl has a series of work [37, 38, 39, 40], partly with V. G. Turaev, on related constructions, but here we concentrate on two methods producing a modular tensor category from a general operator algebraic representation theory. One is within subfactor theory, due to Ocneanu, and presented in this section, and the other is due to Longo, Müger and the author [22], explained in the next section.

We first give a brief explanation on braiding. In a representation theory of a group, two tensor products $\pi \otimes \sigma$ and $\sigma \otimes \pi$ are obviously unitarily equivalent for two representations π and σ, but for two N-N bimodules A, B, we have no reason to expect that $A \otimes_N B$ and $B \otimes_N A$ are equivalent, and they are indeed not equivalent in general. It is, however, possible that for all A and B in a finite system, we have equivalence of $A \otimes_N B$ and $B \otimes_N A$. If we can choose isomorphisms of these two bimodules in a certain compatible way simultaneously for all bimodules in the system, we say that the system has a *braiding*. See [32] for more details, where an equivalent, but slightly different formulation using endomorphisms, rather than bimodules, is presented.

The isomorphism between $A \otimes_N B$ and $B \otimes_N A$ can be graphically represented as an overcrossing of two wires labeled with A and B, respectively. Then the assumption on "compatibility" implies, for example, the Yang-Baxter equation, which represents the Reidemeister move of type III as in Fig. 1, where each crossing represents an isomorphism and each hand side is a composition of three such isomorphisms.

In representation theory of groups, the tensor product operation is trivially commutative in the above sense. This is "too commutative" in the sense that we have no distinction between an overcrossing and an undercrossing in the above graphical representation, and this is not very useful for construction

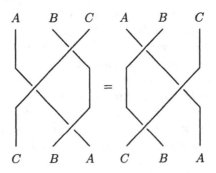

Fig. 1. Yang-Baxter equation

of topological invariants, obviously. So, in order to obtain an interesting topological invariant, an overcrossing and an undercrossing must be "sufficiently different". Such a condition is called nondegeneracy of the braiding. This condition can be also formulated in the language of tensor categories, and then it is called a modularity of the tensor category. A non-degenerate braiding, or a modular tensor category, produces a unitary representation of the modular group $SL(2, \mathbf{Z})$.

We first explain how to obtain such a nondegenerate braiding in subfactor theory. We start with a subfactor $N \subset M$ with finite Jones index and finite depth. Then Ocneanu has found a construction of a new subfactor from this subfactor, which is called the asymptotic inclusion [27]. He realized that the system of bimodules for this new subfactor has a nondegenerate braiding and it can be regarded as the "quantum double" of the original system of N-N (or M-M) bimodules arising from the subfactor $N \subset M$. Note that the original system of N-N bimodules and that of M-M bimodules are not isomorphic in general, but they have the same "quantum double" system of bimodules. Popa has a more general construction of this type, called the symmetric enveloping algebra [31]. Longo-Rehren [25] has found the essentially same construction as the asymptotic inclusion in the setting of algebraic quantum field theory. See [11, Chap. 12] for more details on the asymptotic inclusion and [16, 17] for detailed analysis based on the Longo-Rehren approach.

Suppose we have a nondegenerate braiding. It is also known that such a braiding can arise from quantum groups or conformal field theory. Reshetikhin-Turaev [33] has constructed a topological invariant of 3-dimensional manifold from such a system. First we draw a picture of a link on a plane. This has various overcrossings and undercrossings. We label each connected component with an irreducible bimodule in the system, then each crossing gives an isomorphism arising from the braiding. Then this labeled picture produces a complex number as a composition of these isomorphisms. This is an invariant of "colored links", where coloring means labeling of each component with an irreducible bimodule. Actually, this number is not invariant under the Reidemeister move of type I, and it is invariant under only the Reidemeister moves of type II and type III, so this is not a topological invariant of colored links, but it gives a "regular isotopy" invariant of colored links, for which invariance under the Reidemeister moves of type II and type III is sufficient. Then we sum these complex numbers over all possible colorings, with appropriate normalizing weights arising from dimensions of the bimodules. In this way, we obtain a complex number from a planar picture of a link. There is a method to construct a 3-dimensional oriented closed manifold from such a planar picture of a link, called the *Dehn surgery*. Roughly speaking, we embed a link in the 3-sphere, and remove a tubular neighbourhood, consisting of a disjoint union of solid tori, from the 3-sphere, and then put back the solid tori in a different way. Different links can produce the same 3-dimensional manifolds, but again, it is known that in such a case, the two links can be transformed from one to the other with successive applications of local moves. Such moves are called Kirby moves. Reshetikhin and Turaev have proved that nondegeneracy of the braiding implies invariance of the above complex number, the weighted sum of colored link invariants, under Kirby moves, thus we obtain a topological invariant of 3-dimensional manifolds in this way. Reshetikhin and Turaev considered an example arising from the quantum groups $U_q(sl_2)$ at roots of unity, but the general machinery applies to any nondegenerate braiding. See the book [35] for more details on this construction.

So, starting with a subfactor with finite Jones index and finite depth, we have two topological invariants of 3-dimensional manifolds. One is the Turaev-Viro-Ocneanu invariant arising from the system of N-N bimodules. The other is the Reshetikhin-Turaev invariant of the "quantum double" system of the original system of N-N bimodules. It is quite natural to investigate the relation between these two invariants. Sato, Wakui and the author proved in [23] that these two invariants coincide. Ocneanu [29] has also announced such coincidence and it seems to us that his method is different from ours. Sato and Wakui [34] also made explicit computations of this invariant for various concrete examples of subfactors and manifolds, based on Izumi's explicit computations of the representations of the modular group arising from some subfactors, including the "exotic" one due to Haagerup, in [17].

Another computation of topological invariants arising from subfactors is based on α-induction [3, 4, 5, 25, 41]. This method, in particular, produces subfactors with principal graphs D_{2n}, E_6, and E_8, and the corresponding Turaev-Viro-Ocneanu invariants can be computed once we have a description of the "quantum doubles" by [23], and these quantum doubles were computed in [6]. (Also see [29].) This α-induction is also related to theory of modular invariants [7]. See [3, 4, 5, 20, 21] for more on this topic.

4 Algebraic Quantum Field Theory

Another occurrence of nondegenerate braiding in theory of operator algebras is in algebraic quantum field theory [14], which has its own long history. This theory is an approach to quantum field theory based on operator algebras. That is, in each bounded region on a spacetime, we assign a von Neumann algebra on a fixed Hilbert space. We think that each such von Neumann algebra is generated by observable physical quantities in the bounded region in the spacetime. In this way, we think that this family of von Neumann algebras parametrized by bounded regions gives a mathematical description of a physical theory. We often restrict bounded regions to those of a special form. We impose a physically natural set of axioms on this family of von Neumann algebras and make a mathematical study of such axiomatized systems. A spacetime of any dimension is allowed in this axiomatized approach, and the four dimensional case was studied originally for an obvious physical reason. These studies of Doplicher-Haag-Roberts [8] and Doplicher-Roberts

[9, 10] have been quite successful. Recently, it has been realized that this theory in lower dimensional spacetime has quite interesting mathematical structures. Two-dimensional case has caught much attention in connection to conformal field theory and one-dimensional case also naturally appears in a "chiral" decomposition of a two-dimensional theory. Mathematical structures of one-dimensional theory was studied in [12]. In one-dimensional case, our "spacetime" is simply $\mathbf{R}$ and a bounded region is simply a bounded interval. It is often convenient to compactify the space $\mathbf{R}$ to obtain S^1 and consider "intervals" contained in S^1. In this setting, our mathematical structure is a family of von Neumann algebras on a fixed Hilbert space parameterized by intervals in S^1. We impose a set of axioms. For example, one axiom requires that we have a larger von Neumann algebra for a larger interval. Another axiom requires "covariance" of the theory with respect to a projective unitary representation of a certain group of the "spacetime symmetry". We also have an axiom on "locality" which says if two regions are spacelike separated, then the corresponding von Neumann algebras mutually commute. Another requires existence of a "vacuum" vector in the Hilbert space. Positivity of energy in the sense that a certain self-adjoint operator is positive is also assumed. See [13, 22] for a precise description of the set of axioms. (Actually the main results in [22] hold under a weaker set of axioms, but we do not go into details here.) Under the usual set of axioms, each von Neumann algebra for an interval becomes a factor of type III, so we call such a family a *net of factors*. Now the index set of intervals on the circle S^1 is not directed with respect to inclusions, since the entire circle is not allowed as an interval, so it is not appropriate to call such a family a *net*, but this terminology has been commonly used.

This family is our operator algebraic system and we consider a representation of such a family of von Neumann algebras. Such an idea is due to Doplicher-Haag-Roberts [8] and is called the DHR theory. We have a natural notion of irreducibility, dimensions, and tensor products for such representations. Note that we do not have an obvious definition of tensor products for two representations of such a net of factors. The key idea was that the tensor product operation is realized through compositions of endomorphisms. Also the dimension in the usual sense is always infinite. So it was highly nontrivial to obtain sensible definitions of the tensor product and the dimension. This work is much older than the subfactor theory in the previous section, and its similarity to subfactor theory was soon recognized in [24] in a precise form.

In this way, we have a representation theory for a net of factors. A tensor product operation is "too commutative" for higher dimensional spacetime, but in dimensions one and two, it has an appropriate level of commutativity, and naturally produces a braiding. (See [12] for example.) So we have two problems for getting a modular tensor category from such a representation of a net of factors on S^1. One is whether we have only finitely many equivalence classes of irreducible representations or not. The other is whether the braiding is nondegenerate or not. In [22], Longo, Müger and the author have found a nice operator algebraic condition that implies positive answers to these two problems and we introduced the terminology "complete rationality" for this notion. One of the key conditions for this notion is finiteness of a certain Jones index. Note that in subfactor theory in the previous section, our "family of operator algebras" has only two factors N and M, and its representation theory produced a tensor category, without a braiding in general. Now our "family of operator algebras" is a net of factors and has continuously many factors with more structures, and its representation theory produces a braided tensor category.

Xu has proved in [42] that the $SU(N)_k$-nets corresponding to the WZW-models $SU(N)_k$ are completely rational. Xu worked on coset models in the setting of nets of factors on S^1 in [43], and obtained several interesting examples. He then studied in [44] about topological invariants arising from these nets, which seems to be quite interesting topologically. He also worked on orbifold models in this context in [45]. Finally, we also note that complete rationality is also important in classification theory of nets of factors as in [20, 21].

References

1. M. Asaeda, U. Haagerup: Commun. Math. Phys. **202**, 1–63 (1999).
2. M.F. Atiyah: Publ. Math. I.H.E.S. **68**, 175–186 (1989).
3. J. Böckenhauer, D.E. Evans: Commun. Math. Phys. **197**, 361–386 (1998) II **200**, 57–103 (1999) III **205**, 183–228 (1999).
4. J. Böckenhauer, D.E. Evans, Y. Kawahigashi: Commun. Math. Phys. **208**, 429–487 (1999).
5. J. Böckenhauer, D.E. Evans, Y. Kawahigashi: Commun. Math. Phys. **210**, 733–784 (2000).
6. J. Böckenhauer, D.E. Evans, Y. Kawahigashi: Publ. RIMS, Kyoto Univ. **37**, 1–35 (2001).
7. A. Cappelli, C. Itzykson, J.-B. Zuber: Commun. Math. Phys. **113**, 1–26 (1987)
8. S. Doplicher, R. Haag, J.E. Roberts: I Commun. Math. Phys. **23**, 199–230 (1971) II **35**, 49–85 (1974).
9. S. Doplicher, J.E. Roberts: Ann. Math. **130**, 75–119 (1989).
10. S. Doplicher, J.E. Roberts: Invent. Math. **98**, 157–218 (1989).
11. D.E. Evans, Y. Kawahigashi: *Quantum symmetries on operator algebras*, (Oxford University Press, Oxford, 1998).
12. K. Fredenhagen, K.-H. Rehren, B. Schroer: I. Commun. Math. Phys. **125**, 201–226 (1989) II. Rev. Math. Phys. **Special issue**, 113–157 (1992).
13. D. Guido, R. Longo: Commun. Math. Phys. **181**, 11–35 (1996).
14. R. Haag: *Local Quantum Physics*, (Springer-Verlag, Berlin-Heidelberg-New York, 1996).
15. U. Haagerup: Principal graphs of subfactors in the index range $4 < 3 + \sqrt{2}$. In: *Subfactors*, ed by H. Araki et al. (World Scientific, 1994) pp 1–38.
16. M. Izumi: Commun. Math. Phys. **213**, 127–179 (2000).

17. M. Izumi: Rev. Math. Phys. **13**, 603–674 (2001).
18. V.F.R. Jones: Invent. Math. **72**, 1–25 (1983).
19. V.F.R. Jones: Bull. Amer. Math. Soc. **12**, 103–112 (1985).
20. Y. Kawahigashi, R. Longo: math-ph/0201015, to appear in Ann. Math.
21. Y. Kawahigashi, R. Longo: math-ph/0304022, to appear in Commun. Math. Phys.
22. Y. Kawahigashi, R. Longo, M. Müger: Commun. Math. Phys. **219**, 631–669 (2001).
23. Y. Kawahigashi, N. Sato, M. Wakui: math.OA/0208238.
24. R. Longo: I Commun. Math. Phys. **126**, 217–247 (1989) II Commun. Math. Phys. **130**, 285–309 (1990).
25. R. Longo, K.-H. Rehren: Rev. Math. Phys. **7**, 567–597 (1995).
26. M. Müger: math.CT/0111205.
27. A. Ocneanu: Quantized group, string algebras and Galois theory for algebras. In *Operator algebras and applications, Vol. 2*, ed D. E. Evans and M. Takesaki, (Cambridge University Press, Cambridge, 1988) pp 119–172.
28. A. Ocneanu: Chirality for operator algebras. In: *Subfactors*, ed by H. Araki et al. (World Scientific, 1994) pp 39–63.
29. A. Ocneanu: Operator algebras, topology and subgroups of quantum symmetry – construction of subgroups of quantum groups – (written by S. Goto and N. Sato). In: *Taniguchi Conference in Mathematics Nara '98* Adv. Stud. Pure Math. **31**, (Math. Soc. Japan, 2000) pp 235–263.
30. S. Popa: Correspondences, preprint 1986.
31. S. Popa: Math. Res. Lett. **1**, 409–425 (1994).
32. K.-H. Rehren: Braid group statistics and their superselection rules. In: *The algebraic theory of superselection sectors, Palermo, 1989*, World Scientific Publishing (1990) pp 333–355.
33. N. Reshetikhin, V.G. Turaev: Invent. Math. **103**, 547–597 (1991).
34. N. Sato and M. Wakui: math.OA/0208242, to appear in J. Knot Theory Ramif.
35. V. G. Turaev, *Quantum Invariants of Knots and 3-manifolds*, (Walter de Gruyter, 1994).
36. V.G. Turaev, O. Ya Viro: Topology **31**, 865–902 (1992).
37. V.G. Turaev, H. Wenzl; Internat. J. Math. **4**, 323–358 (1993).
38. V.G. Turaev, H. Wenzl: Math. Ann. **309**, 411–461 (1997).
39. H. Wenzl: Invent. Math. **114**, 235–275 (1993).
40. H. Wenzl: J. Amer. Math. Soc. **11**, 261–282 (1998).
41. F. Xu: Commun. Math. Phys. **192**, 347–403 (1998).
42. F. Xu: Commun. Contemp. Math. **2**, 307–347 (2000).
43. F. Xu: Commun. Math. Phys. **211**, 1–44 (2000).
44. F. Xu: math.GT/9907077.
45. F. Xu: Proc. Nat. Acad. Sci. U.S.A. **97**, 14069–14073 (2000).

Topological Quantum Field Theory and Algebraic Structures[*]

T. Kimura[**]

Department of Mathematics and Statistics, Boston University, 111 Cummington Street, Boston, MA 02215, USA, and Institut des Hautes Études Scientifiques, Le Bois-Marie, 35, routes de Chartres, 91440 Bures-sur-Yvette, FRANCE

Summary. These notes are from lectures given at the *Quantum field theory and noncommutative geometry workshop* at Tohoku University in Sendai, Japan from November 24–30, 2002. We give a survey of operads and their relationship to topological quantum field theories (TQFT). We give simple examples of operads, particularly those arising as moduli spaces of decorated oriented 2-spheres, describe the notion of algebras over them, and we study their "higher loop" generalizations. We then focus upon the example of the moduli space of stable curves and its relationship to cohomological field theories, in the sense of Kontsevich-Manin. The latter can be regarded as a generalization of a TQFT which is relevant to quantum cohomology and to higher KdV integrable hierarchies.

1 Introduction

These lectures provide a survey, from the perspective of someone interested in $(1 + 1)$-dimensional topological quantum field theories (TQFTs) and their generalizations, in which we explain how the notion of an operad and algebras over them can be used to analyze such theories. We have by no means attempted to provide a comprehensive account of this very rich subject. On the contrary, the subject can often appear to be almost too rich, overwhelming the uninitiated with long, unmotivated definitions and constructions. Instead, we have attempted, when possible, to give a "hands-on" approach to several key points relevant to TQFTs through a study of simple examples, many of which admit interesting and nontrivial generalizations.

Many operadic notions involve constructions which should be already familiar to mathematical physicists. One such feature is the natural appearance of trees in the theory of operads which suggests an appealing interpretation in terms of tree-level Feynman diagrams. These trees turn out to be a universal feature of operads and algebras over them. Allowing higher loop (or genus) diagrams correspond to a generalization of the notion of an operad,

---

[*] 2000 *Mathematics Subject Classification*, Primary: 18D50, 81T45. Secondary: 14D21, 53D45, 14N35.

[**] Partially supported by NSF grant DMS-0204824.

called a modular operad, due to Getzler-Kapranov [13]. This interpretation in terms of Feynman diagrams becomes particularly compelling in the study of cohomological field theories, in the sense of Kontsevich-Manin [22], where the underlying (modular) operad is the moduli space of stable curves and the graphs appear as dual graphs associated to a stable curve. Here, it becomes useful to introduce a potential function from which one obtains the Feynman rules in the usual fashion. The relations in the underlying operad imply that the potential function must satisfy certain differential equations. Sometimes, these differential equations are powerful enough to determine this potential function. These techniques are used in the study of Gromov-Witten invariants (or quantum cohomology) of a smooth, projective variety and in the study of the spin-CohFT.

We hope that the interested reader will use this article as a springboard from which to pursue the subject further on his own. We refer the reader to [4] to read further about TQFTs and their cousins. We refer the reader to [25] to read further about operads, homotopy algebras, and their generalizations. Finally, we refer the reader to [24] to read further about cohomological field theories and quantum cohomology.

The paper is structured as follows. In Sect. 2, we review the notion of a topological quantum field theory as described by Atiyah as a functor from a geometric cobordism category to the category of vector spaces. We will be primarily interested in the case of a $(1 + 1)$ dimensional TQFT and we recall this is nothing more than a Frobenius algebra, i.e. a commutative, associative algebra with an identity element and an invariant metric.

In the Sect. 3, we recall the definition of an operad and algebras over them. We describe the endomorphism operad and various operads in the category of topological spaces such as the little intervals and the (framed) little disks operad. We then study the operad of their homology groups and describe the algebraic structures for which they are responsible. We then explain the consequences for a refinement of a TQFT called a topological conformal field theory (TCFT).

In the Sect. 4, we study the moduli space of stable curves from an operadic perspective. This can be thought of as a compactification of the moduli space of Riemann surfaces with marked points. The compactification is realized by allowing the Riemann surfaces to degenerate by acquiring nodes. We can "glue" such surfaces by attaching them along two marked points to obtain a node. This becomes the model example of a modular operad. Restricting to genus zero only and regarding one marked point to be outgoing, one obtains a (usual) operad in the category of complex algebraic varieties. A cohomological field theory (CohFT) in the sense of Kontsevich-Manin is nothing more than an algebra over this modular operad. In particular, a CohFT contains a Frobenius algebra by considering only the suboperad $H_0(\overline{\mathcal{M}}_{0,n})$ regarded as a (cyclic) operad. Therefore, a CohFT can be regarded as a generalization of a TQFT. The genus zero part of a CohFT admits a beautiful description in terms of potential functions associated to tree-level Feynman diagrams.

In the Sect. 5, we provide two constructions of CohFTs. The first is the Gromov-Witten invariants of a smooth, projective variety and involves the moduli space of stable maps. The second is called a spin-CohFT and involves the moduli space of higher spin curves. Constructions of CohFTs do not appear to be very easy to come by at the moment.

Remark 1 Henceforth, we assume that all ground rings are over $\mathbb{C}$ unless otherwise stated.

2 Topological Quantum Field Theory

Our starting point is the notion of a $(d+1)$-dimensional topological quantum field theory (or TQFT) due Atiyah which, roughly speaking, axiomatizes the locality (or gluing) properties associated to the Feynman path integral in the cases where the observables in the associated quantum field theory depend only upon discrete geometric (hence, "topological") data. The idea is to view a TQFT as a functor from a category of geometric objects depending on only discrete geometric data to the category of vector spaces.

This definition provides a model for more complicated quantum field theories such as conformal field theories [28] or cohomological field theories [22] which are obtained by enriching the geometric and linear categories.

Roughly speaking, operads appear in a $(d + 1)$-dimensional TQFT as a certain subcategory of the geometric category consisting of those objects with n inputs and 1 output. An algebra over this operad is then obtained by restriction.

Remark 2 Throughout the remainder of this section, we will assume that all manifolds are oriented and that all homeomorphisms between them preserve orientation. Furthermore, if N is an oriented manifold then N^* is the same underlying manifold N but with its orientation reversed.

Definition 1 A $(d + 1)$-*dimensional topological quantum field theory (or TQFT)* is a collection of the following data:

State Spaces : To each closed d-dimensional manifold N, we assign a finite dimensional vector space $Z(N)$.

Correlators : To each $(d + 1)$-dimensional manifold, M, we assign a vector $Z(M) \in Z(\partial M)$.

There are also the following functorial isomorphisms:

Naturality 1 : Any homeomorphism $f : N \to N'$ of d-manifolds induces an isomorphism $f_* : Z(N) \to Z(N')$.

Duality : An isomorphism $Z(N^*) \to Z(N)^*$.

Empty Set : An isomorphism $Z(\phi) \to \mathbb{C}$

Multiplicativity : An isomorphism $Z(N \sqcup N') \to Z(N) \otimes Z(N')$.

These must satisfy the following axioms.

Naturality 2 : If $f : M \to M'$ is a homeomorphism of $(d+1)$-manifolds
then
$$(f_{\partial M})_*(Z(M)) = Z(M') \,.$$

where $f_{\partial M}$ denotes the restriction of f to ∂M.

Gluing : Let M be a $(d+1)$-manifold such that $\partial M = N_1 \sqcup N_2 \sqcup N_3$ and
let $f : N_1 \to N_2^*$ be a homeomorphism. Let M' be the result of gluing
together M using f then the composition

$$Z(N_1) \otimes Z(N_2) \otimes Z(N_3) \to Z(N_2)^* \otimes Z(N_2) \otimes Z(N_3) \to Z(N_3)$$

is equal to $Z(M')$ where the first map is induced by f_* and the second is
contraction.

Identity Map : Let I be the oriented interval and N be a d-manifold such
that $\partial I \times N = N \sqcup N^*$ then we require that $Z(I \times N)$ is equal to the
image of the identity map on $Z(N)$ in $Z(N^*) \otimes Z(N) \simeq Z(N)^* \otimes Z(N)$.

We will primarily be interested in the case where $d = 1$ for the purposes
of these lectures where things are quite simple.

Definition 2 A *Frobenius algebra (with unit)* is a finite-dimensional commutative, associative algebra (over $\mathbb{C}$) with an identity element and an invariant metric.

In other words, it is a tuple $(\mathcal{H}, \eta, \cdot, \mathbf{1})$ where $\mathcal{H}$ is a finite dimensional
vector space with a metric η together with a multiplication $\cdot$ with identity
element $\mathbf{1}$ such that
$$\eta(a \cdot b, c) = \eta(a, b \cdot c) \tag{1}$$

for all a, b, c in $\mathcal{H}$.

Theorem 1 $(1+1)$-*dimensional topological quantum field theories are in one
to one correspondence with Frobenius algebras.*

Proof. Suppose we are given a $(d+1)$-dimensional TQFT. Let us now construct its associated Frobenius algebra.

Choose, once and for all, a particular circle from the set of all closed,
connected 1-manifolds and call it S^1. Any closed 1-manifold N is isomorphic
to the finite disjoint union of circles. By the multiplicativity , duality, and
the 1st naturality axiom, $Z(S)$ for any closed 1-manifold S is determined by
$\mathcal{H} := Z(S^1)$.

Any 2-manifold M has a finite number of connected components. A connected must be a genus g surface (possibly) with boundary. The key observation is that any such surface can be obtained by gluing together a finite number of disks, cylinders, or pairs of pants along their boundaries. (A disk is a sphere with one boundary component, a cylinder is a sphere with two boundary components and a pair of pants is a sphere with three boundary components.) Choose a particular disk, cylinder, and pair of pants, once and for all, and call them respectively D, C, and P. In each case, let us assume that the orientation assigned to their boundaries agrees with that induced from the surface.

One can view $Z(D)$ as a linear map $\mathbb{C} \to \mathcal{H}$. This map is determined by the image of the unit element in $\mathbb{C}$. Call this image $\mathbf{1}$ in $\mathcal{H}$. It will be the unit element.

Let C' be the exactly the same as C but where its two boundary components have been assigned an orientation opposite that induced from the surface then denote the map $Z(C') : \mathcal{H}^{\otimes 2} \to \mathbb{C}$ by η. This will be the metric.

Let P' be exactly the same as P but where two boundary components are assigned an orientation opposite from the one induced from P while the remaining one is assigned an orientation induced from P so that we have $Z(P') : \mathcal{H}^{\otimes 2} \to \mathcal{H}$. This will be the multiplication.

η is commutative, by the 1st naturality axiom, because there is an isomorphism taking C' to itself which takes one boundary component to the other. The multiplication is commutative for similar reasons.

$\mathbf{1}$ is the unit element in $\mathcal{H}$ because gluing a D to one of the boundary components of C' whose orientation is the opposite of the induced orientation yields a cylinder. One boundary component of the cylinder agrees with the induced orientation and the other does not. Call this cylinder C''. $Z(C'')$ is, by the identity map axiom, equal to the identity map $\mathcal{H} \to \mathcal{H}$. Therefore by the gluing axiom, it follows that $\mathbf{1}$ is the unit element of $\mathcal{H}$.

Let $\eta^{-1} := Z(C) : \mathbb{C} \to \mathcal{H}^{\otimes 2}$. We can see that η^{-1} is the inverse of η as follows. Pick a boundary component of C and of C' and then glue them together. Again, we obtain C'' but $Z(C'')$ is the identity map on $\mathcal{H}$ by the identity axiom. Therefore, η^{-1} is the inverse of η.

Associativity of the product follows by considering a sphere with 4 boundary components, 3 components whose orientations agree with the induced one and the last component whose orientation disagrees with the induced one. Such a surface may be obtained (up to an isomorphism) by gluing together P' in several different ways. This yields the associativity of the multiplication.

Equation (1), the invariance of the metric, follows from a similar argument except that now one glues together a C' with P' instead.

Therefore, $(\mathcal{H}, \eta, \cdot)$ is a Frobenius algebra. It remains to prove that there are no other relations other than those that we have described above. This is the most subtle part of this proof and we the refer the interested reader to [4] for details.

The converse also follows from these arguments.   □

Suppose we now consider (1+1)-dimensional TQFTs but where we restrict to only 2-manifolds, M, which are spheres with $(n+1)$ boundary components where $n \geq 0$ such that n of the boundary components have orientation disagreeing with the induced orientations and the remaining one has an orientation agreeing with the induced orientation so that $Z(M) : \mathcal{H}^{\otimes n} \to \mathcal{H}$. This is essentially the structure map of an algebra over the operad of the moduli space of oriented genus zero surfaces with boundary.

3 Operads and Algebras Over Them

3.1 Some Useful Examples

Operads appear quite naturally in various settings. Rather than diving straight into its definition, which can be somewhat daunting for the uninitiated, let us begin with some key examples.

Example 1 (The Endomorphism Operad) Let $\mathcal{H} = \mathcal{H}_0 \oplus \mathcal{H}_1$ be a $\mathbb{Z}_2$-graded vector space. Elements in $\mathcal{H}_0$ are called the *even elements of* $\mathcal{H}$ while the elements in $\mathcal{H}_1$ are called the *odd elements of* $\mathcal{H}$.

Consider the collection $\mathrm{End}^{\mathcal{H}} := \{\mathrm{End}^{\mathcal{H}}(n)\}_{n \geq 1}$ where $\mathrm{End}^{\mathcal{H}}(n) := \mathrm{Hom}(\mathcal{H}^{\otimes n}, \mathcal{H})$. An element f in $\mathrm{End}^{\mathcal{H}}(n)$ is an n-ary operation on $\mathcal{H}$, i.e. f accepts n elements of $\mathcal{H}$ as an input and yields one element of $\mathcal{H}$ for an output.

$\mathrm{End}^{\mathcal{H}}(n)$ has an action of S_n, the symmetric group on n letters, which permutes the n factors of the tensor product, i.e. for all σ in S_n and homogeneous elements $v_1, \ldots, v_n$ in $\mathcal{H}$,

$$(\sigma f)(v_1, \ldots, v_n) := \pm f(v_{\sigma(1)}, \ldots, v_{\sigma(n)}) . \tag{2}$$

where the sign is the usual one from the category of $\mathbb{Z}_2$-graded vector spaces.

Given $f \in \mathrm{End}^{\mathcal{H}}(n)$ and $f' \in \mathrm{End}^{\mathcal{H}}(n')$, we can construct an element $f \circ_i f' \in \mathrm{End}^{\mathcal{H}}(n + n' - 1)$ by composing the output of f' into the i-th input of f, i.e. for all $i = 1, \ldots, n$,

$$(f \circ_i f')(v_1 \otimes \ldots \otimes v_{n+n'-1})$$
$$:= \pm f(v_1 \otimes \ldots \otimes v_{i-1} \otimes f'(v_i \otimes \ldots v_{i+n'-1}) \otimes v_{i+n'} \otimes \cdots \otimes v_{n+n'-1}) \tag{3}$$

for all $v_1, \ldots, v_{n+n'-1}$ in $\mathcal{H}$.

Composing three multilinear operations should be associative, in the appropriate sense. More precisely, suppose we have homogeneous elements f in $\mathrm{End}^{\mathcal{H}}(n)$, f' in $\mathrm{End}^{\mathcal{H}}(n')$, and f'' in $\mathrm{End}^{\mathcal{H}}(n'')$ then there are two types of compositions that one could perform to obtain an element in $\mathrm{End}^{\mathcal{H}}(n+n'-1)$.

In the first case, we can successively plug in the output of f'' into f' and plug in the output of f' into f. In the second case, we can plug in both of the outputs of f' and f'' into f but at different locations. In either case, the result is independent of the order in which the compositions are performed. In other words,

Associativity 1 : if $1 \leq i < j \leq n$, $n', n'' \geq 1$, and $n \geq 2$ then

$$(f \circ_i f') \circ_{j+n'-1} f'' = (-1)^{|f'||f''|}(f \circ_j f'') \circ_i f' \tag{4}$$

Associativity 2 : if $n, n' \geq 1$, $n'' \geq 1$, and $i = 1, \ldots, n$ and $j = 1, \ldots, n'$ then

$$(f \circ_i f') \circ_{i+j-1} f'' = f \circ_i (f' \circ_j f'') . \tag{5}$$

where $|f'|$ and $|f''|$ denotes the $\mathbb{Z}_2$-gradings of f' and f'', respectively.

Furthermore, the composition maps are equivariant under the action of the permutation groups. Finally, there is a unit element I in $\mathrm{End}^{\mathcal{H}}(1)$, which is the identity map, which satisfies for all f in $\mathrm{End}^{\mathcal{H}}(n)$ and $i = 1, \ldots, n$,

Unit :

$$I \circ_1 f = f = f \circ_i I . \tag{6}$$

Thus $\mathrm{End}^{\mathcal{H}}$ together with its compositions, the action of the symmetric group, and the unit element I yields an example of an operad with unit and is called *the endomorphism operad of* $\mathcal{H}$. This is a model example of an operad in the category of $\mathbb{Z}_2$-graded vector spaces.

Observe that $\mathrm{End}^{\mathcal{H}}(1) = \mathrm{End}(\mathcal{H})$, the space of (usual) endomorphisms of $\mathcal{H}$, and the composition map $\mathrm{End}^{\mathcal{H}}(1) \otimes \mathrm{End}^{\mathcal{H}}(1) \to \mathrm{End}^{\mathcal{H}}(1)$ taking $f \otimes f' \mapsto f \circ_1 f'$ is nothing more than the usual associative composition on $\mathrm{End}(\mathcal{H})$ with unit element I.

The above definition can be immediately generalized to the category of differential graded vector spaces (or dg vector spaces). Suppose that $(\mathcal{H}, d)$ is a dg vector space, i.e. $(\mathcal{H}, d)$ consists of a $\mathbb{Z}$-graded vector space $\mathcal{H}$ whose i-th graded subspace is denoted by $\mathcal{H}_i$, linear maps $d : \mathcal{H}_i \to \mathcal{H}_{i-1}$ for all i such that $d^2 = 0$. The endomorphism operad of the complex $(\mathcal{H}, d)$, $\mathrm{End}^{(\mathcal{H}, d)}$, is a dg vector space whose p-th graded subspace consists of those elements in $\mathrm{Hom}(\mathcal{H}^{\otimes n}, \mathcal{H})$ with $\mathbb{Z}$-grading p. The compositions, the action of the symmetric groups, and the unit element are defined in the same way as before. The compositions preserve the $\mathbb{Z}$-grading, S_n preserves the $\mathbb{Z}$-grading, and I has $\mathbb{Z}$-grading 0. We will denote $\mathrm{End}^{(\mathcal{H}, d)}$ by $\mathrm{End}^{\mathcal{H}}$ when there is no possibility of confusion.

Since the compositions and symmetric groups commute with the differentials and I is closed, by taking homology, one obtains a well-defined operad $\mathrm{End}^{H_\bullet(\mathcal{H})}$ in the category of $\mathbb{Z}$-graded vector spaces.

The next example is an operad in the category of topological spaces. Typically, such operads arise as configuration (or moduli) spaces of geometric objects which can be glued together. This will be situation in which we are most interested.

Example 2 (The Framed Little Disks Operad) Consider the collection $\mathcal{F} := \{\mathcal{F}(n)\}_{n \geq 1}$. Let Δ denote the unit disk about the origin in $\mathbb{C}$. Let $F(n)$ consist of configurations of n little disks embedded inside of Δ via a composition of a translation, rotation, and dilation. In other words, $F(n)$ consists of all n-tuples $(x_1, \ldots, x_n)$ consisting of $x_i : \Delta \to \Delta$ where $x_i(z) = a_i z + b_i$, a_i is nonzero element in $\mathbb{C}$ and b_i belongs to $\mathbb{C}$ for all $i = 1, \ldots, n$. Furthermore, we require that for all $i \neq j$, $x_i(\Delta)$ and $x_j(\Delta)$ can only intersect at most along their boundaries. We can depict an element of $F(n)$ as a unit ("big") disk Δ with n little disks drawn inside it whose interiors are pairwise disjoint, together with a marked point along the boundary of each little disk. The marked point on the i-th little disk is the image of 1 in $\mathbb{C}$ under x_i for all $i = 1, \ldots, n$. It is understood that the big disk has a marked point at 1 in $\mathbb{C}$. $\mathcal{F}$ is a collection of topological spaces.

The symmetric group S_n acts upon $F(n)$ by permuting the order of the n little disks. The composition $F(n) \times F(n') \to F(n + n' - 1)$ taking $(f, f') \mapsto f \circ_i f'$ for all $i = 1, \ldots, n$ is obtained by shrinking f' until its big disk is the same size as the i-th little disk of f and then gluing it into the i-th little disk of f after rotating the shrunken disk so that the marked points coincide.

These compositions are continuous and satisfy the associativity conditions, i.e. for all f in $\mathcal{F}(n)$, f' in $\mathcal{F}(n')$ and f'' in $\mathcal{F}(n'')$, Equation (4) and Equation (5) after ignoring the factor of $(-1)^{|f'||f''|}$ in Equation (4). Associativity boils down to the fact that the result of gluing three objects together in a proscribed fashion is independent of the order in which the gluings are performed.

The compositions are equivariant under the action of the permutation groups. There is a unit element in I in $\mathcal{F}(1)$ which consists of the unit map $x_1(z) = z$ which satisfies Equation (6).

Observe, again, that the composition $\circ_1$ on $\mathcal{F}(1)$ is associative, i.e. $\mathcal{F}(1)$ is a semigroup with unit.

The resulting operad, $\mathcal{F}$, is called the *framed little disks operad* and it is an operad in the category of complex manifolds with corners.

Armed with these examples, let us proceed with the definition. An operad is a notion which exists in any symmetric monoidal category, e.g. the category of topological spaces, dg vector spaces, etc.

Definition 3 An *operad (with unit)* $\mathcal{O} = \{\mathcal{O}(n)\}_{n \geq 1}$ is a collection of objects in a symmetric monoidal category where each $\mathcal{O}(n)$ is endowed with a right action of S_n, the permutation group on n letters, and a collection of morphisms for $n \geq 1$ and $1 \leq i \leq n$, $\mathcal{O}(n) \times \mathcal{O}(n') \to \mathcal{O}(n + n' - 1)$ given by $(f, f') \mapsto f \circ_i f'$ satisfying the following properties.

Equivariance : The compositions are equivariant under the action of the permutation groups.

Associativity : For all $f \in \mathcal{O}(n)$, $f' \in \mathcal{O}(n')$, and $f'' \in \mathcal{O}(n'')$, the associativity conditions, Equation (4) and Equation (5), are satisfied (where the factor of $(-1)^{|f'||f''|}$ should be ignored if $\mathcal{O}$ is not an operad of (graded) vector spaces).

Unit : There exists a element I in $\mathcal{O}(1)$ called the *unit* such that Equation (6) holds for all $i = 1, \ldots, n$ and f in $\mathcal{O}(n)$.

Remark 3 It is implicit in the above that if $\mathcal{O}$ is an operad of (graded) vector spaces then the compositions must respect the $\mathbb{Z}$-grading and the action of the permutation must preserve the $\mathbb{Z}$-grading.

Remark 4 An alternate description of the composition maps favored by algebraic topologists is obtained by iterating k composition maps to obtain the morphism $\gamma : \mathcal{O}(k) \times \mathcal{O}(n_1) \times \cdots \times \mathcal{O}(n_k) \to \mathcal{O}(n_1 + \cdots + n_k)$.

Remark 5 Under the composition map $\circ_1$, it follows that $\mathcal{O}(1)$ is an semigroup with unit I. If $\mathcal{O}$ is an operad with unit in the category of dg vector spaces then $\mathcal{O}(1)$ is a differential graded associative algebra with unit.

Remark 6 Some authors extend the definition of an operad $\mathcal{O}$ to include a component $\mathcal{O}(0)$ as well. In the endomorphism operad, $\mathrm{End}^{\mathcal{H}}(0) := \mathcal{H}$.

Let $\mathcal{O}$ be an operad in the category of topological spaces then it automatically gives rise to two associated operads. First of all, one can define the operad of its singular chain complexes $(C_\bullet(\mathcal{O}), \partial)$. This is an operad in the category of differential graded vector spaces where $C_\bullet(\mathcal{O}) := \{ C_\bullet(\mathcal{O}(n)) \}_{n \geq 1}$, $C_\bullet(\mathcal{O}(n))$ is the space of singular chains on $\mathcal{O}(n)$ with differentials $\partial : C_{p+1}(\mathcal{O}(n)) \to C_p(\mathcal{O}(n))$ for all $p \geq 0$ and the compositions $C_p(\mathcal{O}(n)) \otimes C_{p'}(\mathcal{O}(n')) \to C_{p+p'}(\mathcal{O}(n + n' - 1))$ taking $f \otimes f' \mapsto f \circ_i f'$ for all $i = 1, \ldots, n$ are induced by the corresponding compositions in $\mathcal{O}$. The S_n action on $C_\bullet(\mathcal{O}(n))$ and unit I in $C_0(\mathcal{O}(1))$ are induced. Secondly, by taking the homology, one obtains the induced operad $H_\bullet(\mathcal{O}) := \{ H_\bullet(\mathcal{O}(n)) \}_{n \geq 1}$ in the category of $\mathbb{Z}$-graded vector spaces.

3.2 Morphisms of Operads

We now describe morphisms between operads and provide some concrete examples.

Definition 4 Let $\mathcal{O}$ and $\mathcal{O}'$ be operads. A *morphism of operads* is a collection of maps $\mu : \mathcal{O}(n) \to \mathcal{O}'(n)$ for all $n \geq 1$ taking $f \mapsto \mu_f$ such that

Equivariance : For all σ in S_n and f in $\mathcal{O}(n)$,

$$\mu_{f\sigma} = (\mu_f) \circ \sigma$$

where $\circ$ denotes the composition of μ_f and the action of σ on $\mathcal{O}'(n)$.
Compositions : For all f in $\mathcal{O}(n)$ and f' in $\mathcal{O}(n')$ and $i = 1, \ldots, n$,

$$\mu_{f \circ_i f'} = \mu_f \circ_i \mu_{f'}.$$

Unit : If I is the unit in $\mathcal{O}$ and I$'$ is the unit in $\mathcal{O}'$ then

$$\mu_{\mathrm{I}} = \mathrm{I}'.$$

A morphism of operads $\mu : \mathcal{O} \to \mathcal{O}'$ is said to be an *isomorphism* if it has an inverse morphism.

In particular, if $\mu : \mathcal{O} \to \mathcal{O}'$ is a morphism of operads then its restriction morphism $\mu : \mathcal{O}(1) \to \mathcal{O}'(1)$ is a homomorphism of semigroups with unit.

Example 3 (The Little Disks Operad) Let $\mathcal{D} := \{\mathcal{D}(n)\}_{n \geq 1}$ where $\mathcal{D}(n)$ consists of tuples $(x_1, \ldots, x_n)$ in $\mathcal{F}(n)$ such that $x_i(z) = a_i z + b_i$ where $|a_i| = 1$ for all $i = 1, \ldots, n$. This means that each little disk must embed into the big disk by a composition of only a translation and dilation. Pictorially, it consists of only those framed little disks such that each little disk has a boundary with a marked point that is located at the right most point on that disk.

Since $\mathcal{D}$ is closed under the action of the permutation groups and compositions, and contains the unit element, $\mathcal{D}$ is a suboperad of $\mathcal{F}$, i.e. the inclusion maps $\mathcal{D} \to \mathcal{F}$ is an injective morphism of operads.

Example 4 (The Segal Operad) Let $\mathcal{P} := \{\mathcal{P}(n)\}_{n \geq 1}$ be the moduli space of Riemann spheres with $(n + 1)$ holomorphically embedded disks. In other words, an element in $\mathcal{P}(n)$ consists of $\mathbb{C}P^1$ together with a tuple $(z_1, \ldots, z_{n+1})$ where $z_i : \Delta \to \mathbb{C}P^1$ are holomorphic embeddings of the unit disk Δ for all $i = 1, \ldots, (n+1)$ whose pairwise images can only overlap along their boundaries. Two such configurations are isomorphic if they are related by an automorphism of $\mathbb{C}P^1$. Recall that the group of automorphisms of $\mathbb{C}P^1$ is isomorphic to $\mathrm{PGL}(2, \mathbb{C}) \simeq \mathrm{PSL}(2, \mathbb{C})$ where latter acts on $\mathbb{C}P^1$ in a standard chart as

$$z \mapsto \frac{az + b}{cz + d}$$

where a, b, c, d belong to $\mathbb{C}$ such that $ad - bc = 1$.

S_n acts upon $\mathcal{P}(n)$ by permuting the first n disks, leaving the $(n+1)$-st disk alone. The composition $\mathcal{P}(n) \times \mathcal{P}(n') \to \mathcal{P}(n+n'-1)$ taking $(C, C') \mapsto C \circ_i C'$ is obtained by cutting out the i-th disk of C and the $(n' + 1)$st disk of C' and then gluing along their boundaries using $z \mapsto \frac{1}{z}$ for all $i = 1, \ldots, n$. We call the resulting operad $\mathcal{P}$ the *Segal operad*. It can be regarded as an infinite dimensional version of a complex manifold with corners. In particular, it is an operad in the category of topological spaces.

We will now construct an injective morphism of operads $\nu : \mathcal{F} \to \mathcal{P}$. Let $(x_1, \ldots, x_n)$ denote an element of $\mathcal{F}(n)$. Choose on $\mathbb{C}P^1$ one of its standard charts and call it w. One can thus regard $\mathbb{C}P^1$ as $\mathbb{C} \cup \{\infty\}$. Identify the big disk of $\mathcal{F}(n)$ with the unit disk about 0 in $\mathbb{C}P_1$. Let $z_i := x_i$ for all $i = 1, \ldots, n$. Let Δ' denote the closure of the complement of this (big) unit disk in $\mathbb{C}P_1$. Let $z_{n+1} : \Delta \to \mathbb{C}P_1$ be the embedding whose image is Δ' induced by the chart $\frac{1}{w}$ about the point ∞ in $\mathbb{C}P_1$. The tuple $(z_1, \ldots, z_{n+1})$ yields a well-defined element of $\mathcal{P}(n)$ which preserves the action of the permutation groups, compositions, and the unit. Therefore, $\nu : \mathcal{F} \to \mathcal{P}$ is a morphism of operads.

Suppose there are two elements in $\mathcal{F}(n)$, $(x_1, \ldots, x_n)$ and $(x'_1, \ldots, x'_n)$, such that their images under ν in $\mathcal{P}(n)$ are equal. Let $(z_1, \ldots, z_{n+1})$ and $(z'_1, \ldots, z'_{n+1})$ denote the holomorphically embedded disks in $\mathbb{C}P^1$ associated to $(x_1, \ldots, x_n)$ and $(x'_1, \ldots, x'_n)$, respectively. Notice that $z_{n+1} = z'_{n+1}$. There must be an element of $PGL(2, \mathbb{C})$ which takes $(z_1, \ldots, z_{n+1})$ to $(z'_1, \ldots, z'_{n+1})$. However, the only element in $PGL(2, \mathbb{C})$ which fixes the embedding z_{n+1} is the identity element. Therefore, μ is injective.

Of course, μ cannot be surjective since $\mathcal{F}(n)$ is finite dimensional while $\mathcal{P}(n)$ is infinite dimensional.

3.3 Algebras Over Operads

Let A be an associative algebra with unit. A representation of A is a vector space $\mathcal{H}$ together with a homomorphism of associative algebras $A \to \text{End}(\mathcal{H})$ preserving the unit elements. If $\mathcal{O}$ is an operad, say in the category of vector spaces, then $\mathcal{O}(1)$ is an associative algebra with unit. Does there exist a notion of a "representation" of the operad $\mathcal{O}$ on a vector space $\mathcal{H}$ which when restricted to $\mathcal{O}(1)$ reduces to the usual representation of the associative algebra $\mathcal{O}(1)$? The answer is yes, however, the standard terminology used is that $\mathcal{H}$ is an algebra over the operad $\mathcal{O}$ or $\mathcal{H}$ is a $\mathcal{O}$-algebra rather than $\mathcal{H}$ is a representation of $\mathcal{O}$. It is through this notion that the relationship with topological field theories becomes most evident.

Definition 5 Let $(\mathcal{H}, d)$ be a dg vector space, $\mathcal{O}$ be an operad in the category of dg vector spaces, together with a morphism of operads $\mu : \mathcal{O} \to \text{End}^{\mathcal{H}}$.

Or, let $\mathcal{O}$ be an operad in the category of topological spaces, $\mathcal{H}$ be a topological vector space, and $\mu : \mathcal{O} \to \text{End}^{\mathcal{H}}$ be a morphism of operads.

In either case, we say that $\mathcal{H}$ is an $\mathcal{O}$-*algebra* or is an *algebra over* $\mathcal{O}$.

Remark 7 If $\mathcal{O}$ is an operad in the category of smooth (resp. complex) manifolds then one may wish to impose that the structure morphism $\mu : \mathcal{O} \to \text{End}^{\mathcal{H}}$ be smooth (resp. holomorphic).

Example 5 (The Little Intervals Operad and the Associative Operad) We now introduce a version of the little disks operad but where instead of two dimensional disks, we consider a closed interval. Let $J := [0,1]$ be the unit interval oriented with an arrow pointing towards the right.

Let $\mathfrak{I} := \{\, \mathfrak{I}(n)\,\}_{n \geq 1}$ where $\mathfrak{I}(n)$ consists of configurations of n subintervals in the interval J, i.e. $\mathfrak{I}(n)$ consists of n-tuples $(x_1, \ldots, x_n)$ where $x_i : J \to J$ such that $x_i(z) := a_i z + b_i$ where a_i is a positive real number and b_i belongs to $\mathbb{R}$. We also require that for all $i \neq j$, $x_i(J)$ and $x_j(J)$ can only intersect at most along their boundaries. The composition map $\mathfrak{I}(n) \times \mathfrak{I}(n') \to \mathfrak{I}(n + n' - 1)$ taking $(C, C') \mapsto C \circ_i C'$ arises by inserting the interval C' into the i-th interval of C in an orientation preserving fashion. S_n acts on $\mathfrak{I}(n)$ by reordering the subintervals. The unit I in $\mathfrak{I}(1)$ is the identity map $x_1 : J \to J$. $\mathfrak{I}$ is an operad in the category of topological spaces and is called the *little intervals operad*.

Let's apply the homology functor and consider the operad $H_\bullet(\mathfrak{I})$ in the category of $\mathbb{Z}$-graded vector spaces. Can we characterize algebras over this operad?

Proposition 1 *$\mathcal{H}$ is algebra over $H_\bullet(\mathfrak{I})$ iff $\mathcal{H}$ is an associative algebra.*

Proof. The space $\mathfrak{I}(n)$ contains a connected component $\mathfrak{I}'(n)$ whose n subintervals have an ordering which increases as one goes from left to right. S_n acts simply, transitively on the space of connected components of $\mathfrak{I}(n)$. Therefore, one can identify the set of connected components of $\mathfrak{I}(n)$ with the elements in S_n as an S_n set where S_n acts on itself by right multiplication.

The connected component $\mathfrak{I}'(n)$ is contractible. (Indeed, it is homotopic to the configuration of n ordered points in the oriented interval since each subinterval can be shrunk to its midpoint and the latter is contractible.) Therefore, $H_p(\mathfrak{I}(n))$ vanishes unless $p = 0$ and $H_0(\mathfrak{I}(n))$ is canonically isomorphic (as a S_n-module) to the right regular representation $\mathbb{C}[S_n]$.

Since the compositions $\circ_i$ preserve the subcollection $\{\, \mathfrak{I}'(n)\,\}_{n \geq 1}$ for all $i = 1, \ldots, n$, if $\mathbf{1}_n$ denotes the identity element in S_n then under the identification of $H_0(\mathfrak{I}(n))$ with $\mathbb{C}[S_n]$, we have

$$\mathbf{1}_n \circ_i \mathbf{1}_{n'} = \mathbf{1}_{n+n'-1} \tag{7}$$

for all $i = 1, \ldots, n$.

$H_0(\mathfrak{I}(1))$ contains only multiples of the unit element $I = \mathbf{1}_1$. The image of I must map to the identity map on $\mathcal{H}$.

Given a morphism $\mu : H_0(\mathfrak{I}) \to \mathrm{End}^{\mathcal{H}}$, let $c := \mathbf{1}_2$ be the class in $H_0(\mathfrak{I}(2)) \simeq \mathbb{C}[S_2]$ which is represented by a point in $\mathfrak{I}'(2)$. Call $m := \mu_c : \mathcal{H} \otimes \mathcal{H} \to \mathcal{H}$ the multiplication operation. Equation (7) implies, in particular, that

$$c \circ_1 c = c \circ_2 c \tag{8}$$

in $H_0(\mathfrak{I}'(3))$. Consequently, $m \circ_1 m = \mu_{c \circ_1 c} = \mu_{c \circ_2 c} = m \circ_2 m$. Or, in other words, m is an associative multiplication on $\mathcal{H}$.

$H_0(\mathcal{I}(2))$ contains another class $b\sigma$, where σ is the transposition in S_2, obtained from b by transposing the ordering of the two intervals. The associated binary operation $\mu_{b\sigma} = m \circ \sigma$ is the opposite multiplication which takes $v_1 \otimes v_2 \mapsto (-1)^{|v_1||v_2|} m(v_2, v_1)$. However, it is not another basic operation in the sense that it is completely determined by m and the action of the permutation group.

Since all other classes in $H_0(\mathcal{I}(n))$ for $n \geq 3$ can be generated by iterated compositions of the class b in $H_0(\mathcal{I}(2))$ together with the action of the permutation groups, there are no n-ary operations which cannot be obtained from the multiplication m under successive compositions and the permutation group actions.

Finally, it follows from Equation (7) that there are no relations in $H_0(\mathcal{I}(n))$ aside from those generated by Equation (8).

To prove the converse, given an associative algebra $\mathcal{H}$, we need to construct a morphism of operads $\mu : H_\bullet(\mathcal{O}) \to \mathrm{End}^{\mathcal{H}}$. We define μ_c for $c = 1_2$ to be the given multiplication on $\mathcal{H}$ and then extend μ to all of $H_\bullet(\mathcal{O})$ in a way compatible with the compositions using Equation (7) and by using the equivariance of μ under the permutation group. $\square$

In light of the previous proposition, the operad $H_\bullet(\mathcal{I})$ can be called the *associative operad* since algebras over it are nothing more than associative algebras.

The importance of operads is that they are gadgets which parameterize algebraic structures of a given type just as the associative operad, above, governs associative algebras. Common algebraic structures (for example, associative algebras, commutative associative algebras, Lie algebras, Poisson algebras, A_∞ algebras, etc) consisting of a collection of n-ary operations for $n \geq 1$ have an underlying operad responsible for them.

Definition 6 Let $\mathcal{H}$ be a $\mathbb{Z}$-graded vector space together with two binary operations, a *multiplication* $a \otimes b \mapsto a \cdot b$ of degree 0 and a *bracket* $a \otimes b \mapsto [a, b]$ of degree 1, such that for all homogeneous elements a, b, c in $\mathcal{H}$,

Commutativity of the Product : $a \cdot b = (-1)^{|a||b|} b \cdot a$,
Associativity of the Product : $(a \cdot b) \cdot c = a \cdot (b \cdot c)$,
"Skew"-Symmetry of the Bracket : $[a, b] = -(-1)^{(|a|-1)(|b|-1)}[b, a]$,
Jacobi Identity : $[a, [b, c]] = [[a, b], c] + (-1)^{(|a|-1)(|b|-1)}[b, [a, c]]$, and
Leibnitz Rule : $[a, b \cdot c] = [a, b]c + (-1)^{|b|(|a|-1)} b[a, c]$.

Such a triple $(\mathcal{H}, \cdot, [,])$ is called a *Gerstenhaber algebra*.

A Gerstenhaber algebra is basically a kind of Poisson (super)algebra but where the bracket has degree 1. We will now find the operad responsible for it.

Theorem 2 (F. Cohen [5], V.I. Arnold [2]) $\mathcal{H}$ *is an algebra over* $H_\bullet(\mathcal{D})$ *if and only if* $\mathcal{H}$ *is a Gerstenhaber algebra. Thus,* $H_\bullet(\mathcal{D})$ *is called the Gerstenhaber operad.*

Proof. The operad $\mathcal{D}(n)$ is homotopy equivalent to $C_n^{\mathbb{C}}$, the configuration of n points in $\mathbb{C}$, which consists of pairwise distinct n-tuples $(p_1, \ldots, p_n)$ in $\mathbb{C}$. The equivalence is obtained by shrinking each little disk to its center point.

Let $\mu : H_\bullet(\mathcal{D}) \to \text{End}^{\mathcal{H}}$ be a morphism of operads. $D(1)$ is contractible thus it contains only multiples of the unit I. $D(2)$ is homotopic to S^1. Therefore, $H_\bullet(\mathcal{D}(2)) \simeq H_\bullet(S^1)$. Let β denote the class of a point in $H_0(D(2))$. Let γ denote the class in $H_1(\mathcal{D}(2))$ corresponding to the fundamental class of S^1. One can view γ as being represented by an embedded circle in $D(2)$ where the second disk completes one counterclockwise revolution around the first disk. Thus, μ_β is a binary operation of degree 0, which we'll denote by $a \otimes b \mapsto a \cdot b$, and μ_γ is a binary operation of degree 1. Denote the map $a \otimes b \mapsto [a, b]$ by $\mu_\gamma(a \otimes b)(-1)^{|a|}$ for all homogeneous a, b in $\mathcal{H}$.

Since S^1 is connected, $\beta = \beta\sigma$ where σ is the transposition in S_2. Therefore, $\mu_\beta = \mu_{\beta\sigma} = \mu_\beta \circ \sigma$ or, equivalently, for all homogeneous a, b in $\mathcal{H}$, $a \cdot b = (-1)^{|a||b|} b \cdot a$. Similarly, $\gamma\sigma = \gamma$ implies $[a, b] = -(-1)^{(|a|-1)(|b|-1)}[b, a]$.

The Leibnitz rule follows from an identity between compositions of β and γ in three different ways in $H_1(\mathcal{D}(3))$ while the Jacobi identity follows from an identity between compositions of β in three different ways in $H_2(\mathcal{D}(3))$.

To prove that there are no other basic operations and no other basic relations it is useful to use a presentation of the cohomology groups of configuration spaces due to Arnold [2]. $\square$

Remark 8 $H_\bullet(\mathcal{D})$ has several important suboperads, one responsible for the multiplication only, and the other responsible for the Lie bracket only.

The suboperad $H_0(\mathcal{D})$ of $H_\bullet(\mathcal{D})$ is the operad responsible for commutative, associative algebras. Thus it is may be called the *commutative operad*. Observe that $H_0(\mathcal{D}(n)) \simeq \mathbb{C}$ for all n.

The suboperad $\mathcal{L} := \{\mathcal{L}(n)\}_{n \geq 1}$ of $H_\bullet(\mathcal{D})$ where $\mathcal{L}(n) := H_{n-1}(\mathcal{D}(n))$ is the operad responsible for the Lie algebra structure where the Lie bracket has degree 1. However, after shifting the grading of $\mathcal{H}$ by 1, one may regard the bracket as usual Lie algebra with a degree zero bracket. Hence, $\mathcal{L}$ is often called the *Lie operad*.

3.4 Topological Conformal Field Theory

Let us begin with a variant of topological field theory called a conformal field theory whose axiomatization is due to Segal [28].

Definition 7 A *(g = 0, c = 0) conformal field theory (CFT)* is a topological vector space $\mathcal{H}$ together with a smooth morphism of operads $\mu : \mathcal{P} \to \text{End}^{\mathcal{H}}$. $\mathcal{H}$ is called the *state space*.

The $g = 0$ corresponds to the fact that $\mathcal{P}$ only features Riemann surfaces of genus 0. The $c = 0$ means that μ is a morphism of operads. In a general CFT, μ would only preserve the compositions up to a certain projective factor. In particular, restricting to the semigroup $\mathcal{P}(1)$, $\mathcal{H}$ would be a projective representation of the semigroup $\mathcal{P}(1)$. This is the essentially the geometric origin of the central extension of the Virasoro algebra which is usually denoted by c.

We also remark that Segal's definition of a CFT contains more features than those presented above such as an inner product, decompositions into left ("holomorphic") and right movers ("antiholomorphic") and so forth but the above will suffice for our purposes.

It is not easy to characterize the algebraic structure on $\mathcal{H}$. However, there is a result by Huang which shows that a vertex operator algebra is a CFT where the μ are holomorphic. Indeed, to study the connection with vertex algebras, one must enlarge the operad $\mathcal{P}$ to a larger space, $\mathcal{P}'$, in which the holomorphically embedded disks are now allowed to pairwise overlap provided that their centers do not coincide. The compositions are then analytically continued from $\mathcal{P}$ to a maximal subspace of $\mathcal{P}'$. The result is that $\mathcal{P}'$ is no longer an operad but a partial operad (or perhaps "operadoid"[1]) which means that compositions are not always defined. Vertex operator algebras are essentially an algebra over $\mathcal{P}'$ (up to a projective factor). We refer to [14] and the references therein for more details to this approach.

A purely algebraic construction of the BV algebra structure in terms of the closely related algebraic framework of topological chiral algebras appears in [23].

While a general CFT is a rather complicated object, there is a variant of a CFT called a topological conformal field theory that has a sector which is easier to handle.

Definition 8 A $(g = 0)$ *topological conformal field theory (TCFT)* is a topological dg vector space $\mathcal{H}$ together with a morphism of operads $\mu : C_{\bullet}(\mathcal{P}) \to \mathrm{End}^{\mathcal{H}}$.

Again, we have simplified the definition of a TCFT [29] as one usually assumes in addition, that the morphism μ arises by integrating smooth chains on $\mathcal{P}(n)$ over differential forms on $\mathcal{P}(n)$ with values in $\mathrm{End}^{\mathcal{H}}(n)$. In this situation, a TCFT is a special kind of CFT.

Remark 9 In the physics literature, the differential $d : \mathcal{H}_{p+1} \to \mathcal{H}_p$ associated to a TCFT is sometimes called a BRST operator and the grading p of $\mathcal{H}_p$ is (the negative of) the ghost number.

---

[1] We are pleased to be able to use include this term in light of some of the other topics presented at this conference.

Definition 9 Let $\mathcal{H}$ be a Gerstenhaber algebra together with an operation $\Delta : \mathcal{H} \to \mathcal{H}$ of degree 1 such that $\Delta^2 = 0$ and

$$[a, b] = (-1)^{|a|}(\Delta(a \cdot b) - (\Delta a) \cdot b - (-1)^{|a|} a \cdot (\Delta b)) \tag{9}$$

for all homogeneous a and b in $\mathcal{H}$ then we say that $\mathcal{H}$ is a *Batalin-Vilkovisky* (or *BV*) *algebra*.

The following theorem and lemma are due to Getzler [10].

Theorem 3 *Let* $(\mathcal{H}, d)$ *be a* $(g = 0)$ *TCFT then* $H_\bullet(\mathcal{H})$ *is a BV algebra.*

Proof. The proof follows from the following Lemma and the observation that the morphism of operads $\mathcal{F} \to \mathcal{P}$ is a homotopy equivalence. $\square$

Lemma 1 $\mathcal{H}$ *is a* $H_\bullet(\mathcal{F})$-*algebra if and only if it is a BV algebra.*

Proof. Since $\mathcal{F}(n)$ is a trivial $(S^1)^n$-bundle over $\mathcal{D}(n)$, its homology groups can be readily computed.

Let $\mu : H_\bullet(\mathcal{F}) \to \text{End}^{\mathcal{H}}$ be a morphism of operads. Since $\mathcal{D} \to \mathcal{F}$ is a morphism of operads, it induces a morphism $H_\bullet(\mathcal{D}) \to H_\bullet(\mathcal{F})$. Thus, $\mathcal{H}$ is a $H_\bullet(\mathcal{D})$-algebra which, by Theorem 2, implies that $\mathcal{H}$ is a Gerstenhaber algebra.

Observe that $\mathcal{F}(1)$ is homotopic to (an oriented) S^1 corresponding to the direction of the marked point on the boundary of the little disk. Let c be the element in $H_1(\mathcal{F}(1))$ corresponding to the fundamental class of S^1. Call $\Delta := \mu_c$.

It is clear that $c \circ_1 c = 0$ since $H_2(S^1) = 0$, therefore, $\Delta^2 = 0$. Equation (9) can be verified directly.

The computation of the homology groups implies that there are no other basic operations or relations. $\square$

One can also prove the existence of homotopy algebraic structures associated to a TCFT. For example, Kimura-Stasheff-Voronov [20] proved that a certain relative subcomplex of a TCFT whose structure morphisms arise from integration of smooth differential forms yields a homotopy Lie (or L_∞) algebra. The operad responsible for this structure arises from a certain "real" compactification of the moduli space of Riemann spheres with n marked points. This structure was also been observed in the physics literature by Zweibach [32] in the context of closed string field theory.

In the previous examples, we treated the $(n+1)$st embedded disk of $\mathcal{P}(n)$ differently from the first n embedded disks. The $(n + 1)$st disk is a kind of "output" while the first n disks are "inputs." However, from a geometric point of view, it is reasonable to put all such disks on the same footing, i.e. allow the permutation group S_{n+1} to act upon $\mathcal{P}(n)$ by permuting all disks and to allow sewings between arbitrary pairs of disks on two different spheres. Formalizing this structure gives rise to the notion a cyclic operad [12]. Algebras over cyclic operads typically correspond to algebraic structures (like associative algebras or Lie algebras) with an invariant metric.

By allowing moduli spaces of higher genus Riemann surfaces as well and allowing sewings (including sewing two disks on the same surface to obtain a new surface of higher genus), one obtains an example of a modular operad [13]. Rather than describing such technical notions in complete generality, we will now focus upon the model example of a modular operad, namely the moduli space of stable curves, and study algebras over them. Such algebras are often referred to as a cohomological field theory in the sense of Kontsevich-Manin.

4 The Moduli Space of Stable Curves and Cohomological Field Theories

4.1 The Cyclic Operad of Stable Trees

Let us begin with a useful toy model.

Definition 10 Let $[n]$ denote the set $\{1, \ldots, n\}$ for all $n \geq 1$.

Definition 11 Let I be a finite set such that $|I| \geq 4$. A *stable partition of I* is a partition of I, $I_+ \sqcup I_- = I$, such that $|I_\pm| \geq 2$.

Definition 12 Let I be a finite set such that $|I| \geq 3$. A *stable I-tree*, Γ, is a connected graph containing no circuits (i.e. a tree) together with a bijection from I to its set of tails $T(\Gamma)$ whose vertices are of valence at least 3. Let $T(\Gamma)$, $V(\Gamma)$, and $E(\Gamma)$, respectively, denote the set of tails, the set of vertices, and the set of edges of Γ. A *stable n-tree* when $n \geq 3$ is a stable $[n]$-tree. The *n-corolla*, δ_n, is the unique stable n-tree with exactly one vertex.

Let $\mathcal{T} := \{\mathcal{T}_n\}_{n \geq 2}$ where $\mathcal{T}_n$ is the free $\mathbb{C}$-span of the set of stable n-trees. T_n is an S_n-module by permutation of the labels on its tails.

Let $n \geq 4$ be an integer and consider a partition $N \sqcup N' := [n]$ where $|N| = n$ and $|N'| = n'$ such that $n, n' \geq 2$. Let $N = \{i_1, \ldots, i_n\}$ and $N' = \{j_1, \ldots, j_{n'}\}$ where $i_1 < i_2 < \cdots < i_n$ and $j_1 < j_2 < \cdots < j_{n'}$.

Define the composition maps $\mathcal{T}_{n+1} \times \mathcal{T}_{n'+1} \to \mathcal{T}_{n+n'}$ taking $(\Gamma, \Gamma') \mapsto \Gamma \circ_{(N,N')} \Gamma'$ defined by gluing the $(n+1)$st tail of Γ to the 1st tail of Γ' and then relabeling the remaining tails so that the k-th tail of Γ is labeled by i_k for all $k = 1, \ldots, n$ and the p-th tail of Γ' is labeled by j_{p-1} for all $p = 2, \ldots, n'+1$.

It is probably easiest to visualize the composition by reversing the arrow of time. That is, given a stable n''-tree, Γ'', choose an edge e in Γ'' and cut it. Choose one of the resulting trees and call it Γ and let Γ' denote the other.

Let $n + 1$ denote the number of tails of Γ and $n' + 1$ denote the number of tails of Γ'. Let t and t' denote the new tails on Γ and Γ' respectively obtained by cutting e. Let N denote the set of labels on Γ and N' denote the set of labels on Γ'. Clearly, N and N' form a stable partition of $[n + n']$ where $|N| = n$ and $|N'| = n'$. Let $N = \{i_1, \ldots, i_n\}$ and $N' = \{j_1, \ldots, j_{n'}\}$ where $i_1 < \cdots < i_n$ and $j_1 < \cdots < j_{n'}$. Relabel Γ so that the tail with the label i_k is labeled instead by k for all $k = 1, \ldots, n$ and then label the tail t by $n + 1$. Similarly, relabel Γ' so that the tail with label j_p is labeled instead with $p + 1$ for all $p = 1, \ldots, n'$ and then label the tail t' by 1. With these new labels for Γ' and Γ'', we have $\Gamma \circ_{(N,N')} \Gamma' = \Gamma''$.

Associativity can then be understood, under time reversal, as taking a tree Γ'', and choosing two distinct edges e and e'. Cutting Γ'' at e followed by cutting the result at e' and repeating the process in the last paragraph, one obtains 3 stable trees which when glued back together yield Γ''. On the other hand, cutting Γ'' along e' followed by cutting the result at e and repeating the process above, one obtains 3 stable trees which when glued together in the other order yields Γ''.

These composition maps are equivariant under the action of the permutation groups.

$\mathcal{T}$ contains the structure of an operad.[2] Let $\mathcal{T}(n) := \mathcal{T}_{n+1}$ for all $n \geq 2$. The operadic composition $\mathcal{T}(n) \times \mathcal{T}(n') \to \mathcal{T}(n + n' - 1)$ taking $(\Gamma, \Gamma') \mapsto \Gamma \circ_i \Gamma'$ for all $i = 1, \ldots, n$ is defined by $\Gamma \circ_i \Gamma' = \Gamma' \circ_{(N',N)} \Gamma$ where

$$N = [n] \quad \text{and} \quad N' = \{n + 1, \ldots, n + n'\} . \tag{10}$$

Example 6 Let $(\mathcal{H}, \eta)$ be a finite dimensional vector space $\mathcal{H}$ with a symmetric, nondegenerate, bilinear form. Let $\mathcal{E}nd_n^{\mathcal{H}} := (\mathcal{H}^*)^{\otimes(n)}$ where $\mathcal{H}^*$ denotes the dual vector space.

Given f in $\mathcal{E}nd_{n+1}^{\mathcal{H}}$ and f' in $\mathcal{E}nd_{n'+1}^{\mathcal{H}}$ and a stable partition of $N \sqcup N'$ of $[n + n']$ where $N = \{i_1, \ldots, i_n\}$ and $N' = \{j_1, \ldots, j_{n'}\}$ such that $i_1 < \cdots < i_n$ and $j_1 < \cdots < j_{n'}$ then

$$(f \circ_{(N,N')} f')(v_1, \ldots, v_{n+n'}) := f(v_{i_1}, \ldots, v_{i_n}, e_\alpha)\eta^{\alpha\beta} f'(e_\beta, v_{j_1}, \ldots, v_{j_{n'}})$$

for all $v_1, \ldots, v_{n+n'}$ where $\{e_\alpha\}$ is a basis for $\mathcal{H}$, $\eta^{\alpha\beta}$ is the inverse metric in this basis, and the summation convention has been used. Furthermore, $\mathcal{E}nd_n^{\mathcal{H}}$ has a natural action of S_n.

One can recover the usual endomorphism operad of $\mathcal{H}$ as follows. Using the metric to identify $\mathcal{E}nd_{n+1}^{\mathcal{H}}$ with $\mathrm{End}^{\mathcal{H}}(n)$, one can recover the S_n action on the latter by restriction to the subgroup S_n in S_{n+1}. The operadic compositions in $\mathrm{End}^{\mathcal{H}}$ may be recovered by considering the partitions in Equation (10).

---

[2] Notice that since $\mathcal{T}(1)$ does not exist, this is an operad without unit. However, if one desires, one may adjoin a unit by hand.

4.2 Tree Level Feynman Diagrams

Let us consider the analog of an algebra over an operad in this setting. Let $(\mathcal{H}, \eta)$ be a finite dimensional vector space with a symmetric, nondegenerate bilinear form. Consider a morphism $\mu : \mathcal{T} \to \mathcal{E}nd^{\mathcal{H}}$ preserving the compositions and equivariant under the action of the permutation groups. This means that we must associate to each n-tree in $\mathcal{T}_n$ an element in $(\mathcal{H}^*)^{\otimes n}$ compatible with compositions and the permutation groups.

Specifying μ is equivalent to a choice of an element $\Phi^{(n)}$ in the n-fold symmetric product $S^n \mathcal{H}^*$ for all $n \geq 3$. This is because μ_{δ_n}, where δ_n is the n-corolla, is an element in the n-fold symmetric product $S^n \mathcal{H}^*$. It is symmetric because the tree δ_n is invariant under S_n. Define $\mu_{\delta_n} := \Phi^{(n)}$. Since any tree in $\mathcal{T}$ is obtained by gluing together corollas, μ is completely determined. Furthermore, $\Phi^{(n)}$ can be arbitrary for all $n \geq 3$ because $\mathcal{T}$ is freely generated, with respect to the composition, by the set of corollas.

This is a situation very familiar to physicists. If Γ is an n-tree in $\mathcal{T}(n)$ then associate to each valence n vertex a totally symmetric tensor $\Phi^{(n)}$ ("an interaction vertex") and to each internal edge the inverse metric ("a propagator") contracting in the manner indicated by the graph. In other words, specifying μ is nothing more than a choice of Feynman rules on $\mathcal{H}$. For example, let $\{e_\alpha\}$ be a basis for $\mathcal{H}$, $\{e^\alpha\}$ denote its dual basis in $\mathcal{H}^*$, $\eta_{\alpha\beta}$ be the metric, $\eta^{\alpha\beta}$ be the inverse metric, and $\Phi_{\alpha_1,\dots,\alpha_n}$ denote the symmetric n tensor $\Phi^{(n)}$ with respect to this basis.

For example, the following stable 7-tree

$$\includegraphics{tree} \tag{11}$$

has the associated tensor in $(\mathcal{H}^*)^{\otimes 7}$

$$\Phi_{\alpha_2 \alpha_5 \alpha_6 \beta} \eta^{\beta\gamma} \Phi_{\gamma\alpha_1\delta} \eta^{\delta\epsilon} \Phi_{\epsilon\alpha_3\alpha_4\alpha_7} e^{\alpha_1} \otimes e^{\alpha_2} \otimes \cdots \otimes e^{\alpha_7} \tag{12}$$

where the summation convention has been used.

These Feynman rules are captured by a "Lagrangian" whose "kinetic term" is

$$\frac{1}{2}\eta(x,x) \tag{13}$$

and whose "potential function" is

$$\Phi(x) := \sum_{n=3}^{\infty} \frac{1}{n!} \Phi^{(n)}(x, x, \dots, x) \tag{14}$$

where x belongs to $\mathcal{H}$ and $\Phi(x)$ is regarded as a formal power series on $\mathcal{H}$. However, in the definition of μ, only tree level terms in the usual perturbative expansion are relevant. We'll have use for the higher loop graphs later.

Let us now consider an important special case. Let $\mathcal{T}' := \{\mathcal{T}'_n\}_{n \geq 2}$ denote the subspace in $\mathcal{T}_n$ spanned by trivalent, stable n-trees. This subspace is closed under compositions and the permutation group actions. A morphism $\mu : \mathcal{T}' \to \mathcal{E}nd^{\mathcal{H}}$ is equivalent to Feynman rules associated only to trivalent graphs. Thus, μ is completely characterized, since we keep the same kinetic term, by the (cubic) potential function $\Phi(x) = \frac{1}{3!}\Phi^{(3)}(x, x, x)$.

4.3 The Moduli Space of Stable Curves

Let $\mathcal{M}_{g,n}$ denote the moduli space of smooth, stable curves of genus g with n marked points. This consists of tuples $(C; p_1, \ldots, p_n)$ where C is a smooth, connected, complex algebraic curve of genus g together with pairwise distinct marked points $p_1, \ldots, p_n$ on C where any two such configurations $(C; p_1, \ldots, p_n)$ and $(C'; p'_1, \ldots, p'_n)$ are isomorphic if and only if there is an isomorphism between the curves $\phi : C \to C'$ such that $\phi(p_i) = p'_i$ for all $i = 1, \ldots, n$. (If one prefers, one may also regard this as the moduli space of Riemann surfaces of genus g with n marked points.) Furthermore, we require that the pointed curve $(C; p_1, \ldots, p_n)$ be stable, meaning that it has no infinitesimal automorphisms. This means that $2g - 2 + n > 0$ must hold or, equivalently, if $g = 0$ then $n \geq 3$, if $g = 1$ then $n \geq 1$, and if $g \geq 2$ then n is arbitrary.

$\mathcal{M}_{g,n}$ is a smooth, complex orbifold of complex dimension $3g - 3 + n$. While these spaces are noncompact, $\mathcal{M}_{g,n}$ has a natural compactification $\overline{\mathcal{M}}_{g,n}$ due to Deligne-Mumford called the *moduli space of stable curves of genus g with n marked points*. It consists of tuples $(C; p_1, \ldots, p_n)$ where C is a complex curve with (at worst) nodal singularities, the pairwise distinct points $p_1, \ldots, p_n$ cannot be at a node, and $(C; p_1, \ldots, p_n)$ must have no infinitesimal automorphisms. $\overline{\mathcal{M}}_{g,n}$ is a smooth, connected, compact, complex orbifold of complex dimension $3g - 3 + n$ which contains $\mathcal{M}_{g,n}$ as a dense open subset. This means that a nodal curve with n marked points is stable if and only if each irreducible component of genus 0 has at least 3 special points and each irreducible component of genus 1 has at least 1 special point where by special point we mean either a node or a marked point.

Associated to each partition (N, N') of $[n + n']$, where $|N| = n$ and $|N'| = n'$, compatible with stability, there is a composition morphism $\circ_{(N,N')} : \overline{\mathcal{M}}_{g,n+1} \times \overline{\mathcal{M}}_{g',n'+1} \to \overline{\mathcal{M}}_{g+g',n+n'}$ which takes the curve C of genus g, the curve C' of genus g' and attaches the $(n+1)$st marked point of C with the 1st point of C' thereby creating a node. The remaining $(n + n')$ marked points are then reordered using the partition (N, N') as before. In addition, for each distinct pair i and j in $[n + 2]$, we obtain another gluing morphism $\overline{\mathcal{M}}_{g,n+2} \to \overline{\mathcal{M}}_{g+1,n}$ which glues the curve together along its i-th and j-th marked points to create a node. The collection $\{\overline{\mathcal{M}}_{g,n}\}$ together with these gluing morphisms (which are equivariant under the action of the permutation groups) is the model example of a *modular operad* in the category of smooth, complex orbifolds.

Let us now associate to each stable curve of genus g with n marked points a graph called its *dual graph* as follows. Collapse each irreducible component of genus g to a point to obtain a vertex and label the vertex with g. For each special point on that component, draw a half edge from the vertex. If two irreducible components meet at a node then connect their corresponding half edges together to form and edge. Finally, for each remaining half-edge (or tail) we give it the same label in $[n]$ as its associated marked point. The resulting connected graph is called a *stable graph of genus g with n tails*. An n-corolla of genus g, $\delta_{g,n}$, is a stable graph of genus g with n tails with exactly one vertex and it corresponds to a smooth curve of genus g with n marked points subject to the stability condition that if $g = 0$ then $n \geq 3$ and if $g = 1$ then $n \geq 1$.

Tracing through the definitions, a stable graph of genus g with n tails is a graph, together with a bijection between $[n]$ and its set of tails, which is either $\delta_{g,n}$ or it can be obtained by from gluing together a finite number of stable corollas such that the *genus of Γ*, $g(\Gamma)$ defined by $g(\Gamma) := b_1(\Gamma) + \sum_{v \in V(\Gamma)} g(v)$, is equal to g where b_1 denotes the 1st Betti number, $V(\Gamma)$ denotes the set of vertices, and $g(v)$ denotes the genus label at vertex v.

A stable n-tree can be regarded as a genus 0 stable graph with n tails by assigning each vertex a genus of 0. Conversely, any stable graph of genus 0 with n tails is a stable n-tree by erasing the genus 0 label at each vertex. We shall henceforth make this identification.

To each stable graph, one can associate an orbifold. Given a stable graph Γ of genus g with n tails, we can associate an orbifold $\overline{\mathcal{M}}_\Gamma$ consisting of the closure of the locus of curves in $\overline{\mathcal{M}}_{g,n}$ whose dual graph is Γ. $\overline{\mathcal{M}}_\Gamma$ is a compact orbifold whose complex codimension is equal $|E(\Gamma)|$ where $E(\Gamma)$ is the set of (internal) edges of the graph. In particular, $\overline{\mathcal{M}}_{\delta_{g,n}} = \overline{\mathcal{M}}_{g,n}$. Furthermore, the inclusion map $\overline{\mathcal{M}}_\Gamma \to \overline{\mathcal{M}}_{g,n}$ is a regular embedding.

4.4 $g = 0$ Cohomological Field Theories and Feynman Diagrams

Our goal in this subsection is to consider only genus 0 moduli spaces of stable curves, take their homology groups, and to characterize algebras over them.

Let us restrict to composition morphisms associated to a stable partitions (N, N') of $[n + n']$ where $|N| = n$ and $|N'| = n'$ taking $\overline{\mathcal{M}}_{0,n+1} \times \overline{\mathcal{M}}_{0,n'+1} \to \overline{\mathcal{M}}_{0,n+n'}$. This induces composition morphisms $H_\bullet(\overline{\mathcal{M}}_{0,n+1}) \times H_\bullet(\overline{\mathcal{M}}_{0,n'+1}) \to H_\bullet(\overline{\mathcal{M}}_{0,n+n'})$.

Definition 13 A $g = 0$ *cohomological field theory (CohFT)* is a vector space with metric $(\mathcal{H}, \eta)$ together with maps for all $n \geq 3$, $\mu : H_\bullet(\overline{\mathcal{M}}_{0,n}) \to \mathcal{H}^{*\otimes n}$, which preserve compositions and are S_n equivariant.

Remark 10 Here, we are regarding $\mathcal{H}^{*\otimes n}$ as $\mathcal{E}nd_n^{\mathcal{H}}$.

Remark 11 There is also a flat identity element in the definition of a CohFT which is a generalization of the identity element in a Frobenius algebra but we will ignore it for the purposes of these lectures.

Remark 12 Henceforth, unless otherwise stated, we will assume that $\mathcal{H}$ is an ungraded vector space in order to avoid being distracted by signs.

Let Γ be a stable n-tree. Given any stable n-tree Γ, we can associate the homology class $[\overline{M}_\Gamma]$ in $H_p(\overline{M}_{0,n})$ where $p = 2(n - 3 - |E(\Gamma)|)$. This induces a map $f : \mathcal{T}_n \to H_\bullet(\overline{M}_{0,n})$ for all $n \geq 3$ which is S_n equivariant and preserves compositions. Given any $g = 0$ CohFT $\mu : H_\bullet(\overline{M}_{0,n}) \to \mathcal{H}^{*\otimes n}$, composition with f yields a morphism $\nu : \mathcal{T}_n \to \mathcal{H}^{*\otimes n}$. By the analysis in Sect. 4.2, ν is completely determined by tree level Feynman diagrams associated to a potential function $\Phi(x)$ of the form in Equation (14). However, it's not true that an arbitrary potential function $\Phi(x)$ can arise from the morphism ν. This is because the morphism $\mathcal{T}_n \to H_\bullet(\overline{M}_{0,n})$ is not injective, in general, i.e. there could be two different linear combination of n-trees whose associated cycles are homologous. This means that the result of applying the Feynman rules to these linear combinations of graphs must yield the same element in $\mathcal{H}^{*\otimes n}$. This imposes constraints on the potential function Φ. What are these constraints, exactly?

In order to get an idea of how to think about these relations, let's consider the special case of trivalent trees. If Γ belongs to $\mathcal{T}'_n$ then $[\overline{M}_\Gamma]$ belongs to $H_0(\overline{M}_{0,n})$ since amongst all stable n-trees, a trivalent n-tree has the maximal number of (internal) edges. Therefore, f restricts to a morphism $f' : \mathcal{T}'_n \to H_0(\overline{M}_{0,n})$ for all $n \geq 3$. We recall that since $\overline{M}_{g,n}$ is connected, the kernel of $f' : \mathcal{T}'_n \to \mathcal{H}^{*\otimes n}$, call it $\mathcal{R}_n$ for the space of relations, is quite large in general. However, we seek a more diagrammatic characterization of $\mathcal{R}_n$.

There is only one stable 3-tree, the corolla δ_3. Also, $\overline{M}_{0,3}$ is a point since one can always use the automorphism group, $\mathrm{PGL}(2, \mathbb{C})$, of $\mathbb{C}P_1$ to move the 3 marked points to $0, 1, \infty$, for example. Therefore, $f' : \mathcal{T}'_3 \to H_0(\overline{M}_{0,3})$ mapping δ_3 to $[\overline{M}_{\delta_3}] = [\overline{M}_{0,3}]$ is an isomorphism.

The open moduli space $M_{0,4}$ is a $\mathbb{C}P_1$ together with distinct marked points (p_1, p_2, p_3, p_4). Using $\mathrm{PGL}(2, \mathbb{C})$, one can move p_1 to 0, p_2 to 1, p_3 to ∞ and p_4 can be anywhere else. Therefore, $M_{0,4} \simeq \mathbb{C}P_1 - \{0, 1, \infty\}$. The compactification $\overline{M}_{0,4}$ introduces nodal curves corresponding to the limits where the 4-th point approaches either the 1st, 2nd or 3rd marked points. These three nodal curves have the following dual graphs.

$$(15)$$

Call the above graphs S, T, and U, respectively. Note that U has only two vertices but is drawn in this suggestive fashion for the sequel.

Adding in these three points, we see that $\overline{M}_{0,4}$ is isomorphic to $\mathbb{C}P_1$. Since $\mathbb{C}P_1$ is connected, we have the equality

$$[\overline{\mathcal{M}}_S] = [\overline{\mathcal{M}}_T] = [\overline{\mathcal{M}}_U] \tag{16}$$

in $H_0(\overline{\mathcal{M}}_{0,4})$.[3]

Furthermore, the identity in Equation (16) generates all elements in the kernel of $f' : \mathcal{T}'_n \to H_0(\overline{\mathcal{M}}_{0,n})$ for all $n \geq 4$ in the following sense. Let Γ be a trivalent, stable n-tree where $n \geq 4$. Choose a subgraph of Γ which can be identified with S and then replace it by either a T or a U graph. We call this operation a move on Γ. We can now define an equivalence relation on the space of trivalent n-trees, namely, two stable trivalent n-trees Γ and Γ' are said to be equivalent if and only if there exists a sequence of moves taking Γ to Γ'. It is easy to check that any two trivalent n-trees are equivalent to each other. Therefore, the basic relation in Equation (16) generates the entire kernel of f'.

If $\mu' : H_0(\overline{\mathcal{M}}_{0,n}) \to \mathcal{H}^{*\otimes n}$ is a morphism preserving compositions and permutation group actions then its composition with f', $\nu' : \mathcal{T}'_n \to \mathcal{H}^{*\otimes n}$ for all $n \geq 2$, corresponds to precisely those Feynman rules which satisfy the identity

$$\nu_S = \nu_T = \nu_U . \tag{17}$$

Let $\Phi^{(3)}$ in $S^3\mathcal{H}^*$ be the 3-tensor associated to each trivalent vertex then one can define an operation $m : \mathcal{H}^{\otimes 2} \to \mathcal{H}$ via

$$\eta(m(a,b),c) = \Phi^{(3)}(a,b,c)$$

for all a, b, c in $\mathcal{H}$. Equation (17) is equivalent to saying that $(\mathcal{H}, \eta, m)$ is a commutative, associative algebra with invariant inner product, i.e. a Frobenius algebra although possibly without an identity element.

We have just proven the following.

Proposition 2 *The morphism* $\mu' : H_0(\overline{\mathcal{M}}_{0,n}) \to \mathcal{H}^{*\otimes n}$ *is equivalent to endowing* $(\mathcal{H}, \eta)$ *with the structure of a Frobenius algebra (possibly without an identity element).*

An equivalent way of saying the same thing is to require that the potential function $\Phi(x) := \frac{1}{3!}\Phi^{(3)}(x, x, x)$ satisfy the WDVV (Witten-Dijkgraaf-Verlinde-Verlinde) equation for all a, b, c, d:

$$(\partial_a\partial_b\partial_e\Phi)\,\eta^{ef}\,(\partial_f\partial_c\partial_d\Phi) = (\partial_b\partial_c\partial_e\Phi)\,\eta^{ef}\,(\partial_f\partial_a\partial_d\Phi) , \tag{18}$$

where $\eta_{ab} := \eta(e_a, e_b)$, η^{ab} is in inverse matrix to the metric η_{ab}, ∂_a is derivative with respect to x^a, and the summation convention has been used.

It is this last formulation that turns out to be the most useful in the general case when we allow trees with vertices of arbitrary valences. This meant the potential function is no longer cubic.

---

[3] To a physicist, the choice of names STU for these graphs should be familiar given the previous equality.

Theorem 4 (Kontsevich-Manin) *Let $(\mathcal{H}, \eta)$ be a $g = 0$ CohFT. Let Φ be its associated potential function, as defined by Equation (14), then it must satisfy Equation (18), the WDVV equation. Conversely, any function Φ of the form in Equation (14) satisfying the WDVV equation is the potential function of some CohFT.*

Proof. The idea of this proof is to observe that there are forgetful morphisms $\pi : \overline{\mathcal{M}}_{0,n+4} \to \overline{\mathcal{M}}_{0,4}$ which "forgets" n points. If two algebraic cycles on $\overline{\mathcal{M}}_{0,4}$ yield the same homology class then their flat pullbacks via π must also be in the same homology class of $\overline{\mathcal{M}}_{0,n+4}$. Therefore, pulling back the cycles $\overline{\mathcal{M}}_S$, $\overline{\mathcal{M}}_T$, and $\overline{\mathcal{M}}_U$, one obtains relations in $H_\bullet(\overline{\mathcal{M}}_{0,n+4})$. Furthermore, the pullback of $\overline{\mathcal{M}}_\Gamma$ for $\Gamma = S, T, U$ can be written explicitly in terms of a linear combination of $(n + 4)$-trees. These relations are encoded in the WDVV equation. See [24] for details. $\square$

Remark 13 If $\mathcal{H}, \eta$ is a graded vector space with degree 0 metric η then the same theorem holds except that a factor of $(-1)^{|x_a|(|x_b|+|x_c|)}$ should be inserted to the immediate right of the equal sign.

Remark 14 The triple $(\mathcal{H}, \eta, \Phi)$ together with the flat identity element, which we have suppressed, forms an example of a formal Frobenius manifold, a generalization of a Frobenius algebra. This is a formal version of the notion of a Frobenius manifold due to Dubrovin [8].

Theorem 4 is important because in many cases, the potential function is sufficiently constrained by the WDVV equation so that it can be solved.

4.5 Cohomological Field Theories

To summarize so far, the collection $\{H_0(\overline{\mathcal{M}}_{0,n})\}$ is responsible for Frobenius algebras. The collection $\{H_\bullet(\overline{\mathcal{M}}_{0,n})\}$ is responsible for (formal) Frobenius manifolds. Let us now study CohFTs in all genera.

Definition 14 A *cohomological field theory (CohFT)*, $(\mathcal{H}, \eta, \Lambda)$, is a finite dimensional vector space $\mathcal{H}$ with a nondegenerate, symmetric, bilinear form η together with a collection $\Lambda := \{\Lambda_{g,n}\}$ where for all (g, n), $\Lambda_{g,n}$ belong to $H^\bullet(\overline{\mathcal{M}}_{g,n}) \otimes \mathcal{H}^{*\otimes n}$. Let $\{e_a\}$ be a basis for $\mathcal{H}$ and let $\eta_{ab} = \eta(e_a, e_b)$ and η^{ab} be the inverse. The following conditions must be satisfied for all $v_1, \ldots, v_n$ in $\mathcal{H}$:

Invariance : $\Lambda_{g,n}$ is invariant under the diagonal action of S_n on the space $H^\bullet(\overline{\mathcal{M}}_{g,n}) \otimes \mathcal{H}^{*\otimes n}$.

Splitting : For any $g', g'', g' + g'' = g$, and any partition $N' \sqcup N'' = 1 \ldots, n$ such that $|N'| = n'$ and $|N''| = n''$, let $\rho_{(N',N'')} : \overline{\mathcal{M}}_{g',n'+1} \times \overline{\mathcal{M}}_{g'',n''+1} \to \overline{\mathcal{M}}_{g'+g'',n'+n''}$ where ρ is equal to the gluing morphism $\circ_{(N',N'')}$ then for all $v_1, \ldots, v_n$ in $\mathcal{H}$,

$$(\rho^*_{(N',N'')}\Lambda_{g,n})(v_1,\ldots,v_n) = \Lambda_{g',n'+1}(v_{i_1},\ldots,v_{i'_n},e_a)\eta^{ab}\Lambda_{g'',n''+1}$$
$$(e_b, v_{j_1},\ldots,v_{j''_n}) \tag{19}$$

where $N' = \{i_1,\ldots,i_{n'}\}$ and $N'' = \{j_1,\ldots,j_{n''}\}$ such that $i_1 < i_2 < \cdots < i_{n'}$ and $j_1 < j_2 < \cdots < j_{n''}$ and the summation convention has been used.

Self-Sewing : Let $\rho_{n+1,n+2} : \overline{\mathcal{M}}_{g,n+2} \to \overline{\mathcal{M}}_{g+1,n}$ be the gluing morphism which glues together the $(n+1)$st and $(n+2)$nd marked points then for all $v_1,\ldots,v_n$ in $\mathcal{H}$,

$$(\rho^*_{n+1,n+2}\Lambda_{g+1,n})(v_1,\ldots,v_n) = \Lambda_{g,n+2}(v_1,\ldots,v_n,e_a,e_b)\eta^{ab} . \tag{20}$$

The classes $\Lambda_{g,n}$ are called the genus g, n-point (cohomological) correlators of the CohFT.

Remark 15 The above definition is dual to the earlier notion of CohFT, in some sense. Given a CohFT as above, one can define $\mu : H_\bullet(\overline{\mathcal{M}}_{g,n}) \to \mathcal{H}^{*\otimes n}$ via $c \mapsto \mu_c$. Therefore, by restricting to $H_\bullet(\overline{\mathcal{M}}_{0,n})$ for all n, one recovers the definition of a $g = 0$ CohFT which was introduced earlier.

Let us define a generalization of the previous notion of potential function.

Definition 15 Let $(\mathcal{H},\eta,\Lambda)$ be a CohFT. The (small phase space) potential function of the CohFT, Φ, belongs to $\lambda^{-2}\mathbb{C}[[\mathcal{H},\lambda^2]]$ where λ is a formal variable. It is defined by $\Phi(x) := \sum_{g=0}^{\infty} \lambda^{2g-2}\Phi_g(x)$, $\Phi_g(x) := \sum_n \frac{1}{n!}\Phi^{(g,n)}(x)$, and $\Phi^{(g,n)}(x) := \int_{[\overline{\mathcal{M}}_{g,n}]} \Lambda_{g,n}(x,x,\ldots,x)$.

Remark 16 The genus zero part of Φ, $\Phi_0(x)$, agrees with $\Phi(x)$ from Equation (14). This is just a matter of unraveling definitions. More generally, $\Phi^{(g,n)}(x)$ provides the Feynman rules associated to a genus g vertex of valence n.

The formal parameter λ may be regarded as a kind of coupling constant associated to the genus expansion. λ^2 plays a role analogous to that of $\hbar$ in the usual loop expansion in the usual Feynman diagrams.

It would be nice to have an analog of Theorem 4 in general genus now that we have defined a potential function which contains information about the higher genus moduli spaces. Unfortunately, it's need not be true that a CohFT (in all genera) is completely determined by its potential function $\Phi(x)$.

This is for the following reason. Let us replace $\mathcal{T}_n$ by $\mathcal{T}_{g,n}$, the free linear space of stable graphs of genus g with n tails. Of course, $\mathcal{T}_n = \mathcal{T}_{0,n}$. A morphism $\mathcal{T}_{g,n} \to \mathcal{H}^{*\otimes n}$ is still determined by Feynman rules except that now there are more corollas and loops are now allowed.

One can also still define the morphism $\mathcal{T}_{g,n} \to H_\bullet(\overline{\mathcal{M}}_{g,n})$ in analogy with the previous case. Unfortunately, in general, there will be classes for $g \geq 1$

$H_\bullet(\overline{\mathcal{M}}_{g,n})$ which need not lie in the image of the morphism $\mathcal{T}_{g,n} \to H_\bullet(\overline{\mathcal{M}}_{g,n})$. For example, $H_\bullet(\overline{\mathcal{M}}_{g,n})$ may have odd dimensional cohomology classes when $g \geq 1$. Since our algebraic cycles $\overline{\mathcal{M}}_\Gamma$ are complex, their associated homology classes are necessarily even dimensional. There will also be even dimensional classes in $H_\bullet(\overline{\mathcal{M}}_{g,n})$ in general which do not lie in the image. This means that there will be operations on $\mathcal{H}$ which do not come from applying the Feynman rules to any linear combination of stable graphs. Therefore, there are additional basic operations in a CohFT which are not captured by the Feynman rules in the present form.

Another problem is that the groups $H_\bullet(\overline{\mathcal{M}}_{g,n})$ are unknown in general although in low genera, some facts are known. For example, there is an analog of the WDVV equation but in genus 1 due to Getzler [11].

5 Construction of Cohomological Field Theories

Because of our incomplete knowledge about $H_\bullet(\overline{\mathcal{M}}_{g,n})$, examples of CohFTs (in all genera) appear to be difficult to come by. To my knowledge, there are only two main sources of examples at the moment. One comes from the Gromov-Witten invariants and the second comes from spin CohFTs. However, they share many common features. They both involve moduli spaces of curves with additional data. Forgetting the additional decorations, one obtains the moduli space of stable curves. Construction of a CohFT involves pushing down canonical cohomology classes on these decorated moduli spaces via this forgetful morphism.

5.1 Gromov-Witten Invariants and Quantum Cohomology

Let V be a smooth, projective variety. There is an associated CohFT such that $\mathcal{H} = H^\bullet(V)$ and η is the Poincaré pairing. We need only the classes $\Lambda_{g,n}$ to define the CohFT.

Remark 17 In order to notational clutter due to signs, let us assume that $\mathcal{H} := H^{2\bullet}(V)$, the even dimensional classes on V.

Definition 16 The decorated moduli space in this case is $\overline{\mathcal{M}}_{g,n}(V)$, the *moduli space of stable maps into V*. It consists of tuples $(f : C \to V; p_1, \ldots, p_n)$ up to isomorphism where C is a genus g (at worst) nodal curve with pairwise distinct marked points $p_1, \ldots, p_n$ on C away from its nodes, and $f : C \to V$ is a morphism. (One can think of f as a holomorphic map.) An isomorphism between $(f : C \to V; p_1, \ldots, p_n)$ to $(f' : C' \to V'; p_1', \ldots, p_n')$ is an isomorphism of the underlying curves which takes p_i to p_i' for all i and which takes f to f'. The tuple $(f : C \to V; p_1, \ldots, p_n)$ is said to be *stable* if it has no infinitesimal automorphisms. If β belongs to $H_2(V; \mathbb{Z})$ then $\overline{\mathcal{M}}_{g,n}(V, \beta)$ consists of those tuples $(f : C \to V; p_1, \ldots, p_n)$ in $\overline{\mathcal{M}}_{g,n}(V)$ such that class $[f(C)] = \beta$.

Because of the different definition of stability, $\overline{\mathcal{M}}_{g,n}(V)$ is usually defined even when $2g - 2 + n$ is not positive.

$\overline{\mathcal{M}}_{g,n}(V)$ is a kind of orbifold (a Deligne-Mumford stack). There are evaluation maps $ev_i : \overline{\mathcal{M}}_{g,n}(V) \to V$ taking $(f : C \to V; p_1, \ldots, p_n) \mapsto f(p_i)$ for all $i = 1, \ldots, n$.

There is also a morphism st : $\overline{\mathcal{M}}_{g,n}(V) \to \overline{\mathcal{M}}_{g,n}$ which "forgets" the stable map f, i.e. it forgets f and then collapses any irreducible components of the curve which consequently have become unstable. In particular, when V is a point then st is an isomorphism.

The following theorem follows from the work of many people. Please see [24] and the references therein.

Theorem 5 *For all $v_1, \ldots, v_n$ in $\mathcal{H}$, and g, n, let*

$$\Lambda_{g,n}(v_1, \ldots, v_n) := \text{st}_*((\cup_{i=1}^n ev_i^* v_i) \cap [\overline{\mathcal{M}}_{g,n}(V)]^{\text{vir}}) , \qquad (21)$$

and $\Lambda := \{\Lambda_{g,n}\}$ then $(\mathcal{H}, \eta, \Lambda)$ is a CohFT.

Remark 18 To be precise, the ground ring of $\mathcal{H}$ should be enlarged to include the formal parameter q^β where β belongs to $H_2(V; \mathbb{Z})$. We suppress this in order to avoid notational clutter.

Notice that we did not define $[\overline{\mathcal{M}}_{g,n}(V)]^{\text{vir}}$. It is called the *virtual fundamental class of* $\overline{\mathcal{M}}_{g,n}(V)$ and its definition is very technical. However, the idea is that $\overline{\mathcal{M}}_{g,n}(V)$ has an expected dimension which comes from deformation theory. The virtual fundamental class is a homology class of dimension equal to the expected dimension and it plays the role of the fundamental class for the purposes of this construction. There are cases when the virtual fundamental class is the usual fundamental class, e.g. when $g = 0$ and V is a projective space or flag variety.

The associated potential function Φ has a genus g, n-ary piece, $\Phi^{(g,n)}$. When $\Phi^{(g,n)}$ is regarded as an element of the symmetric product $S^n(\mathcal{H}^*)$ then it is called *the genus g, n-point Gromov-Witten invariants of V*.

Remark 19 The definition of the potential function can be extended to all (g, n) through the formula $\Phi^{(g,n)}(x) := \frac{1}{n!} \int_{[\overline{\mathcal{M}}_{g,n}(V)]^{\text{vir}}} ev_1^* x \cup \cdots \cup ev_n^* x$ for all x in $\mathcal{H}$.

In particular, if we restrict to only genus 0 then the above boils down to a potential function Φ_0 satisfying the WDVV equation. If we restrict to only to the cubic terms in Φ_0 then this contains the information about the Frobenius algebra structure. $(\mathcal{H}, \eta)$ together with the multiplication (called the *quantum multiplication*) of this Frobenius algebra structure is called the *quantum cohomology of V*. It is called the quantum cohomology of V because the cubic terms in Φ_0 are integrals over $\overline{\mathcal{M}}_{0,3}(V) = \sqcup_\beta \overline{\mathcal{M}}_{0,3}(V, \beta)$ but if we ignore all contributions except for those from $\overline{\mathcal{M}}_{0,3}(V, 0)$ then the resulting

Frobenius algebra $(\mathcal{H}, \eta)$ has a multiplication which reduces to the usual cup product. Therefore, the quantum multiplication is a deformation of the usual cup product and one regards the usual cup product as a kind of classical multiplication.

5.2 Spin Cohomological Field Theories

In this section, we will review the construction of spin CohFTs. There is a spin CohFT associated to every integer $r \geq 2$.

The role of the moduli space of stable maps $\overline{\mathcal{M}}_{g,n}(V)$ is played by the *moduli space of r-spin curves* $\overline{\mathcal{M}}_{g,n}^{1/r}$ where $r \geq 2$ is an integer. The moduli space $\overline{\mathcal{M}}_{g,n}^{1/r} = \sqcup_{\mathbf{m}} \overline{\mathcal{M}}_{g,n}^{1/r}(\mathbf{m})$ where the disjoint union is over $\mathbf{m} := (m_1, \ldots, m_n)$, $m_i = 0, \ldots r - 1$, and $i = 1, \ldots, n$.

We will now give a quick overview of r-spin curves. We refer the interested reader to [18] for details. For now, let us assume that r is prime where things are simpler.

Let $(C; p_1, \ldots, p_n)$ be a smooth stable curve of genus g with n marked points. Let $\mathbf{m} = (m_1, \ldots, m_n)$ where $m_i = 0, \ldots, r - 1$ for all i. Let $\omega(-\sum_{i=1}^n m_i p_i) \to C$ denote the holomorphic line bundle $\omega(-\sum_{i=1}^n m_i p_i)$ where $\omega(-\sum_{i=1}^n m_i p_i) := \omega \otimes \mathcal{O}(-\sum_{i=1}^n m_i p_i)$, $\mathcal{O}$ is the structure sheaf, and ω is the dualizing sheaf. (In this smooth case, one can think of ω as the holomorphic cotangent bundle of C.) An *r-spin structure on* $(C; p_1, \ldots, p_n)$ *of type* $\mathbf{m}$ consists of a line bundle $\mathcal{L} \to C$ together with an isomorphism $b : \mathcal{L}^{\otimes r} \to \omega(-\sum_{i=1}^n m_i p_i)$. In other words, $\mathcal{L}$ is an r-th root of $\omega(-\sum_{i=1}^n m_i p_i)$. For reasons of degree, an r-spin structure $\mathcal{L}$ will exist if and only if $2g - 2 - \sum_{i=1}^n m_i$ is divisible by r. When this condition is fulfilled, there are r^{2g} choices. Let $\mathcal{M}_{g,n}^{1/r}(\mathbf{m})$ denote the moduli space of smooth r-spin structures of type $\mathbf{m}$. It will be empty unless the degree condition is satisfied.

Since the $r = 2$ case may be thought of as a kind of holomorphic spin structure, we sometimes call $(\mathcal{L} \to C; p_1, \ldots, p_n; b)$ a *higher spin structure or a higher spin curve*.

Jarvis [15] introduced a smooth compactification $\overline{\mathcal{M}}_{g,n}^{1/r}(\mathbf{m})$ of $\mathcal{M}_{g,n}^{1/r}(\mathbf{m})$. When a smooth curve C degenerates to acquire a node then the r-spin structure $\mathcal{L} \to C$ can either stay locally free at the node (called a *Ramond node*) or it can fail to be locally free at the node (called a *Neveu-Schwarz node*).[4] The morphism b fails to be an isomorphism at the node in the latter case, however, it will be an isomorphism wherever $\mathcal{L}$ is locally free. $\overline{\mathcal{M}}_{g,n}^{1/r}$ is a smooth, compact, complex orbifold of complex dimension $3g - 3 + n$. The forgetful morphism $\mathrm{st} : \overline{\mathcal{M}}_{g,n}^{1/r} \to \overline{\mathcal{M}}_{g,n}$ which forgets the r-spin structure is a branched cover.

When r is not prime then one must include not only an r-th root $\mathcal{L}$ but also a d-th root for all d which divides r together with a collection of isomorphisms

---

[4] The physical terminology associated to the nodes is due to Witten.[31]

generalizing b. These extra roots are necessary for smoothness of the moduli spaces but will otherwise not play much of a role in our discussion.

Let $\mathcal{H}'$ be a vector space with a basis $\{e_0, \ldots, e_{r-1}\}$. Let η' be a metric on $\mathcal{H}'$ such that $\eta'(e_a, e_b) = 1$ if $a + b \equiv (r-2) \mod r$. An important subspace will be $\mathcal{H}$ which is spanned by $\{e_0, \ldots, e_{r-2}\}$ and let η be the restriction of η' to $\mathcal{H}$.

In [18], axioms are stated for a cohomology class (called the *virtual class*) $c^{1/r}$ in $H^{2D}(\overline{\mathcal{M}}_{g,n}^{1/r}(\mathbf{m}))$ where

$$D = \frac{1}{r}\left((r-2)(g-1) + \sum_{i=1}^{n} m_i \right) \tag{22}$$

such that it gives rise to a CohFT. The number D appears in the following manner. Let $\pi : \mathcal{C} \to \overline{\mathcal{M}}_{g,n}^{1/r}$ be the universal curve and let $\mathcal{L} \to \mathcal{C}$ be the universal r-spin structure. Let $R^\bullet \pi_*(\mathcal{L}) \to \overline{\mathcal{M}}_{g,n}^{1/r}$ be its K-theoretic pushforward. Recall that for $i = 0, 1$, the fiber of $R^i \pi_*(\mathcal{L})$ over a point in $\overline{\mathcal{M}}_{g,n}^{1/r}$ consisting of a curve C is $H^i(C; \mathcal{L})$. The Euler characteristic $\dim_{\mathbb{C}} R^0 \pi_*(\mathcal{L}) - \dim_{\mathbb{C}} R^1 \pi_*(\mathcal{L}) =: -D$. When $g = 0$, $R^0 \pi_*(\mathcal{L})$ vanishes for degree reasons. Therefore, $R^1 \pi_*(\mathcal{L})^* \to C$ is a vector bundle of rank D. The class $c^{1/r}$ in $H^{2D}(\overline{\mathcal{M}}_{0,n}^{1/r}(\mathbf{m}))$ is then its top Chern class.

Construction of the virtual class in higher genera has been carried out in [27] but it is rather involved since it is not the top Chern class of the index bundle.

Theorem 6 *[18] Let $(\mathcal{H}', \eta')$ and $(\mathcal{H}, \eta)$ be defined as above. Let $c^{1/r}(\mathbf{m})$ denote the virtual class on $\overline{\mathcal{M}}_{g,n}^{1/r}(\mathbf{m})$, let $\Lambda'_{g,n}$ in $H^\bullet(\overline{\mathcal{M}}_{g,n}) \otimes (\mathcal{H}')^{*\otimes n}$ be defined via*

$$\Lambda'_{g,n}(e_{m_1}, \ldots, e_{m_n}) := r^{1-g} \mathrm{st}_* c^{1/r}(\mathbf{m})$$

for all $m_1, \ldots, m_n = 0, \ldots, r-1$, and $\Lambda' := \{\Lambda'_{g,n}\}$. Let $\Lambda_{g,n}$ in $H^\bullet(\overline{\mathcal{M}}_{g,n}) \otimes \mathcal{H}^{*\otimes n}$ be given by the restriction of $\Lambda'_{g,n}$ to $\mathcal{H}$ and $\Lambda := \{\Lambda_{g,n}\}$ then $(\mathcal{H}, \eta, \Lambda)$ and $(\mathcal{H}', \eta', \Lambda')$ are CohFTs. Furthermore,

$$\Lambda'_{g,n}(e_{m_1}, \ldots, e_{m_n}) = 0 \tag{23}$$

if $m_i = r - 1$ for some i.

In particular, the CohFT associated to $r = 2$ is isomorphic to the the CohFT associated to the Gromov-Witten invariants of a point.

Remark 20 The reason that we make a distinction between the two CohFTs above is that e_0 plays the role of an identity element, which we have been suppressing, of the Frobenius algebra associated to the CohFT $(\mathcal{H}, \eta, \Lambda)$. This is analogous to the role in Gromov-Witten theory played by the usual identity element in $H^0(V)$. However, because of Equation (23), e_0 cannot be the identity element in larger space $(\mathcal{H}', \eta')$. In fact, the one dimensional vector space spanned by e_{r-1} decouples from the rest of the theory. There does not appear to be a counterpart to this phenomenon in Gromov-Witten theory.

Following Witten's argument, the $g = 0$ potential function can be calculated [18] using WDVV and an identity between divisor classes on $\overline{\mathcal{M}}_{0,n}^{1/r}$.

In the previous, we have always restricted the m_i's to lie in the range $0, \ldots, r-1$. It is reasonable to ask what happens if we allow arbitrary positive values instead. Let δ_i be an n-tuple with a 1 in the i-th slot and 0's everywhere else. It is easy to see that $\overline{\mathcal{M}}_{g,n}^{1/r}(\mathbf{m})$ is canonically isomorphic to $\overline{\mathcal{M}}_{g,n}^{1/r}(\mathbf{m}+r\delta_i)$ for all $\mathbf{m}$ with nonnegative components by taking the tensor product of the higher spin structure with $\mathcal{O}(-p_i)$. However, the class $c^{1/r}(\mathbf{m} + r\delta_i)$ has a different dimension from $c^{1/r}(\mathbf{m})$. It turns out that these classes satisfy [19] the equation

$$c^{1/r}(\mathbf{m} + r\delta_i) = -\frac{m_i + 1}{r} \, \psi_i \, c^{1/r}(\mathbf{m})$$

for all $i = 1, \ldots, n$ where ψ_i in $H^2(\overline{\mathcal{M}}_{g,n}^{1/r})$ is the first Chern class of the tautological line bundle associated to the i-th marked point. This complex line bundle over $\overline{\mathcal{M}}_{g,n}$ has a fiber over $(C; p_1, \ldots, p_n)$ which is the complex tangent bundle $T_{x_i}^* C$. The classes ψ_i are called *gravitational descendants*.

If one computes the potential function but where we now allow all nonnegative values of m's then one obtains integrals of products of these ψ classes with the virtual class $c^{1/r}$. The resulting potential function Ψ is called the *large phase space potential*. Witten had a very interesting conjecture about this coming from physics.

Conjecture 1 (The Generalized Witten Conjecture) *The large phase space potential function associated to the integer $r \geq 2$, Ψ, is the unique solution of the r-th KdV integrable hierarchy satisfying one more equation called the string equation.*

When $r = 2$, this conjecture can be shown to reduce to the original Witten conjecture which was proven by Kontsevich [21] using transcendental techniques and, more recently, by Okounkov-Pandharipande [26] using algebraic techniques.

Theorem 7 *[18] The generalized Witten conjecture holds in genus 0.*

The above follows from a calculation of the genus 0 potential function. While some progress has been made towards this conjecture in low genera, it is still open.

5.3 Some Generalizations

Here, we mention a few generalizations to the above. The construction of Gromov-Witten invariants has been extended to the case where the target V is a smooth orbifold by Chen-Ruan [6, 7]. An algebraic version of this theory appears in [3]. In this theory, the moduli space of stable maps must be enlarged to include curves with an orbifold structure at its marked points and nodes. These curves are responsible for what physicists call the *twisted sectors of the theory*. Tantalizingly enough, the notion of r-spin curves admits a clean description in terms of bundles over curves with orbifold structure [1]. There are thus many fascinating connections between the two theories.

The notion of CohFT itself has recently been generalized to a G-equivariant CohFT (or G-CohFT) when G is a finite group [17]. This has the property that when G is the trivial group then one recovers the usual notion of a CohFT. Furthermore, a G-CohFT is related to a G-equivariant version of a Frobenius algebra in precisely the same way that a CohFT is related to a Frobenius algebra. A G-equivariant Frobenius algebra, on the other hand, has been given an equivariant topological field theory interpretation by [30]. The relevant moduli space necessary to define a G-CohFT is called the moduli spaces of pointed, admissible G-covers over a genus g curve with n marked points $\overline{\mathcal{M}}_{g,n}^G$. A pointed admissible G-cover over a stable curve is a generalization of a principal G-bundle over a curve but where the group action on the total space may fail to be free over the marked points and nodes. The associated moduli space $\overline{\mathcal{M}}_{g,n}^G$ forms a special kind of partial operad called a colored operad. A G-equivariant CohFT is defined in terms of $\overline{\mathcal{M}}_{g,n}^G$ in a way analogous to the way that a usual CohFT is defined in terms of $\overline{\mathcal{M}}_{g,n}$ except that G-equivariance is maintained throughout the construction. Furthermore, we prove that after taking an appropriate "quotient" by G, one obtains a (usual) CohFT but with an additional grading by conjugacy classes of G corresponding to the twisted sectors. An example of a G-equivariant CohFT can be constructed from a smooth, projective variety X with a finite group action by introducing the appropriate G-equivariant version of stable maps into X. Restricting to only degree zero stable maps, we recover a G-equivariant theory due to [9]. Taking the appropriate "quotient" by G, we recover the Chen-Ruan orbifold cohomology of the quotient $[X/G]$.

References

1. *Moduli of twisted spin curves*, Proc. Amer. Math. Soc. 131 (2003), no. 3, 685–699
2. Arnold, V.I., *The cohomology ring of the colored braid group*. Mat. Zametki **5**, 227–231 (1969) (English translation: Math. Notes **5**, 138–140 (1969))
3. D. Abramovich, T. Graber, and A. Vistoli, *Algebraic orbifold quantum products*. In A. Adem, J. Morava, and Y. Ruan (eds.), Orbifolds in Mathematics and

Physics, *Contemp. Math.*, Amer. Math. Soc., Providence, RI. **310**, (2002), 1–25. math.AG/0112004.

4. B. Bakalov and A. Kirillov, *Lectures on tensor categories and modular functors.* Univ. Lecture Series **21**, Amer. Math. Soc., Providence, RI, 2001.

5. F.R. Cohen, *Artin's braid groups, classical homotopy theory and sundry other curiosities.* Contemp. Math. **78**, 167–206 (1988)

6. W. Chen and Y. Ruan, *A new cohomology theory for orbifold.* math.AG/0004129.

7. _____, *Orbifold Gromov-Witten theory.* In A. Adem, J. Morava, and Y. Ruan (eds.), Orbifolds in Mathematics and Physics, *Contemp. Math.*, Amer. Math. Soc., Providence, RI. **310**, (2002), 25–85. math.AG/0103156.

8. B. Dubrovin, *Geometry of 2D topological field theories*, in Integrable systems and Quantum Groups, Lecture Notes in Math. 1620, Springer, Berlin, 1996, pp. 120–348.

9. B. Fantechi and L. Göttsche, *Orbifold cohomology for global quotients.* math.AG/0104207.

10. E. Getzler, *Batalin-Vilkovisky algebras and two-dimensional topological field theories, Commun. Math. Phys.* **159** (1994) 265–285.

11. *Intersection theory on* $\overline{M}_{1,4}$ *and elliptic Gromov-Witten invariants*, J. Amer. Math. Soc. **10** (1997), no. 4, 973–998.

12. E. Getzler, M. Kapranov, *Cyclic operads and cyclic homology*, in Geometry, topology, & physics, Conf. Proc. Lecture Notes Geom. Topology, VI, Internat. Press, Cambridge, MA, 1995, pp. 167–201.

13. E. Getzler, M. Kapranov, *Modular operads, Compositio Math.* 110 (1998) 65–126.

14. Y.Z. Huang, *Two-dimensional conformal geometry and vertex operator algebras*, Progress in Mathematics, 148. Birkhaüser Boston, Inc., Boston, MA, 1997. xiv+280 pp. ISBN: 0-8176-3829-6

15. T. Jarvis, *Geometry of the moduli of higher spin curves*, Internat. J. Math. **11** (2000), no. 5, 637–663, math.AG/9809138.

16. T. Jarvis, *Torsion-free sheaves and moduli of generalized spin curves*, Compositio Math. **110** (1998) no. 3, 291-333.

17. T. Jarvis, R. Kaufmann, and T. Kimura, *Pointed Admissible G-Covers and G-equivariant Cohomological Field Theories*, math.AG/0302316

18. T. Jarvis, T. Kimura, and A. Vaintrob, *Moduli spaces of higher spin curves and integrable hierarchies*, Compositio Math. **126** (2001), no. 2, 157–212. math.AG/9905034.

19. T. Jarvis, T. Kimura, and A. Vaintrob, *Gravitational descendants and the moduli space of higher spin curves*, Advances in algebraic geometry motivated by physics (Lowell, MA, 2000), 167–177, Contemp. Math., 276, Amer. Math. Soc., Providence, RI, 2001.

20. T. Kimura, J. Stasheff, and A.A. Voronov, *On operad strucures on moduli spaces and string theory, Comm. Math. Phys.*171 (1995), 1–25.

21. Kontsevich, M., *Intersection theory on the moduli space of curves and the matrix Airy function, Comm. Math. Phys.* 147 (1992) 1–23.

22. M. Kontsevich and Yu. Manin, *Gromov-Witten classes, quantum cohomology, and enumerative geometry. Commun. Math. Phys.* **164** (1994), 525–562.

23. B. Lian, and G.J. Zuckerman, *New perspectives on the BRST-algebraic structure of string theory, Comm. Math. Phys.* **154** (1993), 613–646.

24. Yu. Manin, *Frobenius manifolds, quantum cohomology, and moduli spaces.* Colloquium Publ., **47**, Amer. Math. Soc., Providence, RI, 1999.

25. M. Markl, S. Shnider, J. Stasheff, *Operads in algebra, topology and physics*, Mathematical Surveys and Monographs, 96. American Mathematical Society, Providence, RI, 2002. x+349 pp. ISBN: 0-8218-2134-2

26. A. Okounkov and R. Pandharipande, *Gromov-Witten theory, Hurwitz theory, and matrix models, I*, math.AG/0101147

27. A. Polishchuk and A. Vaintrob, *Algebraic construction of Witten's top Chern class.* In E. Previato (ed.), Contemp. Math. **276**, Amer. Math. Soc., (2001), 229–249. math.AG/0011032.

28. G. Segal, *The definition of a conformal field theory*, Preprint. Oxford.

29. G. Segal, *Two dimensional conformal field theories and modular functors*, IXth Int. Congr. on Math. Phys. (Bristol; Philadephia) (B. Simon, A. Gruman, and I.M. Davies, eds.), IOP Publishing Ltd, 1989, 22–37.

30. V. Turaev, *Homotopy field theory in dimension 2 and group-algebras.* math.QA/99100110.

31. E. Witten, *Algebraic geometry associated with matrix models of two dimensional gravity*, Topological models in modern mathematics (Stony Brook, NY, 1991), Publish or Perish, Houston, TX 1993, 235–269.

32. B. Zweibach, *Closed string field theory: Quantum action and the Batalin-Vilkovisky master equation.* Nucl. Phys. B **390** (1993), 33–152.

An Infinite Family of Isospectral Pairs Topological Aspects

N. Iiyori[1], T. Itoh[2], M. Iwami[3], K. Nakada[4], and T. Masuda[5]

[1] Unit of Mathematics and Information Science, Yamaguchi University, Japan
    iiyori@yamaguchi-u.ac.jp
[2] Department of General Education, Kinki University Technical College, Japan
    titoh@ktc.ac.jp
[3] Graduate School of Pure and Applied Sciences, University of Tsukuba, Japan
    maki@math.tsukuba.ac.jp
[4] Department of Pure and Applied Mathematics, General School of Information
    Science and Technology, Osaka University, Japan
    smv182nk@ecs.cmc.osaka-u.ac.jp
[5] Institute of Mathematics, University of Tsukuba, Japan
    tetsuya@math.tsukuba.ac.jp

Summary. We give some basic properties of the isospectral pairs $(\Gamma, \hat{\Gamma})$ of two bipartite graphs Γ and $\hat{\Gamma}$. Then we present a systematic method of constructing an infinite family of isospectral pairs $(\Gamma, \hat{\Gamma})$ satisfying $\Gamma \neq \hat{\Gamma}$ by making use of the pairs of Young diagrams.

1 Introduction

In the theory of von Neumann algebras, the classification of the hyperfinite factor-subfactor pairing $N \subset M$ with its finite Jones index $[M : N] < \infty$ is an important problem. This problem of classifying subfactors was initiated by V.F.R. Jones in his remarkable work [4], where he introduced a real-valued invariant which we call the Jones index, denoted by $[M : N]$, for a factor-subfactor pairing $N \subseteq M$, where N and M are hyperfinite factors of type II_1. One of the most remarkable aspects of this invariant is that this finite number corresponds to the order of the "Galois group", which measures the relative size of N inside M. The Jones index $[M : N]$ can be a fractional number. For the case $[M : N] < 4$, Jones classified all the possible values, which turn out to be the distinguished set $\{4\cos^2(\frac{\pi}{n}) | n = 3, 4, 5, \cdots \}$. All these values $4\cos^2(\frac{\pi}{n})(n = 3, 4, 5, \cdots)$ are realized in terms of the explicitly constructed factor-subfactor pairing $N \subseteq M$, together with the beautiful classification of such factor-subfactor pairing. Now, it is immediately seen that the particular value $[M : N] = 4$ is an accumulating point of the above distinguished set. Then the reasonable classification of the factor-subfactor pairing $N \subseteq M$ is already carried out.

It was A. Ocneanu [7] who introduced the notion of *paragroups* on the basis of the pairs of bipartite graphs $(\Gamma, \hat{\Gamma})$ and an object called the *flat*

N. Iiyori, T. Itoh, M. Iwami, K. Nakada, and T. Masuda: *An Infinite Family of Isospectral Pairs Topological Aspects*, Lect. Notes Phys. **662**, 289–297 (2005)
www.springerlink.com   © Springer-Verlag Berlin Heidelberg 2005

connection, with the aim to classify the hyperfinite factor-subfactor pairing $N \subset M$ with its finite Jones index $[M : N] < \infty$ in terms of combinatorial data.

The notion of isospectral pairs $(\Gamma, \hat{\Gamma})$ corresponding to $[M : N] \leq 4$ comes from (the two copies of) the Dynkin diagrams of type A, D, E, or the extended Dynkin diagrams. As Ocneanu mentions in his publication [7], we are going to consider this new notion of paragroup corresponding to a given factor-subfactor $N \subseteq M$ as a "generalized" Galois group, which measures the relative size of N inside M for $N \subseteq M$ in such a way that the order of the paragroup for $N \subseteq M$ is equal to $[M : N] \in \mathbf{R}^+$.

In this work, we aim to reformulate Ocneanu's notion of paragroups. Our idea is to generalized this concept in such a way that the notion makes sense in a more general context than only in terms of operator algebra language. Here, we regard the notion of paragroups consisting of two concepts. One is the pair $(\Gamma, \hat{\Gamma})$ of bipartite graphs satisfying the combinatorial condition which we call *isospectral pair*. The other is the so-called *flat connection*: In view of the natural correspondence with the theory of solvable lattice models in two dimensions, this flat connection corresponds to the general notion of a Boltzmann weight in mathematical physics. In particular, we aim to develop this theory over more general fields (of characteristic zero). Therefore, we are obliged to look for a different chacterization of "flatness" (not in terms of complex conjugation), such that the orientations of the edges of the graphs are very likely to become important. By this reformulation, we expect that our "flatness" condition becomes more simple than the exression formulated by Ocneanu in [7], which requires a huge amount of both, multiplication and summation for the Boltzmann weight.

In the present theory of paragroups, the Perron-Frobenius eigenvalue β of graph(s) $\Gamma = \hat{\Gamma}$ is of great importance in the sense that the value β^2 is equal to the Jones' index $[M : N]$ for corresponding factor-subfactor pairing $N \subseteq M$ (in the case $\beta^2 \leq 4$). The notion "isospectral pair" for the pairs $(\Gamma, \hat{\Gamma})$ of bibartite graphs that we introduce in this paper corresponds to the set theoretical part of the finite group, with a bit of information concerning the irreducible representations of our "generalized group" and its dual object.

In [8], Ocneanu presented the following pair of bipartite graphs without any comments or discussions.

This was the beginning of our collaboration.

This paper contains two main theorems. After the definition of the isospectral pairs purely in terms of combinatorial descriptions, our first theorem says that we have an infinite family of such pairs, explicitly constructed in terms of the pairs of Young diagrams. Our construction very well explains the above example of Ocneanu in [8]. Then, our second theorem says that, for an isospectral pair $(\Gamma, \hat{\Gamma})$, the two bipartite graphs Γ and $\hat{\Gamma}$ have the same set of eigenvalues except for zero eigenvalue(s). Therefore, in particular, the corresponding Perron-Frobenius eigenvalue β are the same.

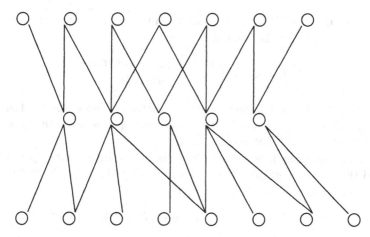

Up to the present, the classification problem of the factor-subfactor for the case of $[M : N] > 4$ in operator algebra seems to have not yet been studied in much detail. Our new family $(\Gamma, \hat{\Gamma})$ of isospectral pairs in terms of Young diagrams provides infinite examples with the property $\beta^2 > 4$ so that our explicit construction could also be useful in such a direction.

What remains to be studied is still a lot. First of all, we haven't touched yet the problem of a good axiomatization of the generalized group structure for a given isospectral pair $(\Gamma, \hat{\Gamma})$, which replaces the flatness condition of Ocneanu. This problem, together with the introduction of a suitable structure(s) for our explicitly constructed examples is left to our future investigations.

2 Definitions and Notations

First of all, we define a generalized notion of graphs, which is called the hyper-graph. Let V and E be finite sets. And let f be a mapping from E to $\mathcal{P}(V)$ (the powerset). The triplet (V, E, f) is called a hyper-graph G over V. An element of V, denoted $V(G)$, is called a vertex and an element of E, denoted $E(G)$, is called an edge. We usually use a symbol G to denote a hyper-graph $G = (V, E, f)$.

For a hyper-graph G, we define a matrix $I(G)$ with its size $|V| \times |E|$ as follows.

$$I(G)(v, e) := \begin{cases} 1 & \text{if } v \in f(e), \\ 0 & \text{othewise,} \end{cases}$$

where $I(G)(v, e)$ is the (v, e)-entry of $I(G)$. This matrix $I(G)$ is called the incident matirx of the hyper-graph G.

We next define an important class of hypergraphs which is called a graph. Let $G = (V, E, f)$ be a hyper-graph over V. Then G is called graph over V if

$$|f(e)| = 2 \quad \text{for} \quad e \in E \quad \text{and for} \quad e, e' \in E \quad f(e) = f(e') \quad \text{implies} \quad e = e'.$$

If G be a graph over V. Then the incident matrix which is square with its size $|V|$ for the graph G is given as follows.

$$A(G)(v, u) := \begin{cases} 1 & \text{if } \{v, u\} \in f(E(G)), \\ 0 & \text{otherwise.} \end{cases}$$

This matrix $A(G)$ is called the adjacency matrix of the graph G. Here we note that this adjency matrix is one-to-one correspondence with the graph. Namely, if $A(\Gamma) = A(\Gamma')$ if and only if $\Gamma = \Gamma'$.

We next define the most important class of graphs for our paper, which is called the bipartite graph. Let $\Gamma = (V \cup U, E, f)$ be a graph over $V \cup U$. This Γ is called a U-bipartite graph over V, if

$$V \cap U = \emptyset \quad \text{and} \quad e \in E \quad \text{implies} \quad f(e) \not\subseteq V, \quad f(e) \not\subseteq U$$

Corollary 1 *Each U-bipartite graph over V is one-to-one corresponding with some hyper-graph over V with U corresponding to the edges of this hyper-graph.*

Now we introduce some notations for the later discussions. The bipartite graph which is corresponding to a hyper-graph G is denoted $\mathrm{Bip}\,(G)$. And the hyper-graph which is corresponding to a bipartite graph Γ is denoted $\mathrm{Hyp}\,(\Gamma)$. And, when Γ is biparite, $I(\mathrm{Hyp}\,(\Gamma))$ is denoted $\mathrm{inc}\,\Gamma$. Now the next theorem between the incidence matrix and the adjacency matrix is the case.

Corollary 2 *If the graph Γ is bipartite with a suitable labelling of its vertices, the adjacency matrix $A(\Gamma)$ is given by the following matrix expression:*

$$A(\Gamma) = \begin{bmatrix} 0 & {}^t(\mathrm{inc}\,\Gamma) \\ \mathrm{inc}\,\Gamma & 0 \end{bmatrix} \qquad \text{where } {}^tA \text{ is the transposition of } A.$$

The above two statements are standard and well-known in the graph theory.

3 Graphs Defined by Young Diagrams

Let Y_n be the set of all Young diagrams whose size is $n \in \mathbf{N}$. Therefore the number $|Y_n|$ is nothing but the partition number of $n \in \mathbf{N}$. Now, we remind that the Young diagrams are defined by decreasing sequence of natural numbers $Y_n \ni \alpha = (\alpha_1, \alpha_2, \cdots, \alpha_\ell)$ with the property that $\alpha_1 + \alpha_2 + \cdots + \alpha_\ell = n$. For two Young diagrams $\alpha = (\alpha_1, \alpha_2, \cdots, \alpha_\ell)$ and $\beta = (\beta_1, \beta_2, \cdots, \beta_m)$, we define the partial order which we denote by $\alpha \leq \beta$ to be α to be $\alpha_j \leq \beta_j$ for $j = 1, \cdots, \max(\ell, m)$. And we define $\mathrm{dist}\,(\alpha, \beta)$ to be:

$$\mathrm{dist}\,(\alpha, \beta) := \sum_{j \in \mathbf{N}} |\beta_j - \alpha_j|.$$

Now, let $\alpha_1, \beta_1, \alpha_2, \beta_2$ be a collection of four Young diagrams. The pair (α_1, β_1) is called a left-successor of the pair (α_2, β_2) if

$$\alpha_1 \geq \alpha_2, \quad \beta_1 = \beta_2 \quad \text{and} \quad \text{dist}\,(\alpha_1, \alpha_2) = 1\,.$$

Similarly, the pair (α_1, β_1) is called a right-successor of the pair (α_2, β_2) if

$$\alpha_1 = \alpha_2, \quad \beta_1 \geq \beta_2 \quad \text{and} \quad \text{dist}\,(\beta_1, \beta_2) = 1\,.$$

Now we use the notational convention $Y_0 := \{\emptyset\}$ and then we define the set $V_{n,m}$ as the following:

$$V_{n,m} := \bigcup_{i=0}^{\min\{m,n\}} \{(\alpha, \beta) \in Y_{n-i} \times Y_{m-i}\}$$

for $n, m \in \mathbf{N}$. For the following discussion, we deal with the two kinds of bipartite graphs by making use of the Young Tableaux. First, we define the $V_{n+1,m}$-bipartite graph over $V_{n,m}$ which is denoted by $V_{n+1,m} - V_{n,m}$ as the following. The vertices are defined by

$$V(V_{n+1,m} - V_{n,m}) := V_{n+1,m} \cup V_{n,m}$$

and the edges are defined by

$$E(V_{n+1,m} - V_{n,m}) := \left\{ \lambda_1 - \lambda_2 \,\middle|\, \begin{array}{l} \lambda_1 \in V_{n+1,m} \text{ is a left-successor of } \lambda_2 \in V_{n,m} \\ \text{or} \\ \lambda_2 \in V_{n,m} \text{ is a right-successor of } \lambda_1 \in V_{n+1,m} \end{array} \right\}$$

One of the simplest examples is given by the following. Since $V_{n,0} = \{(\alpha, \emptyset) | \alpha \in Y_n\}$, by writing α instead of $(\alpha, \emptyset)$, we have:

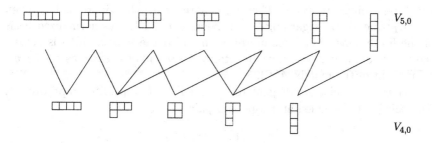

for the graph $V_{5,0} - V_{4,0}$. Actually, this graph is well-known in the representation theory of symmetric groups and this is nothing but the Bratteli diagram for the unital inclusion of the semisimple algebra $\mathbf{C}[S_4]$ into the bigger semisimple algebra $\mathbf{C}[S_5]$.

The next example is rather non-trivial as:

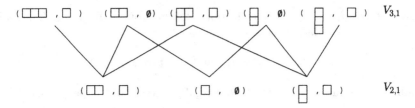

Next we define the $V_{n,m+1}$-bipartite graph over $V_{n,m}$ which is denoted $V_{n,m} - V_{n,m+1}$ as the following. The vertices are defined by

$$V(V_{n,m} - V_{n,m+1}) := V_{n,m} \cup V_{n,m+1}$$

and the edges are defined by

$$E(V_{n,m} - V_{n,m+1}) := \left\{ \lambda_2 - \lambda_3 \left| \begin{array}{l} \lambda_2 \in V_{n,m} \text{ is a left-successor of } \lambda_3 \in V_{n,m+1} \\ \text{or} \\ \lambda_3 \in V_{n,m+1} \text{ is a right-successor of } \lambda_2 \in V_{n,m} \end{array} \right. \right\}$$

By using the simplifications of notations as above, an example of such a graph is given as follows.

4 Isospectral Pairs Coming from Young Diagrams

Let Γ be a bipartite graph. Then we denote by n_Γ the square matrix given by $\mathrm{inc}\,\Gamma^t(\mathrm{inc}\,\Gamma)$. By regarding the elements $u, v \in V(\Gamma)$ as normalized basis of the finite-dimensional vector space on which the matrix $A(\Gamma)$ is acting, the (u, v)-component which we denote by $n_\Gamma(u, v)$ is the same as the (u, v)-component of the matrix $A(\Gamma)$.

Let $\hat{U}$ be a finite set disjoint to V, and $\hat{\Gamma}$ be $\hat{U}$-bipartite graph over V. The pair $(\Gamma, \hat{\Gamma})$ is called an isospectral pair if

$$n_{\hat{\Gamma}}(u, v) = n_\Gamma(u, v)$$

for any $u, v \in V$.

It is straightforward to observe that the next pair of bipartite graphs is isospectral.

This particular pair of bipartite graphs is accidentally given in the lecture note of Ocneanu [7] without any direct relations with the paragroups nor any systematic ways to construct it.

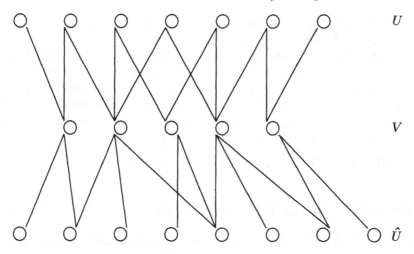

In the following discussions, we give a systematic way of the pairs $(\Gamma, \hat{\Gamma})$ of bipartite graphs Γ and $\hat{\Gamma}$ by making use of the discussions based on the Young diagrams before. Then we see that the above accidental example of Ocneanu is a part of our constructions.

Proposition 1 *Let Γ and $\hat{\Gamma}$ be the bipartite graphs given by $\Gamma := V_{n+1,m} - V_{n,m}$ and $\hat{\Gamma} := V_{n,m} - V_{n,m+1}$ for $n, m \in \mathbf{N}$. Then the pair $(\Gamma, \hat{\Gamma})$ becomes isospectral.*

We denote this isospectral pair by the following diagram as:

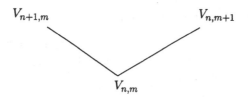

Proof. We put $U_1 := V_{n+1,m}$, $U_2 := V_{n,m+1}$, $V := V_{n,m}$ and we take $u = (\alpha_2, \beta_2)$, $v = (\alpha_2', \beta_2') \in V$. Then we have the following cases.

Case 1 If dist $(\alpha_2, \alpha_2') > 2$ or dist $(\beta_2, \beta_2') > 2$. Then $n_{\Gamma_1}(u, v) = 0$, and $n_{\Gamma_2}(u, v) = 0$. So we have $n_{\Gamma_1}(u, v) = n_{\Gamma_2}(u, v)$.

Case 2 If dist $(\alpha_2, \alpha_2') = 2$ and dist $(\beta_2, \beta_2') = 2$. Then $n_{\Gamma_1}(u, v) = 0$, and $n_{\Gamma_2}(u, v) = 0$. So we have $n_{\Gamma_1}(u, v) = n_{\Gamma_2}(u, v)$.

Case 3 If dist $(\alpha_2, \alpha_2') = 2$ and $\beta_2 = \beta_2'$ or $\alpha_2 = \alpha_2'$ and dist $(\beta_2, \beta_2') = 2$. Then we see that the value $n_{\Gamma_1}(u, v)$ is either 0 or 1. If $n_{\Gamma_1}(u, v) = 0$, then $n_{\Gamma_2}(u, v) = 0$ and if $n_{\Gamma_1}(u, v) = 1$, then $n_{\Gamma_2}(u, v) = 1$. Sowe have $n_{\Gamma_1}(u, v) = n_{\Gamma_2}(u, v)$.

Case 4 If dist $(\alpha_2, \alpha_2') = 1$ and dist $(\beta_2, \beta_2') = 1$. Then $n_{\Gamma_1}(u, v) = 1$ and $n_{\Gamma_2}(u, v) = 1$. So we have $n_{\Gamma_1}(u, v) = n_{\Gamma_2}(u, v)$.

Case 5 Finally we consider the case $u = v$. Let x be the number of left-successors of v and y be the number of Young diagrams with which v is a right-successor. Namely $x + y = n_{\Gamma_1}(v, v)$. Then the number of right-successors of v is given by $y + 1$ and the number of Young diagrams with which v is a left-successor is given by $x - 1$. Therefore we have

$$n_{\Gamma_1}(v, v) = x + y = (x - 1) + (y + 1) = n_{\Gamma_2}(v, v).$$

Hence, this proves that the pair $(\Gamma, \hat{\Gamma})$ is isospectral. $\square$

5 Properties of the Isospectral Pairs

Let Γ be a U-bipartite graph over V and $\hat{\Gamma}$ be a $\hat{U}$-bipartite graph over V. Now we define the two matrices T and $\hat{T}$ by:

$$T := \begin{bmatrix} A(\Gamma) & O \\ O & I_{|\hat{U}|} \end{bmatrix} , \ \hat{T} := \begin{bmatrix} I_{|U|} & O \\ O & A(\hat{\Gamma}) \end{bmatrix}$$

where

$$A(\Gamma) = \begin{bmatrix} 0 & {}^t(\text{inc } _\Gamma) \\ \text{inc } _\Gamma & 0 \end{bmatrix} , \ A(\hat{\Gamma}) = \begin{bmatrix} 0 & \text{inc } _{\hat{\Gamma}} \\ {}^t(\text{inc } _{\hat{\Gamma}}) & 0 \end{bmatrix} .$$

Theorem 1 *The pair of bipartite graphs* $(\Gamma, \hat{\Gamma})$ *is isospectral if and only if* $T\hat{T}T = \hat{T}T\hat{T}$.

Proof. Since we have the two equalities:

$$T\hat{T}T = \begin{bmatrix} O & O & {}^t(\text{inc } _\Gamma)\text{inc } _{\hat{\Gamma}} \\ O & \text{inc } _\Gamma {}^t(\text{inc } _\Gamma) & O \\ {}^t(\text{inc } _{\hat{\Gamma}})\text{inc } _\Gamma & O & O \end{bmatrix} ,$$

$$\hat{T}T\hat{T} = \begin{bmatrix} O & O & {}^t(\text{inc } _\Gamma)\text{inc } _{\hat{\Gamma}} \\ O & \text{inc } _{\hat{\Gamma}} {}^t(\text{inc } _{\hat{\Gamma}}) & O \\ {}^t(\text{inc } _{\hat{\Gamma}})\text{inc } _\Gamma & O & O \end{bmatrix} ,$$

the only non-trivial part is the (u, v)-components of the two matrices $T\hat{T}T$ and $\hat{T}T\hat{T}$ for any $u, v \in V$. Then we have $T\hat{T}T(u, v) = \text{inc } _\Gamma {}^t(\text{inc } _\Gamma)(u, v) = n_\Gamma(u, v)$ and $\hat{T}T\hat{T}(u, v) = \text{inc } _{\hat{\Gamma}} {}^t(\text{inc } _{\hat{\Gamma}})(u, v) = n_{\hat{\Gamma}}(u, v)$. Therefore the pair $(\Gamma, \hat{\Gamma})$ is isospectral if and only if $T\hat{T}T = \hat{T}T\hat{T}$. $\square$

Let Γ be a U-bipartite graph over V. Then we have the equality $|tI_{|V|+|U|} - A(\Gamma)| = t^{|U|-|V|}|t^2 I_{|V|} - n_\Gamma|$. For a hermitian matrix A, we put $\text{Spec}(A) = \{(t, m_t)| \ t \text{ is an eigenvalue of } A \text{ with its multiplicity } m_t.\}$ and $\text{Spec}^*(A) = \{(t, m_t) \in \text{Spec}(A)| \ t \neq 0\}$.

Theorem 2 *If the pair $(\Gamma, \hat{\Gamma})$ is isospectral, then* $\mathrm{Spec}^*(A(\Gamma)) = \mathrm{Spec}^* (A(\hat{\Gamma}))$.

Proof. Let V be the commom vertex set of the pair $(\Gamma, \hat{\Gamma})$. Then $(\Gamma, \hat{\Gamma})$ is isospectral if and only if $n_\Gamma(u,v) = n_{\hat{\Gamma}}(u,v)$ for $u,v \in V$ which implies $\mathrm{Spec}^*(n_\Gamma|_{C_V}) = \mathrm{Spec}^*(n_{\hat{\Gamma}}|_{C_V})$. Hence we obtain $\mathrm{Spec}^*(A(\Gamma)) = \mathrm{Spec}^*(A(\hat{\Gamma}))$. $\square$

6 Conclusion

In this paper, we have obtained the following two theorems:
1) By defining the isospectral pairs in terms of combinatorics, we have an infinite family of such pairs constructed in terms of the pairs of Young diagrams. Even though our present construction is limited to the case of simple edges, our explicit construction gives us the opportunity and motivation to discuss the case $\beta \geq 2$, for which we have now very few informations from the view of both, combinatorics and functional analysis.
2) The sets of all eigenvalues of both Γ and $\hat{\Gamma}$ are just the same including the multiplicities except for zero eigenvalue(s). We believe that the behaviour of the expected "right theory" about to the Perron-Frobenius eigenvalue corresponds to the "Galois conjugate" for factor-subfactor pairing $N \subseteq M$.

We have not touched yet the problem of a good axiomatization of the expected generalized group structure for a given isospectral pair. But we believe that with our approach we have the chance of success to replace the flatness condition of the present Ocneanu's notion of paragroups, and that our new theorems set the stage for it.

References

1. Bratteli, O., *Inductive limits of finite dimensional C^*-algebras,* Trans. Amer. Math. Soc. **171** 195–234 (1972).
2. Evans, D. and Kawahigashi, Y., *Quantum symmetries on operator algebras,* Oxford University press (1998).
3. Goodman, F., de la Harpe, P. and Jones, V.F.R., *Coxeter graphs and towers of algebras,* vol.14, MSRI publications, Springer, 1989.
4. Jones, V.F.R., *Index for subfactors,* Invent. Math. **72**, pp. 1–25 (1983).
5. Jones, V.F.R., *A polynomial invariants for knots via von Neumann algebras,* Bull. Amer. Math. Soc. **12**, pp. 103–111 (1985).
6. Jones, V.F.R., *Hecke algebra representations of braid groups and link polynomials,* Ann. Math. **126**, pp. 335–388 (1987).
7. Ocneanu, A., *Quantized group, string algebras and Galois theory for algebras,* in "Operator algebras and applications, Vol.2 (Warwick, 1987)," London Math. Soc. Lect. Note Series **136**, Cambridge University Press, pp. 119–172 (1988).
8. Ocneanu, A., *Quantum Symmetry, Differential Geometry of Finite Graphs and Classification of Subfactors,* University of Tokyo Seminary Notes **45** (1991).

Lecture Notes in Physics

For information about Vols. 1–615
please contact your bookseller or Springer
LNP Online archive: springerlink.com